International Conference on Extended Defects in Semiconductors 2010

(EDS 2010)

Journal of Physics: Conference Series Volume 281

Brighton, United Kingdom
19-24 September 2010

ISBN: 978-1-61782-765-5
ISSN: 1742-6588

Printed from e-media with permission by:

Curran Associates, Inc.
57 Morehouse Lane
Red Hook, NY 12571

Some format issues inherent in the e-media version may also appear in this print version.

Printed by Curran Associates, Inc. (2011)

For permission requests, please contact the Institute of Physics
at the address below.

Institute of Physics
Dirac House, Temple Back
Bristol BS1 6BE UK

Phone: 44 1 17 929 7481
Fax: 44 1 17 920 0979

techtracking@iop.org

Additional copies of this publication are available from:

Curran Associates, Inc.
57 Morehouse Lane
Red Hook, NY 12571 USA
Phone: 845-758-0400
Fax: 845-758-2634
Email: curran@proceedings.com
Web: www.proceedings.com

TABLE OF CONTENTS

IN MEMORIAM

PAPERS

Journal of Physics: Conference Series **281** (2011) 011001 doi:10.1088/1742-6596/281/1/011001

Preface

The biennial international conference on Extended Defects in Semiconductors started in 1978 with a meeting in Hünfeld, Germany. Subsequent meetings rotated between Poland, France, Great Britain, Germany, Russia and Italy, culminating in EDS2004 in Chernogolovka, EDS2006 in Halle and EDS2008 in Poitiers. EDS2010 was held at the University of Sussex at Brighton, UK from September 19th to 24th. An extension of the tabulation of this history, which first appeared on the EDS2006 website, is given in the appendix. It is with sadness that we note one of the founders of the series, Prof. Dr Helmut Alexander, passed away on 3 December 2009 and we were proud to dedicate EDS2010 to his memory. It has become a tradition to make an award in his name, and this year it was made to Ivan Isacov for his poster "*Electrical levels of dislocation networks in p- and n-type silicon*". A short and warm celebration of Prof. Dr Alexander's life by his friends and colleagues, Prof. Drs Helmut Gottschalk, Eicke Weber and Wolfgang Schröter, is included in this volume.

The conference was a forum for the state-of-the-art of investigation and modelling of extended defects in semiconductors. Scientists from universities, research institutes and industry made contributions to a deeper understanding of extended defects, their interaction with point defects and their role in the development of semiconductor technology.

The remit of the conference included extended defects, nanostructures, nanoparticles, quantum dots and interfaces within semiconducting materials ranging from narrow to wide band gaps, including graphene-derived materials and diamond. Scientific interests range from defect geometry, electronic structure, dynamics, spectroscopy, microscopy, reactions and chemistry to introduction mechanisms, such as implantation and strained layers and the operation of devices such as integrated circuits, heterostructures, and solar cells.

The organisers were confronted with a long period between the afternoon outing to Arundel Castle and dinner in the evening at Wiston House, a mansion of Tudor origin near Steyning, West Sussex. So a short audience-participation seminar was held in the conference room of the manor, covering the history of dislocations and the history of the conference series. We were also able to extend the appreciation of the life of Prof. Yuri Ossipyan (15 Feb 1931 – 10 Sep 2008) briefly given at EDS2008.

EDS2010 continued the drive into graphene-based materials with a session devoted to them, and it gave immense pleasure to many of us who were his former students to dedicate a session to the work of Professor R Jones. We are grateful to his present and former co-workers who came and presented an impressive perspective on their work with him and a vision of a vigorous future for him in his retirement and for AIMPRO, the current Density Functional Theory code that derives from the one he established with his former student, Dr Patrick Briddon.

For EDS2010 we made two minor modifications to the appearance of the conference: a central webpage www.eds-conferences.org, ably managed by our webmaster, Dr Gemma Haffenden, and a Facebook page, "EDS conference series", which Dr Amy Gandy runs enthusiastically. Amongst other things the conference photographs appear here. "I like this". In fact, currently 22 FB-ers "like this" and I am sure it will grow.

Finally, it is a pleasure to acknowledge the significant contributions of Co-Chair, Prof. Marek Skowronski, Conference Manager, Dr Christopher Latham, and the editors of this volume, Drs Jon Goss and Chris Ewels, who in turn wish to thank Dr Alexis Vlandas for his help proof reading the articles.

We all wish the best of luck to Prof. Philomela Khomninou and her team in the organisation of EDS2012.

Malcolm I Heggie

The History of EDS Conferences: dates, locations, titles and organisers		
1978	*Inaugural conference* *"EDS 1"* Hünfeld (Germany)	**Dislocations in Tetrahedrally Coordinated Semiconductors** H. Alexander (Köln), P. Haasen (Göttingen), R. Labusch (Clausthal), W. Schröter (Göttingen)
1980	*"EDS 2"* Krynica (Poland)	**Defect-induced Phenomena in Semiconductors** J. Auleytner (Warszawa)
1983	*"EDS 3"* Aussois (France)	**Properties and Structure of Dislocations in Semiconductors** J. Philibert, B. Sieber, A. Zozime (Meudon-Bellevue)
1986	*"EDS 4"* Zvenigorod (USSR)	**Structure and Properties of Dislocations in Semiconductors** Yu. A. Osip'yan (Chernogolovka)
1988	*"EDS 5"* Szczyrk (Poland)	**Defects in Crystals** J. Auleytner, T. Figielski, E. Kaczmarek, E. Mizera, T.Wosinski (Warszawa)
1989	*"EDS 6"* Oxford (UK)	**Structure and Properties of Dislocations in Semiconductors** P. B. Hirsch, G. R. Booker, D. B. Holt, J. L. Hutchison, S. G. Roberts, P. R. Wilshaw (Oxford)
1992	*EDS 7* Holzhau (Germany)	**Structure and Properties of Extended Defects in Semiconductors** P. Haasen (Göttingen), H. Alexander (Köln), J. Auleytner (Warszawa), J. Heydenreich (Halle), W. Schröter (Göttingen)
1996	*EDS 8* Giens (France)	**Extended Defects in Semiconductors** B. Pichaud (Marseille), S. Pizzini (Milano), A. Cavallini (Bologna), G. Vanderschaeve (Toulouse)
1998	*EDS 9* Jaszowiec (Poland)	**Extended Defects in Semiconductors** T. Figielski (Warszawa), W. Schröter (Göttingen)
2000	*EDS 10* Brighton (UK)	**Extended Defects in Semiconductors** M. Heggie (Sussex)
2002	*EDS 11* Bologna (Italy)	**Extended Defects in Semiconductors** A. Cavallini (Bologna), V. V. Kveder (Chernogolovka), S. Pizzini (Milano)
2004	*EDS 12* Chernogolovka (Russia)	**Extended Defects in Semiconductors** Yu. A. Osip'yan, V. V. Kveder (Chernogolovka)
2006	*EDS 13* Halle (Germany)	**Extended Defects in Semiconductors** H. S. Leipner (Halle), M. Kittler (Frankfurt/Oder)
2008	*EDS 14* Poitiers (France)	**Extended Defects in Semiconductors** J. Rabier, E. Le Bourhis, J.-F. Barbot, M.-F. Beaufort, M.-L. David, J.-L. Demenet, L. Pizzagalli, L. Thilly, C. Tromas, S. Brochard, J. Godet
2010	*EDS 15* Brighton (UK)	**Extended Defects in Semiconductors** M. Heggie (Sussex), M. Skowronski (USA)
2012	*EDS 16* Greece	**Extended Defects in Semiconductors** P. Khomninou

All papers published in this volume of *Journal of Physics: Conference Series* have been peer reviewed through processes administered by the proceedings Editors. Reviews were conducted by expert referees to the professional and scientific standards expected of a proceedings journal published by IOP Publishing.

International Conference on Extended Defects in Semiconductors (EDS 2010) IOP Publishing
Journal of Physics: Conference Series **281** (2011) 012001 doi:10.1088/1742-6596/281/1/012001

Helmut Alexander – A Life for Science – Alive in Science

Wolfgang Schröter[1], Helmut Gottschalk[2] and Eicke R. Weber[3]
[1] IV. Physikalisches Institut der Universität Göttingen, Germany
[2] Abt. Für Metallphysik, II. Physikalisches Institut, Universität zu Köln, Germany
[3] Fraunhofer Institute for Solar Energy Systems ISE and Albert-Ludwigs University, Freiburg, Germany

Abstract. Helmut Alexander was one of the four initiators of the "International Symposium on Dislocations in Tetrahedrally Coordinated Semiconductors" (Hünfeld 1978). This conference was the beginning of a conference series which got its new name EDS (Extended Defects in Semiconductors) first in 1980. He who gave his name to the "Helmut Alexander Award" of the EDS conference died on the third of December, 2009. The international advisory committee of EDS honoured his scientific work in 2002 recognizing him by instituting an award for the best poster presentation for a young scientist at the running conference session. The authors try to review his life with some of his scientific highlights.

Helmut Alexander was born on July 30, 1928 at Mannheim as son of a parish priest. Formative influences on his childhood and youth were his home and the Second World War. His parents strongly supported his special interest in literature and music. In 1944 he was called up by the German Navy (Marine), but soon fell ill with severe tuberculosis. This disease forced him for many years to exercise restraint. He went to grammar school (Gymnasium) in Neustadt/Weinstrasse, from which he graduated with the "Abitur" in 1947. He began to study physics at the University of Mainz in 1949 and obtained his diploma in 1957 at the University of Darmstadt. After a short time at Siemens he began to work for a doctoral thesis at the "Max-Planck-Institut für Metallforschung" in Stuttgart.

1. The beginnings of his scientific career in Göttingen

1.1. Plasticity of germanium compared to that of copper; doctoral thesis

In the first days of 1959, Peter Haasen, just appointed the director of the "Institut für Metallphysik" (IfM) at the university, arrived at Göttingen. He came from the "Max-Planck-Institut für Metallforschung" at Stuttgart and with him came Helmut Alexander, a graduate student busy with a doctoral thesis on Plastic Deformation of Germanium (1961 Plastische Verformung von Germanium *Z. Metallkde.* **52** 344).

The topic of this thesis did not appear to be very exciting. Many materials scientists in leading research centers of the USA, Great Britain, France, Russia and now also at Stuttgart were investigating the problem of plasticity and its interpretation. However, their work concentrated on metals and alloys, not on semiconductors.

Why then germanium? Helmut Alexander's answer to this question is given in his thesis (1960): "... to characterize the deformation mechanism of the diamond lattice by its correspondence and differences with the cubic face-centered lattice by a method, which brought to a large extent a

Journal of Physics: Conference Series **281** (2011) 012001 doi:10.1088/1742-6596/281/1/012001

quantitative understanding for the latter lattice type ...". The method which he refers to is the dynamical tensile test.

For a graduate student at the IfM like Helmut Alexander, the experimental work on germanium started with a piece of raw material, which had to be purified by zone-pulling, then grown with prescribed orientation as a single-crystalline slab, followed by cutting into specimens, mechanical and chemical polishing of their surfaces *etc.*. However, the great challenge on the way towards a tensile test with germanium was to conceive and produce the grips connecting the specimen with the deformation machine. For this problem, until then unsolved, Helmut Alexander found a very elegant solution, which was basic for the success of his work. However, the tensile test remained a delicate procedure. It was later replaced by the compression test, as soon as the equivalence of both tests was shown in the first diploma thesis supervised by Helmut Alexander.

The two main results, which he obtained in his thesis, were starting points for his further research activity in Göttingen and later Cologne.

When a constant elongation velocity is imposed on a copper specimen, the transition from elastic to plastic deformation occurs at a critical (shear) stress τ_c (see Fig. 1), which is weakly temperature-dependent. For germanium, Alexander observed a quite different beginning of plastic deformation: the shear stress runs through a maximum τ_{uy} (upper yield point) and then drops to a minimum τ_{ly} (lower yield point), the whole stage being exponentially dependent on temperature. As a consequence, germanium is brittle at room temperature, while copper is ductile even at 4 K.

To explore the dislocation mechanisms responsible for that yield range of germanium became a key activity of the semiconductor group which was established at the IfM in the following years.

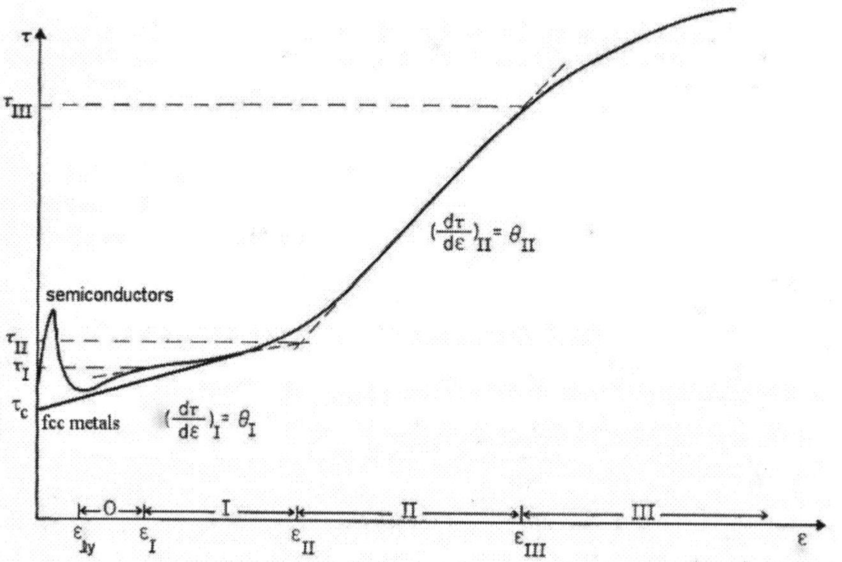

Figure 1. Schematic diagram showing dynamic stress-strain curves (τ versus ε), as measured with specimens uniaxially deformed with constant strain rate, for cubic face-centered (fcc) metals like Cu and elemental semiconductors like Ge and Si.

Concerning the following deformation stages (see Fig.1), Alexander found astonishing similarities between copper and germanium. For both materials, the start of plastic deformation continues with a stage I of easy glide, followed by a stage II with constant and *T*-independent hardening rate $\Theta_{II} = (d\tau/da)_{II}$, and finally leads to a stage III with decreasing hardening rate. In particular, stage II was – and is still – mysterious. As Alexander's work showed, Θ_{II}/G (*G*: shear modulus) has the same value for copper and germanium.

1.2. Dislocation formation and deformation stages in germanium; Venia Legendi

Which dislocation arrangement can produce the same and *T*-independent Θ_{II}/G in two such different materials? At that time, electron microscopy had become available as a tool to make dislocations visible. However, to apply it, thin foils had to be prepared from the material, in which dislocations were exposed to large image forces towards the foil surface. In copper dislocations are highly mobile at room temperature, they react to the image forces and run out of the foil, if not pinned. As one of the experts in the field said, one observes the laggards of an army. On the other hand, in germanium dislocations are immobile at room temperature. Therefore, in specimens cooled fast after deformation under stress applied to room temperature, the electron microscope should give an image of the army.

Figure 2. Start to Helmut Alexander's university career (1960). The photo shows him one day after he handed in his thesis to Peter Haasen and on the day of his engagement with Ingeborg Feld, later to become his wife. (The authors want to thank Ingeborg Alexander for kindly giving us this private photo).

Certainly, Helmut Alexander immediately realized the great chance to solve one of the basic problems of metal plasticity, *i.e.* an interpretation of stage II, by an electron microscopic study of dislocations in germanium. In 1961 the IfM got an electron microscope, and Helmut Alexander decided to work on "Electron microscopy of frozen-in dislocations" for a postdoctoral thesis (Habilitation). A prerequisite for the realization of such a plan was Peter Haasen's offer to him to continue as a "wissenschaftlicher Assistent". It also meant a jump in his private income from zero-payment to a fair salary opening new possibilities also in his private life (see figure 2).

In 1966, he obtained the Habilitation and was awarded by the Faculty of Physics the *Venia Legendi.* From now on he had the right and the duty to teach students.

In 1968, he published the results of his electron microscopic investigations on plastically deformed germanium as two articles (1968 Elektronenmikroskopie eingefrorener Versetzungen *phys. stat. sol.* **26** 725; *phys. stat. sol.* **27** 391). He showed that the dislocations in fact are frozen-in and neither the preparation of TEM foils nor electron irradiation in the microscope under normal imaging conditions nor image forces acting on the dislocations near to the foil surfaces influence the as-deformed dislocation arrangement.

While Helmut Alexander studied frozen-in dislocation arrangements in Ge, scientists in Stuttgart had applied neutron irradiation to pin dislocations in deformed Cu and to observe their arrangements during work hardening by electron microscopy. From a very thorough investigation, Alexander inferred a distinct similarity of dislocation arrangements in Ge to those observed in Cu for ranges I and II (see Fig. 1). This finding not only confirmed the relevance of the observations, made by Essmann (1963 phys. stat. sol. **3** 932; 1965 phys. stat. sol. **12**, 707; 723) in Stuttgart for Cu, but also served to arrive at a fundamental insight into work hardening. In their review of 1968, Alexander and Haasen write: "... It is the slip geometry , interaction between slip systems, dislocations reactions, *etc.,* that determine dislocation structure and work hardening during uniaxial deformation, not dislocation mobility, point defect formation, dislocation core structure, *etc.*. The former are properties of the Bravais lattice only, being the same for fcc and diamond-type crystals; the latter depend on the binding character, crystal unit cell, *etc.* ..."

Furthermore Alexander succeeded in finding the answer to another basic question: are the stages I and II of the deformation curve (see Fig. 1) determined by typical dislocation structures? Varying the deformation temperature from 520^{0}C to 800^{0}C, Alexander generated dislocation patterns that were much more three-dimensional ones with dislocations less fixed to the primary planes than at 520^{0}C. However, the work hardening curve shows the three stages for both temperatures. In their review Alexander and Haasen concluded: "This indicates that there is no typical dislocation structure responsible for the three stages of the stress-strain-curve."

1.3. *A microscopic model on the beginnings of the plasticity in semiconductors; a student group`s work and the Alexander-Haasen theory*

Every year at the beginning of the winter period, the IfM offered to students a practical training consisting of ten experiments within two weeks. For students, who wanted to perform a diploma thesis at the IfM, or elected the subject "Metallphysik" for one of the four oral examinations at the end of the thesis, successful participation in that training was obligatory. As an unusual feature, nearly all scientists at the IfM with their equipment were engaged in that training. The capacity was twenty students, but soon there were many more applicants, and a written examination was introduced to obtain a ranking. The students, who had to work intensively during vacation to prepare for the exam and experiments in rapid succession, got as a benefit in return a picture of the research activities of the IfM, and thus the capability to judge the relevance and difficulties of a diploma thesis at the IfM. This combination of student's practical training with the institute's cutting-edge research possibly explains, why the IfM, which had in 1959 shifted from the faculty of chemistry to that of physics, was able to successfully compete for students among the renowned institutions of Göttingen physics

The IfM also had a small local advantage. Its position inside the historical city walls, opposite to a cinema, a student's theater and restaurant, led the students working at the IfM to move with some part of their private life to the IfM and to combine *e.g.* during evening a long-term experiment with the visit of a movie or restaurant.

One of the two first diploma subjects that Peter Haasen allocated in Göttingen was on the plasticity of InSb and thus developed the nucleus of the semiconductor group. As soon as appointed "wissenschaftlicher Assistent", Helmut Alexander became the leader of this group. While for his post-doctoral thesis he worked on the stages I to III, which were similar in copper and germanium, he directed the research activity of the group towards an understanding of the beginnings of plasticity in germanium. The students began to construct machines for dynamical and static compression, for germanium single crystal growth with low initial dislocation density and low impurity content, methods for oriented polishing of specimens, for the analysis of dislocations by etch-pits, for counting large etch-pit densities using optical and electron microscopy, and for local slip measurements among others. The objective of this work was to model the beginning of plasticity by dislocation properties. One great aim of Helmut Alexander and Peter Haasen was to develop a microscopic theory, based on dislocation (1) multiplication, (2) motion, (3) interaction, and (4) on a statistical model relating microscopic dislocation processes and macroscopic plasticity.

One might follow the progress of the group's work by looking at its first publications (written in German, the titles given here have been translated):

"Plastic deformation of Ge single crystals" by H. Alexander (1961 *Z. Metallkde.* **52** 344, his doctoral thesis);
"Plastic deformation of Ge and InSb in dynamic compression" by S. Schäfer, H. Alexander, and P. Haasen (1964 *phys.stat. sol.* **5** 247);
"Deformation inhomogeneity of Ge within the yield region" by W. Schröter, H. Alexander, and P. Haasen (1964 *phys.stat. sol.* **7** 983);
"Dislocation density and local slip in Ge single crystals" by K. Berner and H. Alexander (1967 *phys. stat. sol.* **15** 963);
"Measurement of the dislocation velocity in Ge" by S. Schäfer (1967 *phys. stat. sol.* **19** 297).

An independent appreciation of this research from outside came 1967, when Alexander and Haasen were invited to write a review for the prestigious book series "Solid State Physics", edited by Seitz, Turnbull and Ehrenreich. In their article on "Dislocations and Plastic Flow in the Diamond Structure" (1968 *Sol. State Phys.* **22** 27-158) the authors gave as the essence of all that work a set of mathematical equations, whose solutions showed an astonishing agreement with measured curves of the initial stage for dynamic as well as for static deformation conditions. The Alexander-Haasen theory was born, until today the basic description of semiconductor plasticity. A key to the success of it was a dislocation multiplication law, discovered by Helmut Alexander and his student Klaus Berner.

With this review and the publication of his habilitation thesis also in 1968, Helmut Alexander became an internationally renowned authority in the field of semiconductor plasticity. He continued working on this field when leaving Göttingen for Cologne.

1.4. Electronic states at dislocations in silicon and germanium; first results and move to the "Universität zu Köln"

There is a further research activity that Alexander started in Göttingen and continued with in Cologne, the results of which brought him and his group a high repute in the international community. William B. Shockley from Bell Laboratories, Nobel Prize winner for the discovery of the transistor, had made in 1953 a very interesting conjecture about the electrical activity of dislocations in silicon and germanium (*Phys. Rev.* **91** 228). He argued, that an edge-type dislocation in those materials should have a one-dimensional arrangement of dangling bonds in the dislocation core at the edge of the inserted half-plane (see Fig. 3).

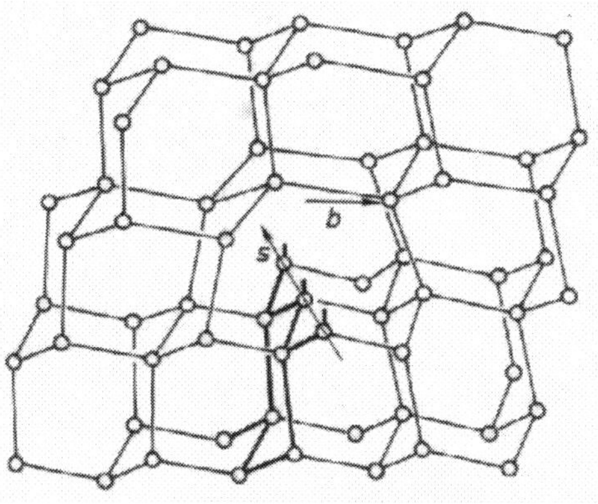

Figure 3. A 60° shuffle dislocation; bold lines indicate the extra half-plane inserted from the bottom.

In the electrically neutral state, they should be occupied by one electron and then be half-filled. With the overlap of neighboring dangling bonds a one dimensional metal should be associated with the dislocation core.

Shockley's conjecture was an issue hotly disputed in the semiconductor group, specifically regarding experiments that could prove or disprove it. Under the guidance of Peter Haasen, one of the authors (Schröter W 1967 *phys. stat. sol.* **21** 211) performed a doctoral thesis on the Hall effect of dislocations in p-type germanium. He showed that dislocations with edge-component are associated with electronic states whose occupation follows Shockley's conjecture.

But how to prove the properties of a one-dimensional metal? This was the point of intensive discussions in particular between Helmut Alexander and Reiner Labusch, a young theoretician educated in the school of Friedrich Hund and now working at the IfM. He had made some calculations and found that electron paramagnetic resonance (EPR) could be a promising method.

However, application of EPR to germanium was difficult because a strong spin-lattice coupling and multiple hyperfine interactions broaden all EPR lines in Ge. One could try silicon, but experiments with plastically deformed silicon were rare, one earlier application of EPR to deformed silicon at Bell Laboratories failed. Nevertheless, Helmut Alexander decided that the experiment should be repeated. Pieces of silicon, obtained from industry, were plastically deformed and measured at a neighboring institute, having the facility to measure EPR down to 20K (liquid hydrogen). Wilhelm Sander, the EPR-expert there indeed detected an EPR-line in plastically deformed silicon, but no fine structure, especially relevant for its interpretation, could be resolved. H. Alexander, R. Labusch, and W. Sander published this first evidence in 1965 (*Solid State Commun.* **3** 357). At the time it was clear that for the differentiation between a one dimensional metal and a row of localized states, the EPR-experiment and the deformation procedure needed further refinements. To arrive at them became one of the major challenges for Alexander's work in Cologne.

1968 was not only the year, in which Helmut Alexander published two of his most important articles, but also brought him the offer of a full professorship in Cologne.

2. Helmut Alexander in Cologne (1968 – 1997)

When Helmut Alexander came to Cologne to build up the "Abteilung für Metallphysik im II. Physikalischen Institut der Universität zu Köln", he had to take over the heritage of an already existing "Metallphysik"- group which at that time was part of the Institute of Theoretical Physics. The main subject of prior research in this group was the investigation of magnetic properties and transport phenomena of metallic alloys. This field of research was very different from his former activities. Nevertheless, he accepted this challenge and decided to lead both this "Magnetiker"-group and the "Germanisten"-group for all the time until he retired in 1993. He focused his interest in this line of research on magnetism of metallic compounds with a ferromagnetic component, with special respect to the liquid phase. One main result was: The liquid phase of Au-Co is not ferromagnetic, as it was suggested in the literature (Busch G, Guentherodt H J 1968 *Phys. Letters* **27A** 110). Magnetic measurements could be conducted with greater precision than conventional, purely thermal analysis, and allowed corrections to phase diagrams, as well as the study of properties of alloys at high temperature, such as precipitation kinetics.

2.1. Electron microscopy and dislocation velocity

In 1968, in Helmut Alexander's new lab in Cologne existed a modern TEM (Siemens Elmiskop IA) facility and a small TEM group consisting of four doctoral students. Soon he continued to investigate the early stages of deformation. The search for dislocation sources by TEM and Lang topography was successful. More and more the activities changed from Ge to Si, being a material better suited to EPR measurements.

When Ray and Cockayne (Ray I L F and Cockayne D J H 1970 *Phil. Mag.* **22** 853; 1971 *Proc. Roy. Soc.* **A325** 543) published the results of their newly developed TEM weak beam technique applied to silicon a fundamental change of dislocation models and theory occurred. Nearly all the dislocations appeared to be dissociated into partial dislocations with a stacking fault ribbon between them. This result was revolutionary as it decided the discussions whether the dislocations were glide or shuffle set ones - until then with a clear tendency to prefer the shuffle set model.

The weak beam technique was applied to III-V compounds to measure the splitting width of dislocations and to calculate the stacking fault energies. Alexander suggested a correlation of the "reduced stacking fault energies γ'" with the ionicities f_i of the compounds where γ' is the stacking fault energy referred to the specific area per atom in the {111}-plane. In a plot of γ' *vs* f_i the values for six III-V compounds fall on a single line. (Fig. 4) (Gottschalk H, Patzer G and Alexander H 1978 *phys. stat.sol. (a)* **45** 207; Fig. 5, p. 216: Copyright Wiley-VCH Verlag GmbH & Co. KGaA, reproduced with permission). A stacking fault represents a plane of hcp-type stacking inserted into the cubic sphalerite lattice. Thus, decreasing stacking fault energies in the cubic lattice point to an increasing preference of the hexagonal structure. Indeed, compound semiconductors with ionicities larger than the cross-over point found in Fig. 4 near f_i = 0.47, such as most of the III-nitride compounds and alloys, preferentially crystallize in the wurtzite structure, showing the relevance of the findings summarized in Fig. 4.

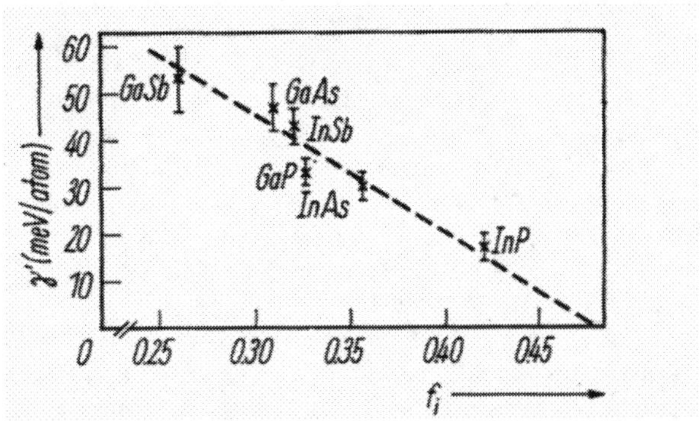

Figure 4. Reduced stacking fault energy γ' *vs* the iconicity f_i of III-V compounds.

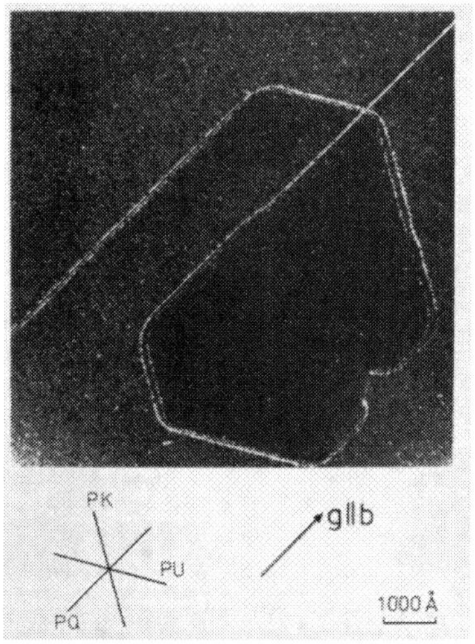

Figure 5. Dislocations in Si after high stress deformation.
PQ = 30/30; PK = 90/30; PU = 30/90 (see text).

One of the milestones of the early Cologne years was the development of a deformation mode to generate long straight dislocation segments of well defined types (Wessel K and Alexander H 1977 *Phil. Mag.* **35** 1523). A Si specimen with the standard stress axis [213] was predeformed at high temperature (750°C) to introduce dislocations, then cooled to 420° C to be deformed for a second time under high stress and finally cooled to room temperature under the full stress applied. As a result the dislocations became aligned straight along the <110>-directions so that three types of dislocations appeared: two kinds of 60°-dislocations, one with a 30° partial moving in front and a 90° partial trailing (30/90) and another one with the 90° partial moving in front followed by the 30° partial (90/30), and screw dislocations (30/30) (Fig. 5). Additionally, a very surprising result was obtained: the measured dissociation widths of the three dislocation types differed significantly from their equilibrium widths measured in annealed specimens.

The evidence for the glide set dislocation model led to the assumption that dislocations move in their split state. Consequently, the mobility of a dislocation should be composed of the mobilities of both partials. From measurements of the splitting widths of the three dislocation types ratios of the mobilities were derived and it was found that a unique velocity *e.g.* for 60° dislocations does not exist. Instead, different mobilities are expected for 30/90 and 90/30 dislocations. A particularity is the behavior of screw dislocations (30/30) with a strongly reduced mobility of the trailing 30° partial. The reason is not fully understood but there are indications for EPR active centers sited on the partial.

An important step forward for this research was achieved when a new TEM microscope was installed (Philips EM 400T) in 1980. At 120 keV much brighter images could be obtained so that the use of the weak beam technique became much easier.

Dislocations in Si at 420°C move very slowly. Though there is indeed some plastic deformation the material behaves on a macroscopic scale like a brittle one and, consequently, a lot of specimens were broken. The fraction of good high stress specimens grew significantly after the heated pistons in the testing machine formerly made of stainless steel were replaced by quartz glass ones with an ingenious device allowing to perfectly align the two opposing surfaces of the pistons (Küsters K H and Alexander H 1983 *Physica B+C* **116** 594).

High-stress deformed specimens with a population of only three types of straight dislocation have been very interesting for a lot of investigations using very different measuring methods. The idea was to ascribe special properties to the different dislocation types. For a while Alexander's lab was worldwide the only one to produce uniaxial high-stress deformed specimens of Si. Helmut Alexander, always open to cooperation, willingly gave specimens to other research groups all over the world.

One of the continuing threads in Helmut Alexander's scientific life had always been the mobility of dislocations. Therefore, beside the microscopic investigations precise macroscopic measurements of dislocation velocities in high stress/low temperature deformed Si were performed by etch pitting (Alexander H, Kisielowski-Kemmerich C, Swalski A T 1987 *phys. stat. sol. (a)* **104** 183). The fact that dislocations at the surfaces bend when the specimen is deformed for a second time was considered. These measurements showed the difference of the velocities of the two 60° dislocations (30/90 and 90/30).

The influence of light on dislocation mobility, the photoplastic effect, was also investigated using the etch pit method (Küsters and Alexander 1983; see above).The surprising result that only the 30/90 and not the 90/30 dislocation is affected by illumination, was further evidence for the special role of the trailing 30° partial.

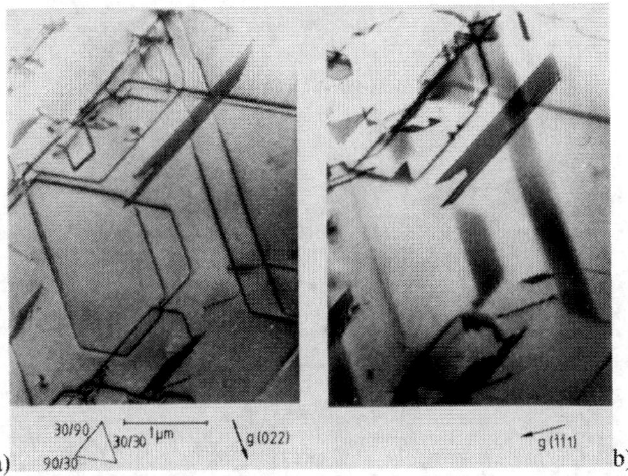

a) b)

Figure 6. Anomalously wide dissociated dislocations in high stress/low temperature deformed Si. Uniaxial stress was applied along [2 1 11] a) g = (022); the partial dislocations in the "main" glide plane are visible. b) g = (-1 -1 1); the stacking faults in the "main" glide plane appear with a dark contrast.

If the stress axis is parallel to <2 1 11>, the forces acting on the partials are directed so as to separate the partials more and more. Thus large stacking faults are generated, mainly at screw dislocations. Such anomalously wide split dislocations are in an extreme non-equilibrium state. The stacking fault tries to pull back the partials. Relaxation experiments by heating in the microscope gave some interesting results (Alexander H, Eppenstein H, Gottschalk H and Wendler S 1979 *J.* Mic. **118** 13; Fig. 6). Surprisingly, not all screw dislocations were extremely widened (left part of fig. 6a).

Helmut Alexander was not only interested in conventional TEM available in his Cologne institute but also in HREM applications. Therefore, he left Cologne in 1985 for a four months stay as a guest scientist at the Department of Physics of the Arizona State University, Tempe, Az., USA. The subject of his research was HREM lattice imaging of split dislocations in Si (Alexander H, Spence J C H, Shindo D, Gottschalk H and Long N 1986 *Phil. Mag. A* **53** 627). John Spence continued this line of research after Helmut Alexander returned to Cologne, and it culminated with an important joint paper (Kolar H R, Spence J C H and Alexander H 1996 *Phys. Rev. Lett.* **77** 4031) in which the activation energy for dislocation motion could be distinguished into the kink formation and migration energies. The results were obtained on widely split dislocations by HREM in situ relaxation experiments at 600°C and at 130°C.

The result W_m = 1.24 eV is in accordance with earlier relaxation experiments performed in Cologne using the conventional TEM with a heating stage at 100°C to 140°C. From the measured displacement changes of the partials the kink migration energy W_m was estimated to be 1.0 – 1.2 eV (Gottschalk H, Alexander H and Dietz V 1987 *Inst. Phys. Conf. Ser.* **No 87 Sect. 5** 339).

Some years later Helmut Alexander saw a chance to measure dislocation velocities in Si on a microscopic scale using a new technique. Dislocations in high temperature deformed and annealed specimens contain mainly edge dislocations which are fairly straight. The dislocations are not split continuously on their whole length. Many constrictions occur which could be correlated to jogs (Tillmann J 1976 Diploma Thesis Univ. Köln; Farber B Ya and Gottschalk H 1991 Springer Proceedings in Physics **54** Polycrystalline Semiconductors II 242; Fig. 7). At low temperatures, for Si about 400°C, jogs act as strong obstacles against the dislocation motion (Fig. 8). The distances the dislocations had moved between the jogs in a subsequent high stress deformation at this low temperature were measured. The jogs marked the starting positions of the dislocation segments (Fig.8). (Gottschalk H, Hiller N, Sauerland S, Specht P, Alexander H 1993 *phys. stat. sol. (a)* **138** 547; p. 548, Figs. 1 and 2: Copyright Wiley-VCH Verlag GmbH & Co. KGaA, reproduced with permission).

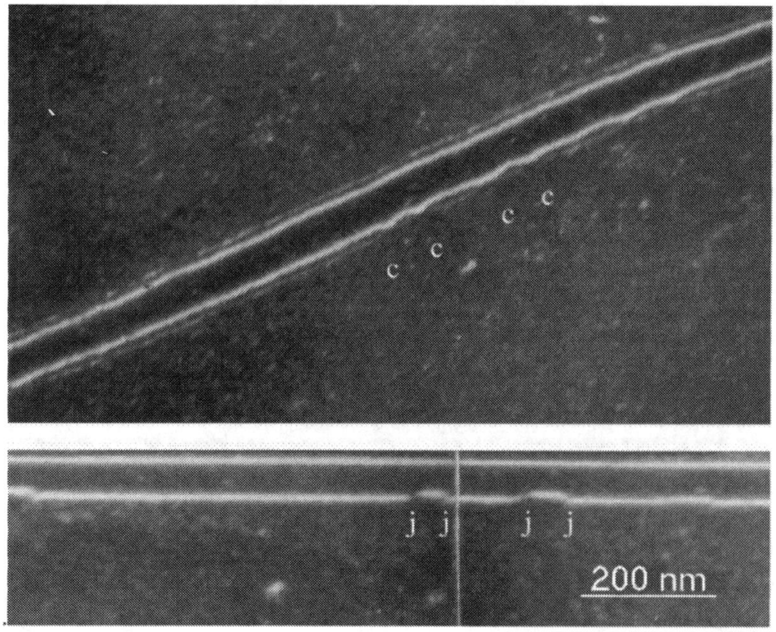

Figure 7. 60°-dislocation dipole in Si (TEM Weak Beam), top: Plane view on the glide plane: constrictions c; bottom: same area; dislocation viewed parallel to the glide plane: jog pairs j. The dislocation segments between the jog pairs are split.

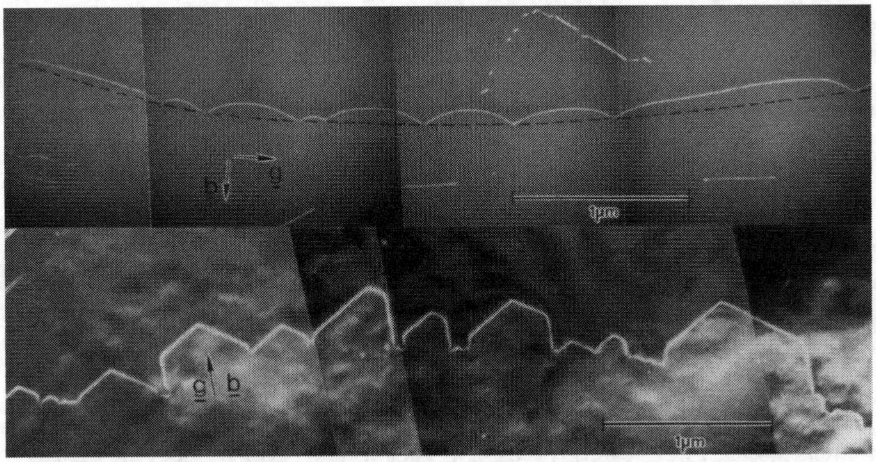

Figure 8. Free segments of near edge dislocations have moved while jogs (the constrictions) are still lying along a smooth line representing the former course of the dislocation (dashed line). Deformation parameters: $T = 370°$ C, $\tau = 50$ MPa, $t = 300$ s (top); $T = 420°$ C, $\tau = 200$ MPa, $t = 600$s (bottom).

In 1985 the scientific equipment in Cologne was completed by a new scanning electron microscope, equipped with an EDX device (energy dispersive X-ray spectroscopy). This microscope added significant possibilities to characterize materials by imaging of electron induced physical effects. Besides SEM, equipment for the measurement of electron beam induced current, light beam induced current, scanning deep level transient spectroscopy and cathodoluminescence was built up.

That time marked an important step in Helmut Alexander's life. Previously he had been a man of pure science, who strictly rejected collaboration with industrial institutes. But when he was asked to collaborate with a group of researchers of eight universities and some solar Si and solar cell producers to help to enhance the efficiency of multicrystalline Si solar cells he was enthusiastic. The ecological aspects of this "DIXSI" project convinced him so much that these activities more and more moved to the center of his scientific interests for the last ten years of his scientific career. The experimental equipment available in his lab was well suited to examine these less pure Si wafers with respect to dislocations, grain boundaries and precipitates. Beside the raw material, fully processed solar cells were made available in this project to measure and to image by SEM methods the electrical properties of defects limiting the cell efficiency.

Helmut Alexander was so much engaged in this project that he developed a model of dislocation generation during the solidification of a cast Si block, which, unfortunately, remained unpublished.

A large part in Helmut Alexander's career is characterized by teaching. He took this task very seriously and it is not exaggerating to call him an exemplary teacher. After having retired, his passion for teaching made him write a textbook on electron microscopy "Physikalische Grundlagen der Elektronenmikroskopie" (Physical Fundamentals of Electron Microscopy), which was published in 1997 (Teubner, Stuttgart). With this book one of his dreams came true.

2.2. *Electron Paramagnetic Resonance and other electrical and optical investigations*

After arriving in Cologne, Helmut Alexander decided to continue the study of dislocation structures in Si by EPR that had led to the very promising first results described in section 1.4. Wilhelm Sander had accepted around the time of Alexander's change to Cologne a chair in the Physics Department of the

"Rheinisch-Westfälische Technische Hochschule Aachen", where he established a modern EPR facility. With Aachen located 70 km from Cologne it was possible to establish a close cooperation. Udo Schmidt, who had started with a Diploma thesis in magnetism, but decided to change topics to investigate the EPR signals of dislocations in silicon, shuttled regularly between Cologne and Aachen to perform EPR measurements in the laboratory of W. Sander with samples prepared in Cologne. In 1970, one of the present authors (ERW) joined this line of research as a graduate student.

The analysis of the EPR signal introduced in silicon by plastic deformation made rapid progress with the sensitive instrumentation available in the Aachen lab, using well-controlled samples uniquely prepared with the dedicated instrumentation in Cologne. A key factor was the availability of low-temperature EPR measurements, generally using liquid hydrogen, an old Göttingen tradition, and for measurements below 20K a quite cumbersome liquid helium evaporation cryostat. A delicate fine structure of the EPR signal of plastically deformed Si was revealed and distinguished into lines near $g = 2.00$ denoted as Si-K1 (K for Köln, *i.e.* Cologne in German) and a complex set of lines appearing pairwise on both sides of the central spectrum denoted Si-K2. Udo Schmidt and ERW succeeded in determining the g-tensors and fine structure tensors of these spectra by analyzing the complex angular dependence upon rotation of the sample in the magnetic field at cryogenic temperature. The g-tensors revealed two lines with $S = ½$ for each glide system, related to the glide systems of the plastic deformation geometry used to introduce the dislocations.

The fine structure tensor of Si-K2 proved to be aligned along the Burgers vector of the primary glide system (Schmidt U, Weber E, Alexander H, and Sander W 1974 Solid *State Commun.* **14** 735) and could be interpreted either as $S = 1$ spectra in a multitude of crystal fields, or as sets of spectra with larger spin, ranging from $S = 1$ to $S = 3$ (Bartelsen L 1977 *phys. stat. sol. (b)* **81** 471). The g-tensor of Si-K2, and of the Si-Y-line, were found to be orthorhombic with C_{2v} symmetry, where the axis with the g-value closest to the free electron value lay along a <110> direction perpendicular to the (total) Burgers vector of the respective slip system. The hyperfine structure of Si-K1 identified this defect as a typical dangling bond defect with an orbital inclined 22° from a <111> bond axis (Weber E and Alexander H *J. de Physique* 1979 **40-C6**, 101). These findings were strong evidence that the paramagnetic centers are located in screw dislocations, were the dislocation line even of the partials runs along the total Burgers vector of the dislocation.

To interpret these findings it is necessary to consider a realistic picture of the geometry of dislocations split into partials. Alexander prepared a now-famous sketch of the possible dangling bond geometries in a dissociated 60° glide-set dislocation, containing a 30° and a 90° partial, see Fig. 9.

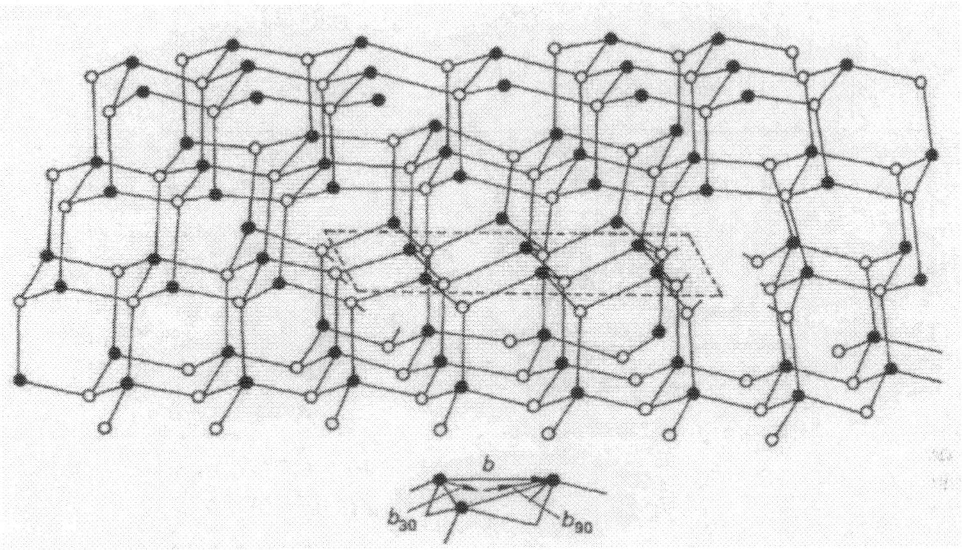

Figure 9: Dangling bonds in a dissociated 60° glide set dislocation. A stacking fault ribbon is bound by the two partials: on the left side the 30° partial, on the right side the 90° partial. (Partials are shown unreconstructed).

A more detailed analysis of the complex *g*-tensor geometries pointed to a vacancy in the core of a 30° partial of a screw dislocation as the most likely site of the unpaired electrons of Si-K1, and possibly the spins creating the Y-line (Kisielowski-Kemmerich C 1990 *phys. stat. sol.* **161** 11). Si-K2 in this model could be caused by rows of such defects along the screw dislocation line direction, but this assignment still awaits further confirmation, especially by theory. Recent calculations (Lehto and Öberg 1997 *Phys. Rev.* **B56** R 12706 and Csanyi G, Ismail-Beigi S and Arias T A 1998 *Phys. Rev. Lett.* **80** 3984) are at variance with the Kisielowski-Kemmerich model.

Upon increasing the measurement temperature above 50K new lines appeared that were identified as three distinguishable spectra labeled Si-K3, K4 and K5 (Alexander H, Kenn M, Nordhofen B, and Weber E 1975 *Inst. Phys. Conf. Ser.* **23** 433). The most intriguing finding was that the *g*-tensors of those spectra showed no relation to the deformation slip systems. A comparison of spectra obtained from samples deformed along <213>, *i.e.* with a primary glide system, and <110> where glide in four glide systems is preferred, clearly reveals this distinction (see Fig. 10): whereas the spectra taken at high temperatures (Si-K3, K4 and K5) are identical for both deformation conditions, those taken at low temperature (Si-K1) are quite different (Weber E and Alexander H *J. de Physique* 1979 **40-C6**, 101).

This finding allowed a very important conclusion: whereas the EPR spectra Si-K1 and Si-K2 are clearly related to the dislocation types present in the sample, Si-K3 to Si-K5 obviously were formed by defects introduced by deformation but left over far away from the dislocation core, interpreted as deformation-induced point defect clusters that are left as a debris from the dislocation motion.

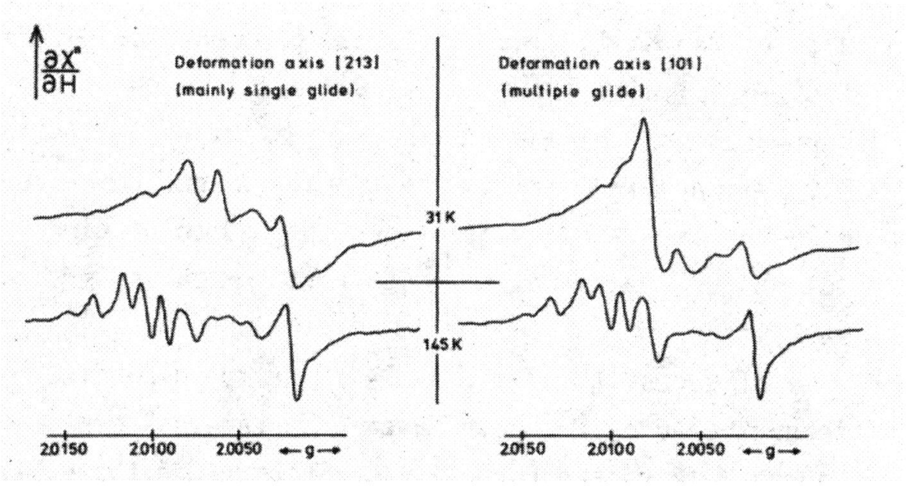

Figure 10. Comparison of EPR spectra of plastically deformed silicon obtained at 31K and 145K for magnetic field along [1 1 -1], using samples prepared with different uniaxial deformation geometries.

All EPR spectra come from dangling-bond type defects, no evidence for one-dimensional conduction along the dislocation line as hypothesized by Shockley was ever found by the Alexander group. A very important finding was obtained by R. Erdmann: Light can transform spectrum Si-K1 into the Si-K2 spectrum of the same glide system (Erdmann R and Alexander H 1979 *Phys. stat. sol. (a)* **55** 251).

Having achieved the differentiation between clearly dislocation-related defects and defects in point-defect clusters produced as debris of dislocation motion a very important issue was the question of the concentration of dislocation-related paramagnetic defects:

The quantitative analysis of the EPR spin density yielded a very low density of well-defined paramagnetic sites, corresponding to one atomic site in a hundred or more along the dislocation line. This allowed to suggest another major conclusion: that dislocation cores in silicon should be predominantly reconstructed, avoiding those rows of dangling bonds depicted in Figs. 3 and 9. Ossip'yan and his coworkers in Chernogolovka as well as Suezawa *et al.* in Sendai, Japan, pointed out that in addition to the Si-K1 to K5 spectra plastic deformation of silicon introduces a broad line called the 'Y-line' (Suezawa M, Sumino K, and Iwaizumi M 1981 *Inst. Phys. Conf. Ser.* **59** 407) that remains upon thermal annealing of the samples in the 600-800°C range when Si-K1 to K5 disappear. However, even when taking into account the spin density of this line the above conclusion holds. Later, theoretical calculations by Bob Jones and his group confirmed this suggestion: perfect dislocations in silicon should show little electrical activity. Lightowlers and Higgs later pointed out that electrical activity (or at least photoluminescence) of dislocations in silicon is dominated by residual metal contamination that is almost unavoidably introduced during high-temperature treatment (Lightowlers E C, V. Higgs V, 1993 *phys. stat. sol. (a)* **138** 665-672) and the latest speculation is that the Si-K1,K2 EPR lines, too, are related to point defect clusters at or near dislocations (Blumenau A T, R. Jones R, Öberg S, *et al.* 2001 *Phys. Rev. Lett.* **87** 187404).

A wide body of literature exists on detailed analysis of the deep energy levels introduced by plastic deformation, to which the Alexander group contributed as well, summarized together with important results on the dislocation structure in a series of excellent reviews by Alexander, see below.

In conclusion, Helmut Alexander made a significant impact in our understanding of dislocations and plasticity in covalent semiconductors. Many modern technological developments profited from the understanding derived by the Alexander group in Göttingen and Cologne, such as the possibility to produce inexpensive solar cells out of silicon with high metal content, and the development of high-efficiency light emitting diodes and lasers in the lattice-mismatched thin film structures of III-nitrides with high dislocation densities. Therefore it is fitting that the International Advisory Board of the EDS conference series decided to dedicate a best-poster award given in his name at each conference to a young scientist in this field.

Figure 11. Helmut Alexander 1993.

More than 120 scientists including the authors of this article were acquainted thoroughly with Helmut Alexander as their academic teacher since the years of their graduate studies. We all appreciated his willingness to discuss actual scientific problems in depth at almost any time and we are grateful for the fruitful cooperation over many years. He will be alive in science by the inheritance of his scientific life – by his numerous publications and by these important reviews:

Alexander H and Haasen P 1968 "Dislocations and Plastic Flow in the Diamond Structure" Ed. Seitz,Turnbull and Ehrenreich *Sol. State Phys.* **22** 27-158

Alexander H 1986; "Dislocations in Covalent Crystals" in: Dislocations in Solids **Vol. 7** Ed. Nabarro F R N (Amsterdam, North-Holland, Elsevier) 114-234

Alexander H and Teichler H 1991 "Dislocations" in: Materials Science and Technology, Ed. Cahn R W, Haasen P and Kramer E J **Vol. 4** "Electronic Structure and Properties of Semiconductors", Ed. Schröter W (Weinheim, VCH) 249-320

Alexander H and Teichler H 2000 "Dislocations" in: Handbook of Semiconductor Technology **Vol. 1**; Ed. Jackson K A and Schröter W (Weinheim, WILEY-VCH) 291-376

Journal of Physics: Conference Series **281** (2011) 012002 doi:10.1088/1742-6596/281/1/012002

Dislocation cores in silicon: new aspects from numerical simulations

Laurent Pizzagalli, Julien Godet, Julien Guénolé and Sandrine Brochard

Department of Physics and Mechanics of Materials, Institut Pprime, CNRS UPR 3346, Université de Poitiers, SP2MI, BP30179, 86962 Futuroscope Chasseneuil, France

E-mail: Laurent.Pizzagalli@univ-poitiers.fr

Abstract. Recent theoretical investigations of the properties of dislocation cores in silicon are reviewed. New results, obtained from numerical simulations for the non-dissociated screw and 60° dislocations, are presented and discussed in relation with experiments.

1. Introduction

The plasticity properties of silicon have been largely investigated since the first studies made sixty years ago. It could therefore be surprising that nowadays there are still several important issues. What appears to be well confirmed is the occurrence of two distinct plastic regimes [1], one at temperatures higher than approximately 900 °C (where Si is ductile), and the other for lower temperatures (where Si is brittle), and the fact that the plasticity is governed by dislocation core properties. This behaviour is not specific to silicon, since other covalent materials with equivalent structures have been shown to exhibit a similar characteristic [2–7]. Open questions depend on the considered plastic regime. Most of the available data refer to the high temperature case, for which the ductility of silicon allows conventional mechanical testing. Although it has first been thought that dislocations in this regime were non-dissociated and located in the shuffle set of $\{111\}$ planes in the diamond cubic structure [8], pioneering microscopy experiments unambiguously showed the occurrence of a weak dissociation. These dislocations are characterised by a $1/2\langle 110\rangle$ Burgers vector with screw and 60° characters, and dissociate into 30° and 90° Shockley partials located in the glide set of $\{111\}$ planes. There have been many recent numerical simulations attempting to determine the core structure of partial dislocations. It is generally agreed that the 30° partial core exhibits a double period reconstruction along the dislocation line. The situation is less clear for the 90° partial since there are two possible structures which are very close in energy [9–11]. There is a general consensus that these dislocations move thanks to the thermally activated formation and migration of kinks along the dislocation line [8], with an activation energy of 2.2 eV for pure silicon. However, the elementary atomistic mechanisms responsible for this displacement are not formally known, because of the large number of possible kink configurations [12], the difficulty to compute or measure formation or migration energies with accuracy [13–19], and the likely role of topological defects along the dislocation line [20,21]. It is therefore difficult to make the connection between kink mechanisms and the mobility law of dislocations [22,23]. Several authors have also hinted the need to take into account the correlated motion of the partials [24,25].

The other plastic regime is observed at temperatures approximately lower than 900 °C. Since silicon is brittle in this temperature range, experiments are usually made using a confining pressure for avoiding cracks, or in a localised area of the sample by nanoindentation [1]. In all cases, high stresses are required for plastically deforming silicon [26]. Overall, there is much less available information regarding this plastic regime, mainly because of experimental limitations. Yet the growing interest for systems with sub-micron or nanometric dimensions calls for a better understanding. In fact, in nanostructures such as nanowires or nanopillars which are often free of defects, plastic deformation has been shown to occur at large stress and low temperature [27–30]. Also, undesirable dislocations could be nucleated in silicon-based devices because of the very high stresses [31–33]. In this low temperature / high stress regime, perfect non-dissociated dislocations with orientations screw, 30° and 41° have been observed depending on the experiments [1]. The required stress for displacing dislocations has been estimated to be about 1.5 GPa [26]. Additional experiments on the physical signatures of these dislocations do not bring much more information [34]. Fortunately, many theoretical studies using numerical simulations have been done during the last decade, leading to a better understanding of dislocation properties in this regime thanks to the investigations of the stability and mobility of the dislocation cores. In this paper, we review and comment these results, which are essentially focused on the screw and 60° dislocation, and we discuss remaining issues.

2. Modelling and numerical simulations

In many materials, the Peierls-Nabarro model allows to determine basic properties of dislocation cores such as dissociation width and Peierls stress, in particular thanks to several recent improvements [35–37]. However, dislocation cores in silicon are often characterised by a complex topology, which may be different than in the bulk due to bond reorientation or reconstruction. As a result, such a model based on an approximate core description is usually not well suited for providing quantitative results [38]. Most of the recent investigations are then made using numerical simulations based on an atomistic description of the dislocations. They provide a better description of the core structure, and almost all related core quantities can be computed. The system can be modelled using semi-empirical inter-atomic potentials, or, at a more fundamental level, with electronic structure approaches such as tight-binding or first principles density functional theory (DFT). Usual potentials for silicon include the Stillinger-Weber [39] and the Tersoff [40] potentials, or the more recent EDIP [41] and MEAM [42]. Nonetheless, not everything in the garden is rosy. In fact, available potentials for silicon are not always accurate, and qualitatively incorrect results may be obtained in specific cases. Since none appears to be superior to the others, the usual remedy is to compare the results from several potentials. Conversely, DFT calculations provide robust and trustable results in the case of silicon. The only drawback is the large need in computational resources, which prevents the investigations of large systems. Several methods have been proposed in order to make an accurate treatment of dislocations in small systems. A descriptive listing is out of the scope of the present paper, but the interested reader will find valuable information in the references [21, 43, 44].

Calculations with atomistic methods usually deal with the determination of the most stable configuration for the dislocations core, or for defects along the dislocation line such as kinks or jogs, yielding structures and energies. The Peierls stress can also be numerically computed by applying a stress on the system encompassing the dislocation. These static calculations can be completed with molecular dynamics simulations which allow to monitor the system evolution at non zero temperature. Because of the severe time limitations for the latter, several methods have also been proposed for computing activated mechanisms using a static formalism. String-based approaches such as the nudged elastic band (NEB) method [45] are increasingly used nowadays. They allow to calculate transition mechanisms and their associated activation energies with accuracy within a reasonable time. Other kinds of transition state search methods have also

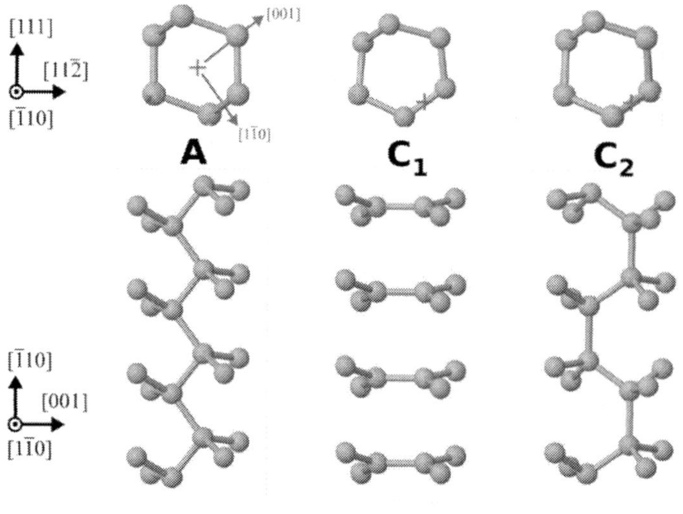

Figure 1. Three possible structures for a screw dislocation in a diamond cubic lattice, viewed along two different directions (top and bottom). Adapted from [48].

been tried [12,46,47].

3. The screw dislocation

We first focus on the non-dissociated screw dislocation. Although this is not always the main observed character in the microstructure, the screw usually plays a special role in the plastic deformation of materials. In 1958, Hornstra proposed two possible core structures for the screw dislocation in the diamond lattice [49]. Both core structures, labelled A and C_1 respectively, are represented in the figure 1. One is obtained when the dislocation line is placed in the centre of one hexagon, as obtained when the diamond structure is viewed along a $\langle 110 \rangle$ direction. The dislocation core therefore belongs to a shuffle $\{111\}$ plane. The second proposed configuration is located in a glide $\{111\}$ plane, and corresponds to a dislocation line passing through the centre of a hexagon small edge. Atomistic calculations using different potentials showed that A is always more stable than C_1 for silicon [50]. A third configuration has been suggested by Koizumi *et al* [51], with a core centred on a hexagon long edge, but it has been revealed to be unstable by first principles calculations [52]. Note that Celli in an early investigation already pointed that this configuration would not be stable [53]. Finally, Wang *et al* recently demonstrated that the energy of the C_1 configuration could be greatly lowered thanks to a double period reconstruction along the dislocation line [54]. This configuration, called C_2 here, is represented in the right side of the figure 1. The activation energy for this reconstruction has been shown to be low [48], which implies that the C_1 core is unlikely to occur. As a consequence, one can say that for a screw dislocation in silicon the two configurations A and C_2 are possible.

The relative energies of A and C_2 configurations have been computed within various methods [1], and it seems that C_2 is always the most stable core structure. Since a large hydrostatic pressure could be present in experiments, the effect of pressure on the stability of both configurations has been recently investigated. Although calculations revealed that the pressure has a noticeable influence on dislocation core energies and tends to favour the A core [55], the C_2 structure remains clearly the lowest energy configuration for a screw dislocation in silicon.

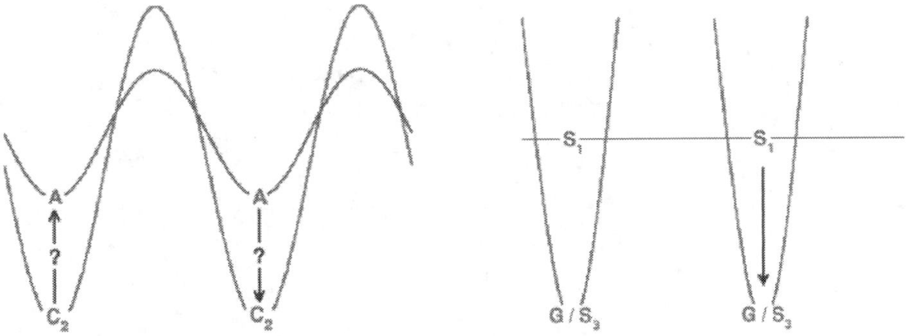

Figure 2. Schematic variation of energies for the core configurations involved in the displacement of the screw (left) and the 60° (right) dislocations.

Preliminary information on the mobility of both configurations can be gained from the value of the Peierls stress. Computed values of 4 GPa [56] and 6 GPa [54] for A and C2 suggest that the former configuration is the more mobile. Nevertheless, it is likely that the displacement of the screw dislocations in the low temperature regime is thermally activated, and activation energies for the formation and migration of kinks should be considered for a finer analysis. These quantities have been computed for the A core using NEB and DFT simulations [57], yielding values of about 1 eV for the formation energy and lower than 50 meV for the migration energy. Using these values, a theoretical comparison of the relative mobilities of the A screw and the partial dislocations showed that the former was more mobile for all stress [58], in contradiction with an early analysis [59]. To our knowledge, kink formation and migration energies are not known for the C_2 core. However, to a first approximation one could consider that the kink pair formation energy is roughly proportional to the Peierls stress magnitude, which would give a kink formation energy in the range 1.4–2.0 eV, thus much larger than values for the A core or than the ones commonly accepted for partials [22]. For the kink migration process, the obvious similarity of the C_2 core with the double period core of the 90° partial dislocation suggests close mechanism and energies for kink migration, i.e. an energy of about 1.2 eV. Therefore the C_2 dislocation core appears to be less mobile than the A core, and even than partial dislocations.

Since A is less stable than C_2 but is likely to be the core involved during the screw dislocation displacement, one can imagine a three-steps mechanism (figure 2). First, C_2 is the core configuration for a screw dislocation at rest, and the combined action of temperature and stress allows the transformation $C_2 \rightarrow A$. Then, the A core moves through the crystal. Finally, the A core is stopped for a time long enough to obtain the reverse transformation $A \rightarrow C_2$. This issue has been recently studied by Guénolé *et al* using NEB calculations and interatomic potentials [48]. It appears that the C_1 configuration is an intermediate step of the $A \rightarrow C_2$ transformation. But more importantly, the energy barrier that must be overcome for the $A \rightarrow C_2$ transformation is in the same range than the energy required for the thermally activated motion of partial dislocations, and it depends only weakly on the applied stress. It is therefore unlikely that such a mechanism could be triggered for the temperatures associated with non-dissociated dislocations.

To summarise, theoretical investigations show that the most stable configuration for a non-dissociated screw in silicon is the C_2 structure. The other possible configuration A is stable too, albeit with a larger core energy. The A screw is mobile in the considered stress / temperature range, which is not the case for the C_2 core. Finally, a $A \rightarrow C_2$ transformation would be obtained

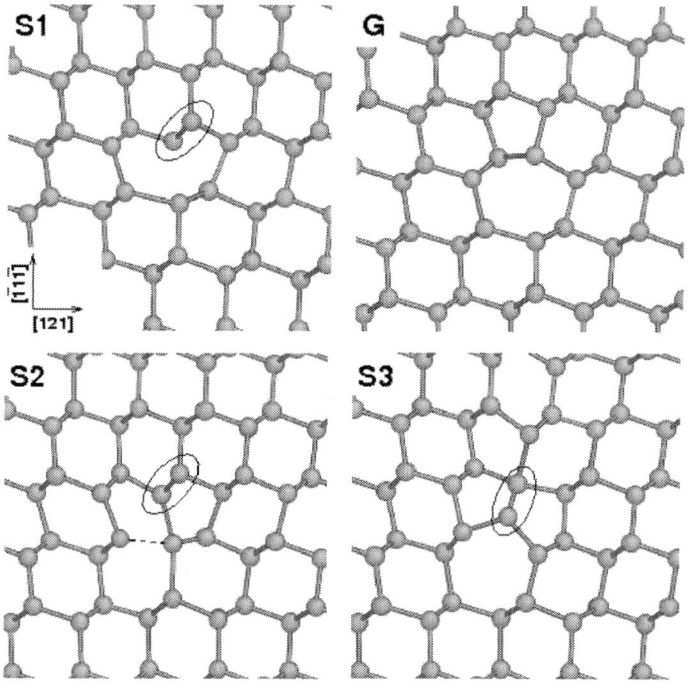

Figure 3. Four possible structures for a 60° dislocation in a diamond cubic lattice [61].

only for high temperatures, for which dissociated dislocations are observed. In a possible scenario, the core A is the likely configuration for the non-dissociated screw, and the C_2 does not play any part in plastic deformation at low temperature. This is supported by theoretical investigations of dislocation nucleation in silicon, which show the formation of the A core of the screw dislocation [60]. However, since C_2 is topologically close to partial dislocation cores, it may occur as an intermediate step during a possible transformation leading to the dissociation of the screw dislocation.

4. The 60° dislocation

The second important dislocation core in the diamond structure corresponds to a 60° orientation, which is dissociated into a 30° and a 90° partial dislocations at high temperature. However, at low temperature, 60° dislocations are not observed in the microstructure of plastically deformed silicon. One explanation is that the non-dissociated 60° dislocation is supposedly relatively mobile, and only dislocation segments for the slowest characters can be clearly observed after deformation. Additional arguments for the existence of non-dissociated 60° dislocation are given by investigations of dislocation nucleation from surfaces [60,62,63], or from interfaces of stressed systems [31,33].

Hornstra suggested one possible configuration for the 60° dislocation [49]. It is obtained when the dislocation core is located in a shuffle {111} plane, and is characterised by an atom with a dangling bond (figure 3). It is labelled S_1 in the following. Another possible core, reported in the Hirth and Lothe textbook [8] and shown in the figure 3, corresponds to a dislocation centred on a glide {111} plane. This G structure has no apparent under-coordination like the S_1 configuration, which would suggest a larger stability. It is worth to mention that this core exhibits a double period reconstruction along the dislocation line [61], a feature initially reported by Blumenau and co-workers in the case of diamond [64].

Until recently, only theoretical investigations using semi-empirical potentials have been performed, and they all confirmed the much larger stability of G over the S_1 core. Still, all studies involving 60° dislocations used the configuration S_1 instead of G for the core structure [65, 66]. Actually, within these potential calculations, S_1 configuration behaves as expected, i.e. it is mobile with a reasonable Peierls stress, whereas the highly stable G configuration appears to be sessile, with a large Peierls stress.

Surprisingly, to our knowledge there have been no first principles calculations of the 60° dislocation in silicon. We have then performed such calculations, which have confirmed the high stability of the G configuration [61]. But the striking result of this work is that the S_1 structure is an unstable configuration, which spontaneously evolves with an energy gain of about 1 eV/b to a so-called S_2 structure, shown in the figure 3. All tests made with different cell sizes, boundary conditions, and calculation methods, confirmed this point. The S_2 geometry is different from the one obtained in the case of diamond, since weak bonds are present into the core. Now, a slight perturbation of the S_2 structure brings the system into another third configuration, unimaginatively called S_3 (figure 3), with an extra 0.8 eV/b energy gain. The final S_3 configuration remains less stable than the G core, but with an energy difference of only 0.6 eV/b. To summarise, it appears that only the configurations G and the new one S_3 are the possible core geometries for the non-dissociated 60° dislocation. Note that this is in apparent contradiction with first-principles investigations of dislocation nucleation from surfaces, showing clearly the formation and propagation of 60° dislocations with the S_1 structure [63].

Additional clues can be obtained by studying the mobility of possible structures. We have performed molecular dynamics calculations using the EDIP and Tersoff potentials, increasing step by step the temperature and the applied stress on both G and S_3 dislocation cores [67]. Both configurations were found to be sessile even at temperatures and stresses larger than 1200 K and 3 GPa, respectively. For higher values, amorphisation occurs in the vicinity of the cores. This aspect is not surprising if one considers that both configurations are very stable, and are characterised by reconstructed bonds (especially S_3). Therefore it appears that none of the available stable configurations for the 60° dislocation are mobile in the low temperature / high stress regime. However, first principles calculations showed that if a suitable stress is applied on an unstable S_1 core, it will move and its geometry is conserved during the displacement [61]. This dazzling result explains why a S_1 dislocation core can be nucleated and displaced in a highly stressed system [63]. A crude calculation of the Peierls stress for the S_1 core yields values ranging from 1 to 2 GPa, which is in relatively good agreement with experimental estimations [26]. The dislocation displacement is obviously not thermally activated and only caused by the stress. Consequently, when the resolved stress on the S_1 dislocation core becomes too low, the $S_1 \rightarrow S_3$ transformation should occur. Since the S_1 configuration only exists in motion, the role of nucleation is essential here. This scenario is sketched in the figure 2.

To summarise these results, first principles calculations suggest that the core of a 60° dislocation in silicon is either stable and sessile, or glissile and exhibiting a transient character. This results highlights the role of nucleation in the low temperature / high stress regime, a feature which could be also dominant in the case of silicon nanostructures. Overall, the occurrence of transient dislocation cores appears to be a new finding, and tends to break the traditional pattern of theoretical investigations, where one looks first for stable core configurations before determining mobility properties. Unfortunately, an experimental confirmation will be difficult to achieve. For instance, the usual post-mortem microscopy investigations are obviously not appropriate for determining a transient dislocation core.

5. Discussion

The previous sections show that our knowledge concerning dislocation cores properties in silicon is not yet complete, despite significant advances during the last decade. In the low temperature / high stress regime, at present we know that there are two possible cores for the screw dislocation, one being more mobile than the other and likely to occur during plastic deformation. Besides, theoretical investigations suggest a glissile transient configuration for the 60° dislocation, which emphasises the role of nucleation in this regime. Little is known for the other dislocations, with 30° and 41° orientations. Geometrical arguments suggest that this original 41° orientation could be obtained by combining short segments of 30° and 60° dislocations [68], but this remains to be confirmed.

Another open question concerns the transition between the high temperature / low stress and the low temperature / high stress regimes. Our last investigations clearly showed that this transition could be triggered by nucleation only [60]. Another possibility is the transformation of non-dissociated dislocations at low temperature into partial dislocations by dissociation [69,70]. To our knowledge, there is no available information on the mechanisms and the associated activation energies for such a process. For the screw dislocation, we already proposed that a process involving first the transformation from the A core to the C_2 core followed by the dissociation of the latter, could occur for temperatures in agreement with the experimental transition.

Finally here we only focused on silicon because of the large amount of available works, and also because it is usually considered as a model material. However, it is not clear whether the results and conclusions reported in this paper can be applied to other covalent systems such as diamond, germanium, or zinc-blende compounds like SiC or GaAs (mentioning only the most common ones), although the occurrence of two plastic regimes seems to be a general feature in these systems. For instance, a tight-binding calculation of the screw calculation in diamond points that the C_1 and C_2 configurations are quasi degenerate in energy, in clear contrast to silicon. Therefore it seems appropriate to conclude this paper by the well-known sentence, saying that additional investigations are required for a better understanding.

References

[1] Rabier J, Pizzagalli L and Demenet J L 2010 Dislocations in silicon at high stress *Dislocation in solids* vol 16 ed Kubin L and Hirth J P (Elsevier) chap 93, p 47
[2] Suzuki T, Nishisako T, Taru T and Yasutomi T 1998 *Philos. Mag. Lett.* **77** 173
[3] Suzuki T, Yasutomi T, Tokuoka T and Yonenaga I 1999 *Philos. Mag. A* **79** 2637
[4] Rabier J, Demenet J L, Denanot M F and Milhet X 2005 *Mater. Sci. Eng. A* **400-401** 97
[5] Demenet J L, Milhet X and Rabier J 2005 *Phys. Stat. Sol. (c)* **2** 1987
[6] Kedjar B, Thilly L, Demenet J L and Rabier J 2010 *Acta Mater.* **58** 1418
[7] Kedjar B, Thilly L, Demenet J L and Rabier J 2010 *Acta Mater.* **58** 1426
[8] Hirth J P and Lothe J 1982 *Theory of dislocations* (New York: Wiley)
[9] Bennetto J, Nunes R W and Vanderbilt D 1997 *Phys. Rev. Lett.* **79** 245
[10] Lehto N and Öberg S 1998 *Phys. Rev. Lett.* **80** 5568
[11] Miranda C R, Nunes R W and Antonelli A 2003 *Phys. Rev. B* **67** 235201
[12] Bulatov V V, Yip S and Argon A S 1995 *Philos. Mag. A* **72** 453
[13] Jones R 1980 *Philos. Mag. B* **42** 213
[14] Bigger J R K, McInnes D A, Sutton A P, Payne M C, Stich I, King-Smith R D, Bird D M and Clarke L J 1992 *Phys. Rev. Lett.* **69** 2224
[15] Huang Y M, Spence J C H and Sankey O F 1995 *Phys. Rev. Lett.* **74** 3392
[16] Öberg S, Sitch P K, Jones R and Heggie M I 1995 *Phys. Rev. B* **51** 13138
[17] Kolar H R, Spence J C H and Alexander H 1996 *Phys. Rev. Lett.* **77** 4031
[18] Valladares A, White J A and Sutton A P 1998 *Phys. Rev. Lett.* **81** 4903
[19] Beckman S P and Chrzan D C 2007 *Phys. Rev. B* **76** 144110
[20] Heggie M and Jones R 1983 *Philos. Mag. B* **48** 365
[21] Lehto N and Heggie M I 1999 Modelling of dislocations in c-si *Properties of Crystalline Silicon* (*EMIS Datareviews* no 20) ed Hull R (London: INSPEC) p 357

[22] Bulatov V V, Justo J F, Cai W, Yip S, Argon A S, Lenosky T, de Koning M and de la Rubia T D 2001 *Philos. Mag. A* **81** 1257
[23] Scarle S, Ewels C, Heggie M and Martsinovich N 2004 *Phys. Rev. B* **69** 075209
[24] Cai W, Bulatov V, Chang J, Li J and Yip S 2004 Dislocation core effects on mobility *Dislocations in Solids* vol 12 ed Nabarro F and Hirth J (North Holland) chap 64, p 1
[25] Vanderschaeve G and Caillard D 2007 *Mater. Sci. Eng. A* **462** 418
[26] Rabier J, Renault P, Eyidi D, Demenet J, Chen J, Couvy H and Wang L 2007 *Phys. Stat. Sol. (c)* **4** 3110
[27] Kizuka T, Takatani Y, Asaka K and Yoshizaki R 2005 *Phys. Rev. B* **72** 035333
[28] Justo J, Menezes R D and Assali L 2007 *Phys. Rev. B* **75** 045303
[29] Han X, Zheng K, Zhang Y, Zhang X, Zhang Z and Wang Z L 2007 *Adv. Mater.* **19** 2112
[30] Kang K and Cai W 2007 *Philos. Mag.* **87** 2169
[31] Marzegalli A, Montalenti F and Miglio L 2005 *Appl. Phys. Lett.* **86** 041912
[32] Izumi S, Miyake T, Sakai S and Ohta H 2005 *Mater. Sci. Eng. A* **395** 62
[33] Izumi S, Ohta H, Takahashi C, Suzuki T and Saka H 2010 *Philos. Mag. Lett.* **90** 707
[34] Pizzini S, Binetti S, Donne A L, Marzegalli A and Rabier J 2006 *Appl. Phys. Lett.* **88** 211910
[35] Lu G, Bulatov V V and Kioussis N 2003 *Philos. Mag.* **83** 3539
[36] Lu G 2005 The peierls-nabarro model of dislocations: a venerable theory and its current development *Handbook of Materials Modeling. Volume I: Methods and Models* 2.20 ed Yip S (Springer) p 1
[37] Denoual C 2007 *Comput. Methods Appl. Mech. Engrg.* **196** 1915
[38] Schoeck G 2005 *Mater. Sci. Eng. A* **400-4001** 7
[39] Stillinger F H and Weber T A 1985 *Phys. Rev. B* **31** 5262
[40] Tersoff J 1989 *Phys. Rev. B* **39** 5566
[41] Justo J F, Bazant M Z, Kaxiras E, Bulatov V V and Yip S 1998 *Phys. Rev. B* **58** 2539
[42] Baskes M, Nelson J and Wright A 1989 *Phys. Rev. B* **40** 6085
[43] Cai W 2001 *Atomistic and Mesoscale Modeling of Dislocation Mobility* Ph.D. thesis MIT
[44] Bulatov V and Cai W 2006 *Computer simulations of dislocations* Oxford Series on Materials Modelling (New York: Oxford University Press)
[45] Jónsson H, Mills G and Jacobsen K W 1998 Nudged elastic band method for finding minimum energy paths of transitions *Classical and Quantum Dynamics in Condensed Phase Simulations* ed Berne B J, Ciccotti G and Coker D F (World Scientific) chap 16, p 385
[46] Ewels C P, Leoni S, Heggie M I, Jemmer P, Hernndez E, Jones R and Briddon P R 2000 *Phys. Rev. Lett.* **84** 690
[47] Pedersen A, Pizzagalli L and Jónsson H 2009 *J. Phys.: Condens. Matter* **21** 084210 a. Pedersen, L. Pizzagalli, and H. Jonssón, J. Phys.: Condens. Matter, in print (2009)
[48] Guénolé J, Godet J and Pizzagalli L 2010 *Modelling Simul. Mater. Sci. Eng.* **18** 065001
[49] Hornstra J 1958 *J. Phys. Chem. Solids* **5** 129
[50] Pizzagalli L, Beauchamp P and Rabier J 2002 *J. Phys.: Condens. Matter* **14** 12681
[51] Koizumi H, Kamimura Y and Suzuki T 2000 *Philos. Mag. A* **80** 609
[52] Pizzagalli L, Beauchamp P and Rabier J 2003 *Philos. Mag. A* **83** 1191
[53] Celli V 1961 *J. Phys. Chem. Solids* **19** 100
[54] Wang C Z, Li J, Ho K M and Yip S 2006 *Appl. Phys. Lett.* **89** 051910
[55] Pizzagalli L, Demenet J L and Rabier J 2009 *Phys. Rev. B* **79** 045203
[56] Pizzagalli L and Beauchamp P 2004 *Philos. Mag. Lett.* **84** 729
[57] Pizzagalli L, Pedersen A, Arnaldsson A, Jónsson H and Beauchamp P 2008 *Phys. Rev. B* **77** 064106
[58] Pizzagalli L and Beauchamp P 2008 *Philos. Mag. Lett.* **88** 421
[59] Duesbery M S and Joós B 1996 *Philos. Mag. Lett.* **74** 253
[60] Godet J, Hirel P, Brochard S and Pizzagalli L 2009 *J. Appl. Phys.* **105** 026104
[61] Pizzagalli L, Godet J and Brochard S 2009 *Phys. Rev. Lett.* **103** 065505
[62] Godet J, Pizzagalli L, Brochard S and Beauchamp P 2004 *Phys. Rev. B* **70** 54109
[63] Godet J, Brochard S, Pizzagalli L, Beauchamp P and Soler J M 2006 *Phys. Rev. B* **73** 092105
[64] Blumenau A T, Heggie M I, Fall C J, Jones R and Frauenheim T 2002 *Phys. Rev. B* **65** 205205
[65] Li C, Meng Q, Zhong K and Wang C 2008 *Phys. Rev. B* **77** 045211
[66] Li C and Meng Q 2009 *Superlatt. Microstruc.* **45** 1
[67] Pizzagalli L (unpublished)
[68] Rabier J 2007 *Phys. Stat. Sol. (a)* **204** 2248
[69] Rabier J and Demenet J L 2005 *Phys. Stat. Sol. (a)* **5** 944
[70] Saka H, Yamamoto K, Arai S and Kuroda K 2006 *Philos. Mag.* **86** 4841

Journal of Physics: Conference Series **281** (2011) 012003 doi:10.1088/1742-6596/281/1/012003

Effect of electronic doping on the plasticity of homoepitaxial 4H-SiC single crystals

J L Demenet[1], M Amer[1], A Mussi[1,2] and J Rabier[1]

[1]Institut PPRIMME, Département de Physique et Mécanique des Matériaux - UPR 3346 CNRS, Université de Poitiers, ENSMA - SP2MI, BP 30179, F-86962 Chasseneuil Futuroscope Cedex, France
[2]now at Unité Matériaux et Transformations, UMR 8207 CNRS - Université Lille 1, F-59655 Villeneuve d'Ascq Cedex, France

jean.luc.demenet@univ-poitiers.fr

Abstract. Instrumented micro-indentations have been performed at room temperature on 4H-SiC homoepitaxial single crystals with different doping. For these experiments, it appears that the pop-in event occurs at the same level of load for intrinsic and n-type SiC and at a higher load level for p-type. Correlation of the pop-in event with dislocation nucleation indicates that doping acts on dislocation nucleation and that p-type doping plays a hardening role on the plastic behaviour of 4H-SiC. This result is confirmed by the conventional measurement of imprint size using scanning electron microscopy.

1. Introduction

It is well known that impurities play an important role in plasticity of materials. They usually act as obstacles to dislocation propagation through direct interaction with dislocations. This metallurgical effect requires generally a noticeable concentration of impurity to be evidenced. However, impurities play also a remarkable role in the plasticity of semiconductors at concentrations much lower than 1%. In this case, doping does not act through direct interaction with dislocations but through electronic effects. Such effects have been put in evidence in several semiconductors. Silicon is the one in which the effect has been observed and the most studied, by direct measurements of dislocation velocity [1], as well as by measurements of the lower yield stress [2] or extension of rosette arms after indentation [3]. As a consequence, electronic doping yields a modification of the Brittle-to-Ductile Transition Temperature (T_{BDT}). It is expected that due to the wide band-gap of SiC (3.2 eV for the 4H-polytype) doping could significantly modify T_{BDT} in this material. However this has not been evidenced yet, although silicon carbide has both important mechanical and electronic applications.

In this paper, some results of microindentations performed at room temperature on intrinsic, n-type and p-type 4H-SiC homoepitaxial layers are presented. Results are compared with the literature for other semiconductors and discussed in the framework of the plasticity of these materials.

2. Experimental

Materials used in this study have been supplied by LETI (CEA-Grenoble, France) and were grown by the Lely method. A 10μm thick layer of different doping intrinsic (N_i ~ $6{\cdot}10^{13}$ cm^{-3}), n-type (N_n ~ $8{\cdot}10^{17}$ N atoms cm^{-3}) and p-type (N_p ~ $4{\cdot}10^{17}$ Al atoms cm^{-3}) was homo-epitaxied at 1550°C on a

4H-SiC substrate oriented 7° off the basal plane. Microindentations using a Fischerscope H100CXYp equipped with a Vickers indenter were performed at room temperature on the (0001) Si-surface up to maximum loads of 50 mN and 300 mN. The maximum load was reached within 20 seconds and dwelt for 5 s. An array of fifty indentations was made on each sample (figure 1). The indentation diagonals were oriented to be parallel to <1120> and <1100> directions (figure 2) determined from back reflection X-ray Laue patterns taken prior to the tests. Scanning Electron microscopy (SEM) was used to measure the size of imprints.

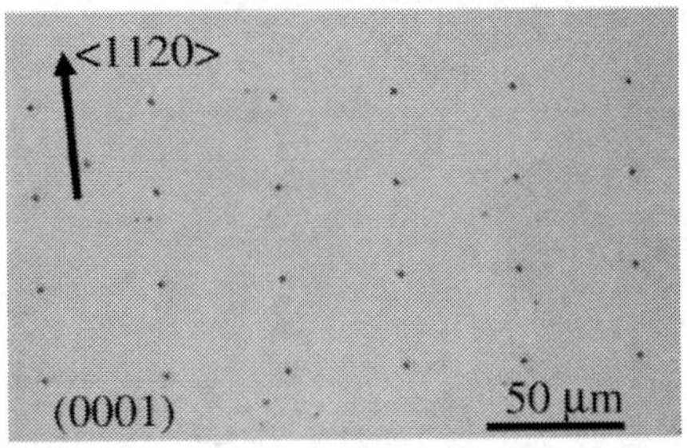

Figure 1. Optical microscopy of a typical array of indentations made on a homoepitaxial layer under a load of 300mN.

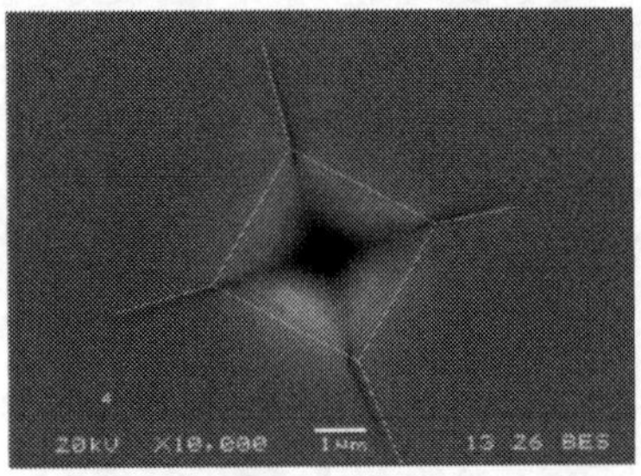

Figure 2. Micrograph of an imprint observed by SEM. Diagonals are along <1100> and <1120> directions.

3. Results

Typical curves under a maximum load of 50 mN obtained on different doped layers are shown in figure 3. The maximum penetration depth is lower than 0.3 µm, which is much smaller than the thickness of the layer. Under these conditions, the influence of the substrate can be neglected and the recorded curves are representative of the behaviour of the material of the layers. A well-defined pop-in event, i.e. a sudden penetration of the indenter at constant load, is observed on each curve. The pop-in is usually associated to the onset of plasticity: the first irreversible events occur at this point, and are usually assumed to correspond to the nucleation of dislocations. The level of pop-in occurs at the same load for intrinsic and n-type material. For the p-type layer, pop-in occurs at a higher load. A statistical analysis has been made from the fifty imprints on each sample. This leads to pop-in levels of 4.7 ± 0.6 mN, 4.2 ± 0.5 mN and 7.9 ± 1.6 mN, for n-type, intrinsic and p-type material, respectively. Unfortunately, from these values it is not possible to get the actual stresses responsible of pop-in events because the actual stress field beneath a Vickers indenter is difficult to sort out. Nevertheless, the link between the pop-in event and the dislocation nucleation indicates that nucleation of dislocations is more difficult when material is p-type doped as compared to intrinsic or n-type doped material.

From the unloading part of the load-penetration curve, a hardness value can be extracted providing that the response of the material during unloading is elastic and the contact surface beneath indenter remains constant. The systematic observation of cracks at the ends of imprints (see figure 2) does not allow such a hypothesis to be valid. This is why we used the classical way to get the hardness value, i.e. by measuring *post-mortem* by SEM the size of the diagonals of the imprints, as it is usually done for bulk materials under high loads. From measurements of imprint size made under a load of 300 mN on intrinsic and p-type (no measurements have been performed on n-type) we obtained Vickers hardness values of 32.6 ± 0.2 GPa and 33.8 ± 1.2 GPa, respectively. Whatever the quite poor accuracy of the hardness value for p-type material, these results are in accordance with the pop-in levels. It can be claimed that p-type doping has a hardening effect on the plasticity of 4H-SiC.

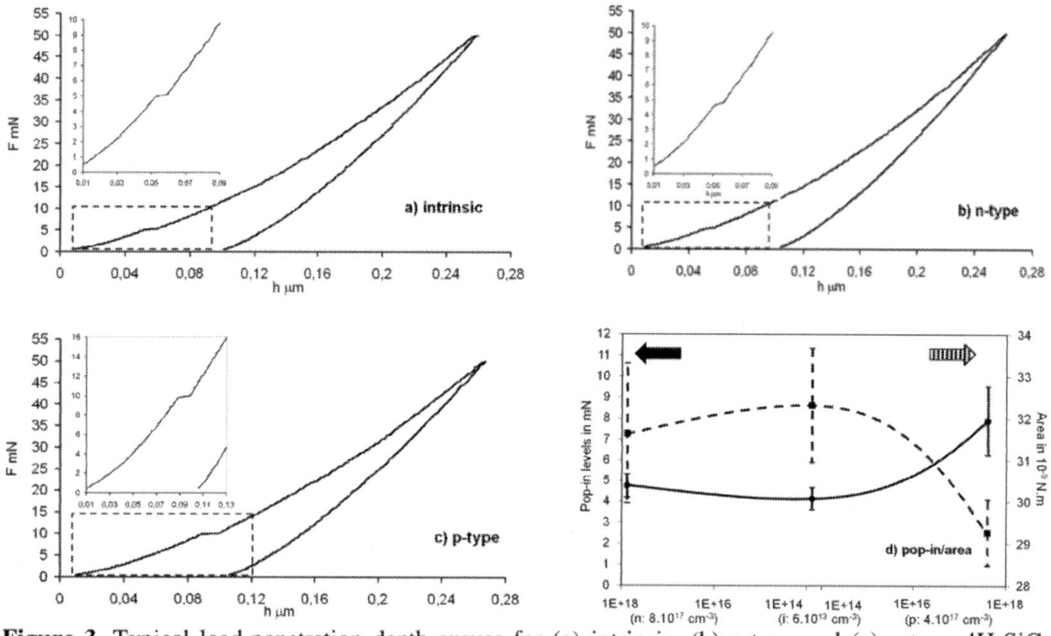

Figure 3. Typical load-penetration depth curves for (a) intrinsic, (b) n-type and (c) p-type 4H-SiC. Pop-in is observed for the three materials. A magnification of the pop-in region is given on each curve. (d) Pop-in level and area of load-penetration curve as a function of doping.

A value of the dissipated energy during the loading-unloading experiment is given by the area of the load-penetration curve. It is assumed that this value gives a good description of the resistance of materials to indentation [4]. The results are presented in figure 3 (d) from experiments made up to the maximum load of 300 mN. The energy dissipated by plasticity and/or fracture is found to be lower in p-type. Qualitatively, this is in accordance with Vickers hardness values and pop-in levels.

4. Discussion

Many studies by micro- and nano-indentation have been conducted in semiconductors, some of them being dedicated to the influence of doping. In the case of silicon, analysis of nanoindentation tests suffers from phase transformations occurring beneath the indenter. Multiple pop-ins and pop-outs have been reported; the cause of pop-ins is not clear at this time, but the pop-outs are assumed to be due to phase transformations [5, 6]. The phase transformations disappear and the sample deforms by classical dislocation plasticity when the sample becomes thin and comparable in width to the size of the nanoindenter, however the influence of doping has not yet been checked in these conditions [7]. Following other authors, pop-in events could be the signature of subsurface cracking or sudden dislocation bursts [8]. Le Bourhis and Patriarche have analyzed the structure of nanoindentations in heavily n- and p-doped GaAs. This material does not suffer phase transformation, and the authors observed that the mechanical response of both types of samples is relatively similar with a level of pop-in which is not affected by doping [9]. In contrast, the indentation rosette structure is different. Perfect dislocations with long screw segments emanate from imprints in n-doped specimens, as well as partial dislocations on both arms of rosettes, whereas p-type specimens show no partial dislocations. From these results, the different behaviour as a function of doping could be related to a difference in mobility of dislocations. This is in accordance with the huge effect of doping observed by plastic deformation under pressure in bulk GaAs [10].

Contrary to silicon and similar to GaAs, it has been shown that the plasticity response of SiC to nanoindentation occurs by dislocation plasticity rather than by densification or phase transformation.

The first indirect evidence of such a behaviour was the observation of the pop-in phenomena in 6H-SiC [11]; TEM observations have demonstrated that dislocations are emitted from imprints and lie in the basal plane, but no precise analysis of dislocations (perfect or partials) was performed [12]. By compression of bulk material [13] and classical indention tests [14] as a function of temperature, a high asymmetry in mobility of partial dislocations has been shown, which is important in that the temperature is low. Unfortunately, all experiments reported above were conducted on as-grown crystals in which doping level was not controlled during the elaboration process; however all silicon carbide samples used had a high content of N atoms. From the results presented in this paper, it appears that doping modifies the mechanical response of 4H-SiC. P-type doping leads to a higher level of pop-in as compared to intrinsic material whereas n-type has no significant effect. Considering that pop-in is related to nucleation of dislocations, our results show that this phenomenon is strongly affected by p-doping, making it more difficult. As a matter of fact it is the first time that the level of pop-in can be related to the doping effect in a semiconductor allowing it to be shown that nucleation of dislocations can be affected by electronic doping as was shown for dislocation velocity. However in this context a point to be questioned is the actual site of dislocation nucleation. The basal plane is the easy glide plane for hexagonal silicon carbide, but from a geometrical point of view, dislocations should be nucleated on other planes to accommodate deformation due to an imprint on the basal plane. A few papers in the past have mentioned the activity of prismatic planes [15, 16]. By indentation on a prismatic plane, Mussi *et al.* [16] have observed short segments of dislocations on other prismatic planes. Those dislocations are perfect dislocations but they cross-slip in the basal plane where they dissociate, indicating the possibility of dislocations to be nucleated on such planes. TEM observations on homoepitaxial layers of different doping have to be done to check dislocation nucleation sites. Moreover, nanoindentation using a spherical indenter would give access to the actual stress corresponding to pop-in events and to dislocation nucleation stress. This work is in progress.

Acknowledgments
This work is partly supported by the "Matinex" joined CEA-CNRS program., The authors would like to thank J-L Santailler (Laboratoire d'Electrotechnique et de Technologie de l'Information, CEA-Grenoble, France) for the supply of high quality 4H-SiC single crystals and T Rouxel (LARMAUR-Rennes, France) for access to the Fischerscope nano-indenter.

References

[1] George A and Champier G 1979 *Phys. Stat. Sol. (a)* **53** 529
[2] Demenet J L, Desoyer J C, Rabier J and Veyssière P 1984 *Scripta Metall.* **18** 41
[3] Roberts S G, Pirouz P and Hirsh P B 1983 *J. de Physique* **44** C4-75
[4] Berriche R 1995 *Scripta Metall et Mater.* **32** 617
[5] Juliano T, Domnich V and Gogotsi Y 2004 *J. Mater. Res.* **19(10)** 3099
[6] Nagy P M, Horvath P, Peto G and Kalman E 2009 *Mater. Sci. Forum* **604-605** 29
[7] Ge D, Minor A M, Stach E A and Morris Jr J W 2006 *Phil. Mag.* **86** 4069
[8] Chang Li and Zhang L C 2009 *Key Engineering Materials* **389-390** 453; 2009 *Mater. Sci. & Eng. A* **506** 125
[9] Le Bourhis E, Patriarche G 2008 *Acta Mater.* **56** 1417
[10] Boivin P, Rabier J, Garem H 1990 *Phil. Mag. A* **61** 619
[11] Page T F, Oliver W C and McHargue C J 1992 *J. Mater. Res.* **7** 450
[12] Page T F, Riester L and Hainsworth S V 1998 *Mater. Res. Soc. Symp. Proc.* **522** 113
[13] Fujita S, Maeda K and Hyodo S 1986 *J. Mater. Sci. Let.* **5** 450
[14] Ning X and Pirouz P 1997 *J. Amer. Ceram. Soc.* **80** 1645
[15] Maeda K, Suzuki K, Fujita S, Ichihara M and Hyodo S 1988 *Phil. Mag. A* **57** 573
[16] Mussi A, Demenet J L and Rabier J 2006 *Phil. Mag. Lett.* **86** 561

Journal of Physics: Conference Series **281** (2011) 012004 doi:10.1088/1742-6596/281/1/012004

Characterisation of Dislocations in ZnO by X –Ray Imaging

N Burle[1], F Baudin[2], A Jonchere[2], P Gergaud[2] and G Feuillet[2]
[1] IM2NP- CNRS 6242, UPCAM, Campus Scientifique St Jérome, case 262, 13397 Marseille cedex 20 France
[2] CEA - LETI, MINATEC, 17 rue des Martyrs, 38054 Grenoble cedex 9, France

e-mail : nelly.burle@im2np.fr ; guy.feuillet@cea.fr

Abstract. Transmission X-Ray topography (X-ray imaging) was used to characterize the density and type of the native dislocations in ZnO substrates. C-oriented ZnO bulk materials from different origins were compared, either high purity Chemical Vapor Transport (CVT) grown or commercial, hydrothermally grown wafers. Elongated dislocations lying within the substrate were found in hydrothermal substrates. In CVT crystals, the density of dislocations was found to be too high for X-ray determination of their Burgers vector. However, in hydrothermal crystals, the density of dislocations were found to be in the range $< 10^4$ / cm² and extend mainly within the substrate. Whereas complete Burgers vector identification is not achieved, two kinds of dislocations have been evidenced : grown-in, swirled dislocations ("high" temperature) and linear, gliding dislocations, most probably developped during the cooling steps in (0 1 -1 0) glide planes.

1. Introduction

Quantum well ZnO/ZnMgO heterostructures are promising materials for the realization of electro-luminescent diodes. As for most optical applications, dislocations must be avoided in the emissive materials, where they would act as recombination centers and limit the optical efficencies of the devices. A strong advantage associated to ZnO is related to the availability of ZnO substrates allowing for homoepitaxial processes. In this case, dislocations could be avoided in the active regions of the devices. But this will be true only if the substrate dislocation density is low enough to prevent replication of threading dislocations in the layers and if no misfit dislocation is developped at the substrate- buffer layer interface due to misfit strain relaxation.

As a substrate for growing emissive structures, the ZnO wafers should combine high chemical purity and good crystalline quality. In this respect, we have compared substrates from two growth techniques : hydrothermal growth which allows to get low dislocation density substrates, with rather high impurity concentrations and Chemical Vapor Transport (CVT) [1] which leads to purer material but with higher dislocation densities.

This paper presents our first results about dislocation imaging in various substrates. High purity CVT wafers were found to exhibit too high dislocation densities to allow their crystallographic characteristics to be determined. However, some hydrothermally grown commercial wafers are of a good enough structural quality to image individual dislocations. The determination of their characteristics could help understanding their origin to prevent them from occurring in CVT crystals, opening the way to the realization of pure and perfectly crystalline wafers for further applications.

2. Experimental results

2.1. Some ZnO Crystallographic data

ZnO exhibits a hexagonal (wurtzite) structure; the lattice constants are a=0.325 nm and c=0.5027 nm; each of the 2 sites (0, 0, 0) and ($^1/_3$, $^2/_3$, $^1/_2$) is occupied by 2 atoms: Zn (0, 0, 0) and O (0, 0, $^3/_8$) (figure 1).

Positions of the compact crystallographic planes expected for dislocations glide are also reported. One must remember that the rule to transform Miller indices between 3 and 4 indices is not the same for planes and for directions : "a" axis is quoted [100] or 1/3[2 -1 -1 0] ; "m" axis (perpendicular to "a" axis) is quoted [1 2 0] or [0 1 -1 0] ; more easily, prismatic (100) plane is also quoted (1 0 -1 0) and prismatic (2 -1 0) plane is also called (2 -1 -1 0).

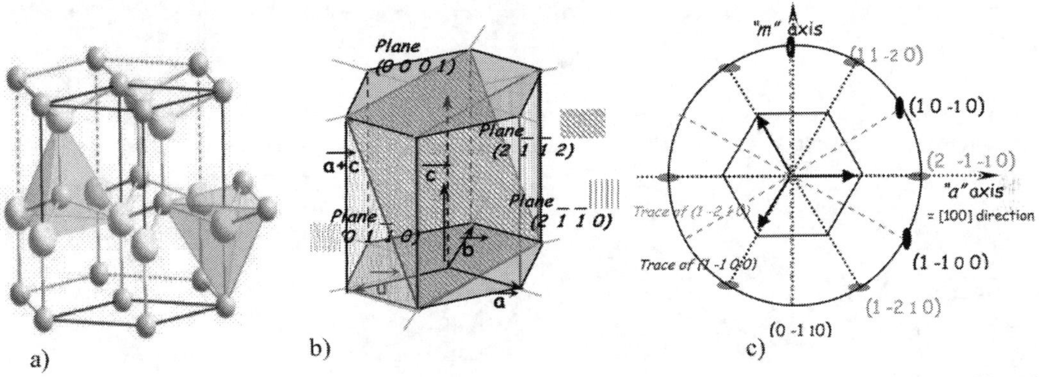

Figure 1. a) Atoms in the hexagonal cell (Largest atoms are Zn, smallest are O); b) Positions of basal plane (0 0 1), prismatics planes (0 1 -1 0) and (2 -1 -1 0), pyramidal plane (2 -1 -1 2) in the hexagon; c) Stereographic projection on (001) : poles of the prismatic planes are plotted by respect with basal vectors ***a***, ***b***, and ***u*** = -(***a***+***b***) and with usual indications of "a" axis and "m" axis.

2.2. Dislocation densities depending on the type of substrate

Samples from various providers have been imaged: CVT wafers grown at LETI, and commercial hydrothermal wafers from CrysTec GmbH (referred to as CT) and Tokyo Denpa (TD). All samples are (001) oriented, their sizes are about 10 x 10 mm^2 and thicknesses between 0.5 mm and 1 mm.

X-ray images have been recorded on Ilford L4 nuclear plates with λ K_{α} $_{Ag}$ provided by a rotating anode Rigaku generator, using Lang (transmission) laboratory setting.

In CVT and CT samples dislocation densities are too high to be measured on X-Ray topograms, which means they are over than 10^4, in accordance with the values estimated from etch pit density counting (EPD): 10^5cm/cm^3 in MOVPE and between 10^4 and 10^5cm/cm^3 in CT samples; in TD samples, the density is found under 10^3cm/cm^3 both by EPD and X-ray imaging (figure 2).

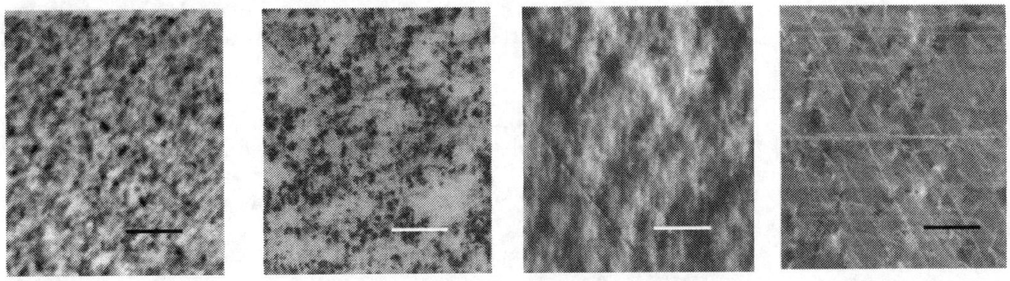

Figure 2. a) CVT sample 77-A, g = (1 0 -1 0); b) CVT sample 79-A, g = (-2 1 1 0); c) sample CT4, g = (-2 1 1 0); d) sample TD 2, g = (-2 1 1 0); scale 0.5 mm.

2.3. Geometrical characteristics of dislocations in TD wafers.

Two kinds of dislocations can be observed in the lowest density samples (Tokyo Denpa, $[D] < 10^3$ cm/cm^3): few very linear lines and some swirled dislocations. Nevertheless the general directions of the dislocation lines are along the crystallographic directions **a**, **b** and **u** = **-(a+b)** (figure 3).

The directions correspond to the trace of the {10-10} type prismatic planes of the hexagonal cell in the c plane. This is in good accordance with previous X-ray or TEM observations [2, 3] where such slip lines in the basal plane are frequently observed even if Burgers vectors are not always determined.

The images obtained with the 3 possible {2 -1 -1 0} diffracting conditions lead to intense, or weak contrasts for the various dislocations. It is obvious from large scale images that the swirled lines follow roughly the same three directions, but seem to be pinned at the emerging points or at several points along the lines; they look like typical "high temperature" dislocation structures observed after cooling in Semiconductors. [4]

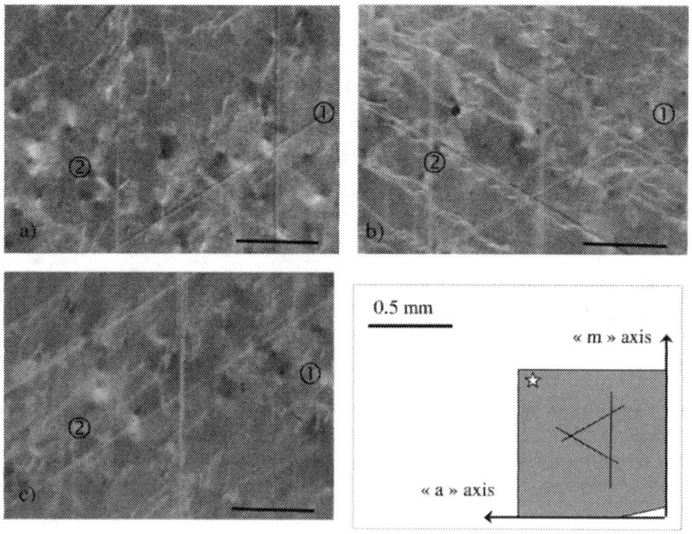

Figure 3 :

Transmission X-Ray topographies of the same area in ZnO sample TD-2

a) g = (2 -1 -1 0)

b) g = (-1 2 -1 0)

c) g = (1 1 -2 0)

$\lambda = K_{\alpha\ Ag} = 0.056$ nm

3. Discussion and conclusion

Observations of dislocation contrasts on several images usually allow one to determine the Burgers vector **B** using the extinction rule "**g**.**B** =0", if two extinctions are obtained; if the glide systems are well known, one single extinction could give a good indication of the vector **B**. Unfortunately, this rule is not always powerful enough since it is only a simplified criterion. It is the case in our samples, as it can be noticed in figure 3.

For example, the straight line ① is visible with g = (2 -1 -1 0) (figure 3a) or (-1 2 -1 0) (figure 3b) and vanishes with g = (1 1 -2 0) (figure 3c); the line direction is u = -(a+b) wich is perpendicular to (11 -2 0) plane, so it would be a pure edge dislocation. Whereas edge growth dislocations have already been observed in some cases [5], such straight lines in semiconducting materials evocate "low" temperature glide along Peierls directions, and would be screw or 60° dislocations, so the expected vector would be one of the basal plane, **a**=[1 0 0] = 1/3[2 -1 -1 0] or **b** or **u**.

The swirled line ② vanishes with **g** = (2 -1 -1 0) or **g** = (1 1 -2 0), which should lead to a Burgers vector **c** perpendicular to the basal plane; if true, this dislocation would also vanish with **g** = (-1 2 -1 0) (figure 3b) where it is in contrast.

These two examples illustrate clearly that in these samples the simplified rule "**g.B**=0" cannot lead to complete identification of Burgers vectors by X-ray imaging. It can be due at least partially, to the diffracting conditions, as the samples are quite thick so Bormann effect is not negligible. As it is well known, the complete extinction rule is "**g.R**=0" where **R** is the displacement vector around the dislocation, and the expression of **g.R** contains several terms including not only **g** and **B** but also the line vector **t**. This line vector can be determined in the case of straight dislocations so the contrast of dislocations could be calculated and compared with images. In the case of the swirled dislocations this would be more problematic.

Moreover, the glide planes of these non linear dislocations are not obvious: primary glide and cross-slip have already been reported in ZnO and similar wurtzite materials depending on the deformation method [6, 7] and the dislocations in ZnO are reported to be easily mobile, with low activation energy [8], and not dissociated, which would favour cross-slip events [9]. Easy slip occurs in basal or prismatic (01 -1 0) planes but (2 -1 -1 0) prismatic planes and various pyramidal planes have been observed in wurtzite semiconducting materials [2, 5, 6, 8, 9].

Optimization of the experimental conditions, as thinning samples to get Bragg diffracting conditions instead of Bormann, will render the images easier to be interpreted; comparison with contrast calculations would lead to an easier determination of Burgers vectors.

It has been shown that the dislocation density in ZnO wafers as well as their characteristics is highly dependant on the growth method. Correlation with EPD and cathodoluminescence measurements is in progress. Controlled, low temperature plastic deformation will also be undertaken in low dislocation density samples to allow glide mechanisms to be precise.

References

[1] Santailler J C G, Audouin C and Chichignoud G 2010 *J. Crystal Growth* to be published
[2] Zheng Y, Boulliard J C, Demaille D, Bernard Y and Petroff J F 2005 *J. Crystal Growth* **274** 156
[3] Yoshino K, Yoneta M and Yonenaga I 2008 *J Mater Sci: Mater Electron.* **19** 199
[4] Pichaud B, Burle-Durbec N, Minari F and Duseaux M 1985 *J. Crystal Growth* **71** 648
[5] Dhanaraj G, Dudley M, Bliss D, Callahan M and Harris M 2006 *J. Crystal Growth* **297** 74
[6] Bradby J E, Kucheyev S O, Williams J S, Jagadish C, Swain M V, Munroe P and Phillips M R 2002 *Appl. Phys. Lett.* **80** (24) 4537
[7] Kucheyev S O, Bradby J E, Williams J S, Jagadish C and Swain M V 2002 *Appl. Phys. Lett.* **80** 956
[8] Yonenaga I, Ohno Y, Taishi T and Tokumoto Y 2009 *Physica B* **404** 4999
[9] Suzuki K, Ichihara M, Takeuchi S, Nagakawa K and Maeda K 1984 *Philos. Mag. A* **49** 451

Investigation of non-decorated glide dislocations by infra-red light scattering tomography

V Monier[1], O Kononchuk[2], L Capello[2] and B Pichaud[1,3]
[1]Aix-Marseille Université, IM2NP CNRS UMR 6242, Faculté des Sciences et Techniques de Saint-Jérôme, Case 142, F-13397 Marseille Cedex, France
[2] R&D; SOITEC S.A., Parc Technologique des Fontaines, F-38190, Bernin, France

E-mail: bernard.pichaud@im2np.fr

Abstract. The laser scattering tomography technique enables the observation of non-decorated dislocations in Si crystals. In polarization and tomography measurements, changes in the dislocation scattering intensity are observed. A study of the light scattering by a dislocation was then developed and described in this paper. Based on theoretical results and experimental observations, slip system of a non-decorated mixed dislocation can be totally determined.

1. Introduction

Laser Scattering Tomography (LST) has been used to study oxygen precipitates in Si wafers for many years [1,2]. This technique is able to measure densities in the range of 10^5–10^{10}cm^{-3}, and sizes down to 20 nm [3]. Such defects fulfill the approximations of Rayleigh scattering (size < wave length) for which the scattered intensity can be easily calculated. The ability to image extended defects by LST has also been demonstrated in various materials (quartz, sapphire, III-V, Si) [1,4-7]. All works on Si has addressed dislocations introduced during long high temperature annealing, which necessarily leads to decoration of dislocations by oxygen atoms and precipitates. No non-decorated dislocations in Si were investigated so far. Dislocations cannot be considered as point-like scatterers and their scattering must be evaluated from the Maxwell equations as done by Moriya [4].

In this paper, we focus on the study of oxygen-free mixed dislocations in Si by the IR-LST technique. After discussing the theory of light scattering by dislocations, we demonstrate the possibility to image non-decorated glide dislocations. Using polarized light, we were able to identify the dislocation slip plane and Burgers vector using scattering simulation developed from [4]. Finally we show that it is possible to discriminate between non-decorated and decorated glide dislocations by LST.

2. Simulation of 90° light scattering from a dislocation in silicon

For non-decorated dislocations, the LST scattering contrast arises from photo-elastic effect [4,6], that is changes in the refractive index originated by the stress field surrounding the dislocation. The electric field scattered by the dislocation is [8]:

$$\vec{E}_s(R) = \frac{\exp(ikR)}{R}\left(\vec{s}\times\vec{s}\times\left[F(\vec{g})\right]\vec{E}_I\right) \quad (1)$$

[3] To whom any correspondence should be addressed.

with R the position where the field is measured, k the wave number, $\vec{g}$ the scattering vector defined as $\vec{g} \equiv \vec{s} - \vec{s}_0$, $\vec{s}_0$ being the direction of the incoming light, $\vec{s}$ the scattering direction and $\vec{E}_I$ the incident electric field in the case of incoming plane wave. The scattering factor $\vec{F}$ is defined as the Fourier transform of the tensor of the inverse dielectric constant variation $\Delta\varepsilon^{-1}(\vec{r})$:

$$\vec{F}(\vec{g}) = FT\left(\Delta\varepsilon^{-1}(\vec{r})\right) \quad \text{where} \quad \Delta\varepsilon^{-1}(\vec{r}) = p \times \vec{S} \quad \text{and thus} \quad \vec{F}(\vec{g}) = p \times \hat{\vec{S}}$$

with $\vec{S}$ the strain tensor of the dislocation, $\hat{\vec{S}}$ the Fourier transform of $\vec{S}$ and p the strain-optic 4 rank tensor with in the case of silicon reduces to only 3 non-zero coefficients [9] (p_{11}, p_{12} and p_{44}).

If one considers dislocations in silicon, provided they are not produced at too high temperature, it is well known that their lines are parallel to the three <110> Peierls valleys in the glide plane. So we will consider a hexagonal dislocation loop in the {111} planes.

The strain field tensor of such a loop is complex and has been simplified as follows:

i.) the loop is considered as the sum of linear straight segments

ii.) the strain field tensor is evaluated in a position close enough to the dislocation segment, so the latter can be considered to have infinite length.

The calculation of the strain field tensor is developed in the dislocation coordinate system (x,y,z). This coordinate system is built to have the dislocation line along the x-axis. The x-y plane corresponds then to the {111} glide plane and the Burgers vector of the dislocation forms an angle φ with the x-axis. By means of φ, the mixed dislocation can be introduced in the calculation. In this coordinate system, the Fourier transform of the strain field components [10] can be expressed as:

$$\hat{S}_{xx} = 0 \qquad \hat{S}_{yy} = \frac{q}{ik\left(g_y^2 + g_z^2\right)^{1/2}} \frac{b \sin\varphi}{4(1-\nu)} \left(\cos 3B + (3 - 4\nu)\cos B\right)$$

$$\hat{S}_{zz} = \frac{q}{ik\left(g_y^2 + g_z^2\right)^{1/2}} \frac{b \sin\varphi}{4(1-\nu)} \left(\cos 3B + (4\nu - 1)\cos B\right) \qquad \hat{S}_{xy} = \frac{q}{ik\left(g_y^2 + g_z^2\right)^{1/2}} \frac{b\cos\varphi}{2} \cos B$$

$$\hat{S}_{yz} = \frac{q}{ik\left(g_y^2 + g_z^2\right)^{1/2}} \frac{b \sin\varphi}{4(1-\nu)} \left(\sin 3B + \sin B\right) \qquad \hat{S}_{xz} = \frac{q}{ik\left(g_y^2 + g_z^2\right)^{1/2}} \frac{b\cos\varphi}{2} \sin B$$

Here, B is defined by $\tan B = -{g_y}/{g_z}$ and $q = \frac{2\sin(kg_x x_0)}{kg_x}$

where b is the modulus of Burgers vector, g_y and g_z are the scattering vector components, ν the Poisson's ratio. Here $2x_0$ is the length of the dislocation line visible by the objective lens taking into account the numerical aperture of the objective.

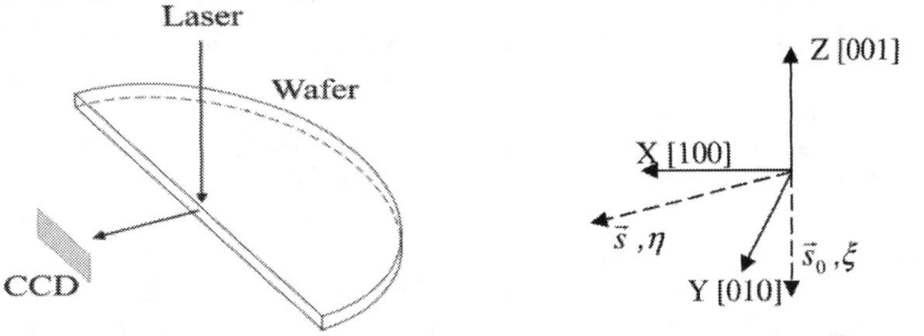

Figure 1. Scheme of the LST geometry and of the relative orientations of coordinate systems attached to Si and observation set-up.

The calculation of the scattering factor in the observation coordinate system (ξ,η,ζ) can then be obtained by an additional matrix transformation. If M is the transformation matrix between the Si coordinate system (X,Y,Z) and the dislocation coordinate system (x,y,z), and P is the matrix transformation between the observation coordinate system (ξ,η,ζ) and the Si coordinate system (X,Y,Z), then one can obtain as a final result:

$$\bar{\bar{S}}_{(X,Y,Z)} = M^{-1} \times \bar{\bar{S}}_{(x,y,z)} \times M \qquad \vec{F}(\vec{g})_{(X,Y,Z)} = p \times \bar{\bar{S}}_{(X,Y,Z)} \qquad \vec{F}(\vec{g})_{(\xi,\eta,\zeta)} = P^{-1} \times \vec{F}(\vec{g})_{(X,Y,Z)} \times P$$

Perfect 90° scattering corresponds to the case where incoming and scattering lights have fixed directions and are strictly perpendicular to each other. Hence, Eq. (1) can be written as:

$$\vec{E}_s(R) = \frac{\exp(ikR)}{R}\begin{pmatrix} 0 & F_{12} & F_{32} \\ 0 & 0 & 0 \\ 0 & F_{13} & F_{33} \end{pmatrix}\vec{E}_I \quad \text{and} \quad \vec{E}_I = \begin{pmatrix} 0 \\ E_\eta \\ E_\zeta \end{pmatrix}$$

F_{ij} being the coefficients of the scattering factor matrix. Introducing the polarization angle β of the incoming light as the projection of incident electric field in the observation coordinate system gives:

$$\vec{E}_s(R) = \frac{\exp(ikR)}{R}\begin{pmatrix} 0 & F_{12}\sin\beta & F_{32}\cos\beta \\ 0 & 0 & 0 \\ 0 & F_{13}\sin\beta & F_{33}\cos\beta \end{pmatrix} E_0 \quad \text{with} \quad \tan\beta = -\frac{E_\eta}{E_\zeta}$$

with E_0 the amplitude of the incident electric field. Finally, the scattered intensity can be written as:

$$I_R = |E_R|^2 / |E_I|^2 = \frac{1}{R^2}\left((F_{12}\sin\beta + F_{32}\cos\beta)^2 + (F_{13}\sin\beta + F_{33}\cos\beta)^2\right)E_0$$

Based on these calculations, the theoretical scattering intensity of a non-decorated glide dislocation in Si can be determined as a function of the polarization angle β, the Burgers vector being introduced via the angle φ and the dislocation line orientation in the coordinate system (x,y,z).

In a real optical system, the 90° scattering equations remain valid but the incoming and scattered lights spread out within a cone which numerical aperture depends on the system. In order to simulate more completely the measured signal, this experimental feature must be included in the calculations. In our specific case, the light is considered to be collected within a cone of 11° surrounding the scattering vector $\vec{g}$ parallel to the [111] direction.

Applying the method detailed above, we implemented in Matlab software a numerical simulation of the intensity scattered by a dislocation described by φ and its coordinates (x,y,z) as the function of the light polarization angle β. This enables a direct comparison between the scattered intensity measured by our LST tool and the calculated intensity.

3. Observation and simulation of scattered intensity from non-decorated dislocation

3.1 Experimental procedure

300 mm wafers free of COP (crystal originated particle, which are in fact voids due to vacancy agglomeration) and without detectable LST defects were annealed at 1150°C for 2 min in H_2 in a clean CVD epitaxial reactor. This setup allows a large temperature gradient between the wafer centre and edge to be introduced intentionally generating glide dislocations. Wafers were analyzed by LST 300A from Semilab using the standard configuration, and the tomography options developed for individual defect measurement and polarization options. Details were described previously [11,12].

In the wafers, we analyzed several dislocation half loops exhibiting various contrasts. Sizes of these half loops vary from 50 to 500 μm and their density can reach 10^7 cm^{-3}. We will show here, as an example, how the full characterization of such a half loop can be obtained from its scattered intensity.

3.2 Glide plane determination

Depending on the orientation of the glide plane, the dislocation half loop can result in different images on the CCD camera (110) plane. In our experimental set-up, if the dislocation lies in (111) or (11-1) planes, as shown in figure 2a, the half loop image is a facetted or curved line. Such an image is shown in figure 2b, where the longest segment (e.g. *b* in figure 2b) is almost parallel to [-110]. Only a short oblique straight line would be visible if the dislocation lies in the (-111) or (1-11) planes. Oblique lines were observed, but here we will concentrate on the first configuration: the half loop.

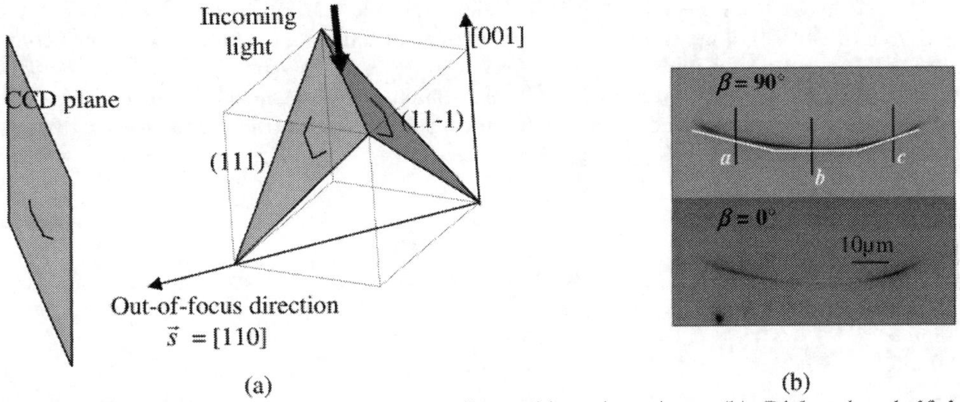

Figure 2. (a) Experimental set-up with crystallographic orientations. (b) Dislocation half loop observed in (110) plane, with polarization direction at 90° and 0°. Three intensity profiles were measured on the left side ***a***, on the centre ***b*** and on the right side ***c*** of the dislocation.

Using the tomography option we investigated the relative position of the maximum of the scattered intensity (MSI) along the [110] direction for the different branches of the image (***a***, ***b*** and ***c*** in figure 2b). It is clear from figure 2a that if the MSI of branch ***b*** appears after the MSI of branches ***a*** and ***c*** by moving the beam in [110] direction, the half loop belongs to the (111) plane and it belongs to (11-1) if the MSI of branch ***b*** arises before that of branches ***a*** and ***c***. The output of such measurement is shown in figure 3, where the relative position of the MSIs corresponds to a dislocation belonging to the (11-1) plane.

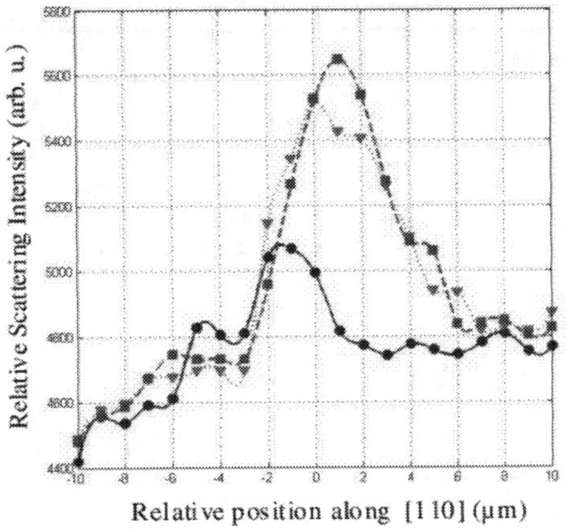

Figure 3. Scattered intensity of ***a*** (full triangles), ***b*** (full circles) and ***c*** (full squares) dislocation segments (along lines indicated in figure 2b) that is as a function of the relative position of the light beam along the [110] direction.

3.3 Burgers vector determination

As shown in figure 2b, the light scattered by dislocations depends on the polarization of incoming light. The outcome of the simulation performed with the method described in section 2 for the three possible Burgers vectors of the theoretical loop in (11-1) plane is shown in figures 4a-c. Figure 4d displays the observed scattered intensities. For the ***a***, ***b***, and ***c*** segments, the MSI positions correspond to different polarization angles: for segment ***b*** the MSI is close to β=85° , for segment ***a*** it is at 50°, while the intensity from segment ***c*** shows basically no polarization dependence. This behavior depends on the orientation of the Burgers vector of the dislocation and its slip plane. Thus, the Burgers vector orientation can be determined by comparing the experimental dependence of scattered light on polarization (figure 4d) of the beam with the simulations of figure 4a-c. Our measured data correspond well to the Burgers vector direction [101]. Accordingly, segment ***a*** being oriented along [011] and segment ***b*** along [-110] both correspond to 60° dislocations while segment ***c*** oriented along [101] is of screw character. The values used in calculation for elasto-optical coefficients are: p_{11} = - 0.094, p_{12} = 0.017, and p_{44} = - 0.051 [13, 14].

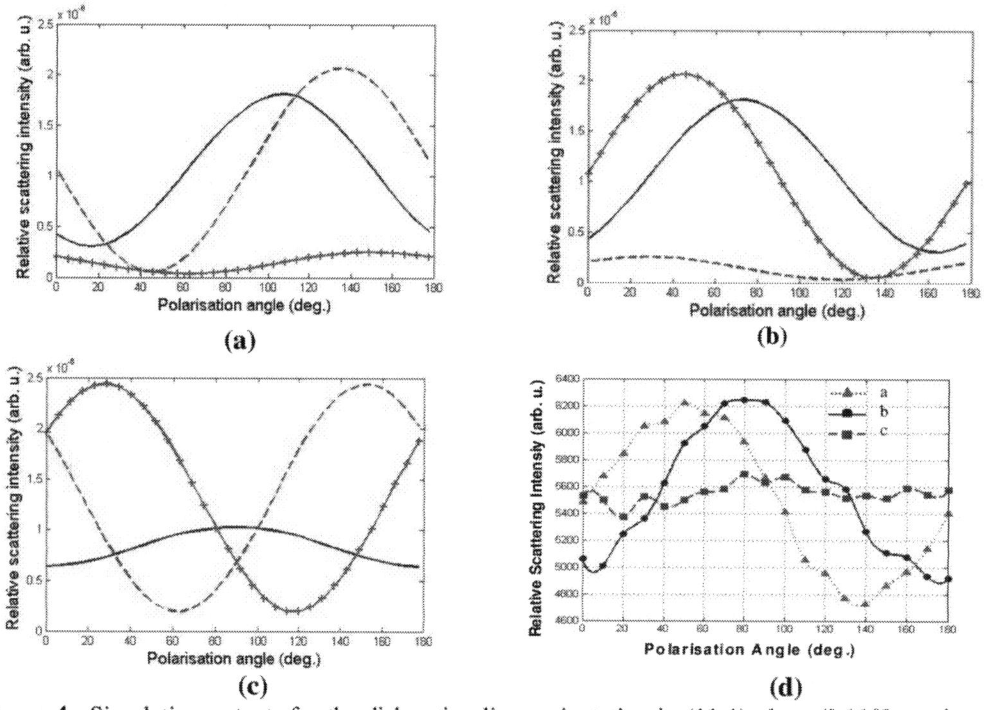

Figure 4. Simulation outputs for the dislocation lines orientation in (11-1) plane ([-110] continuous line, doted line [101] and continuous line with crosses [011]). (a) Burgers vector parallel to [011], (b) Burgers vector parallel to [101], and (c) parallel to [-110]. (d) Scattered intensity of dislocation segments ***a***, ***b*** and ***c*** (along lines indicated in figure 2b) as a function of β.

4. Discrimination between non-decorated and decorated dislocation

Additional information about the morphology of the light scattering defects is provided by the analysis of the scattered intensity dependence on the polarization of the illuminating beam [8]. For small particles, where the Rayleigh approximation works, the scattering distribution of linear polarized light is similar to a dipole radiation pattern. Scattered intensity is proportional to $\sin^2(\beta)$. It means that small defects are visible only when incoming light has a polarization component perpendicular to the observation direction. In practice, the objective lens collects scattered light within an angle of about 11°, which leads to non-zero defect intensity when polarization of the incoming light is parallel to the objective optical axis. Examples of scattered intensity dependence on polarization of the illuminating beam are shown in figure 5a for oxygen precipitates of different sizes.

The wafer in which non-decorated dislocations were developed was annealed at 950°C for 2 hours to obtain oxygen precipitation along the dislocations. Figure 5b shows the scattering behavior of the dislocation half loop analyzed in the previous section, after this anneal. The polarization dependence of scattered intensity has completely changed and is now very close to the one observed for small oxygen precipitates. This experimental result corroborates our hypothesis that the generation and fast movement of dislocations during the first anneal resulted in non-decorated loops. During the second anneal the dislocation did not move and became decorated by oxygen atoms or small precipitates.

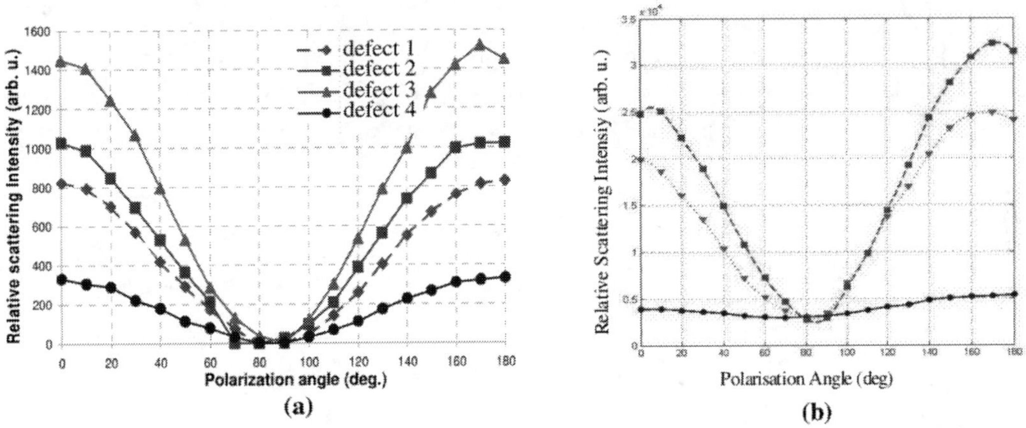

Figure 5. (a) Scattered intensity dependence on polarization of the beam for oxygen precipitates of different sizes with $d << d_{eff}$, and (b) the same profiles as in figure 4d after annealing at 950°C for 2 hrs.

5. Conclusion

Non-decorated glide dislocations in Si have been used as model defects to develop a characterization method based on the tomography and polarization options of LST 300A. The glide plane of the dislocation is determined by tomography, while the polarization dependence of the scattered intensity allows the Burgers vector to be identified. A simulation code calculating the incoming light polarization dependence of scattering by a dislocation has been developed and is able to identify the Burgers vector by comparison with the experiment. LST enables to distinguish between non-decorated and oxygen-decorated dislocations in Si using the polarization dependence of the scattered light. The reported results confirm as well that non-decorated glide dislocations can be created in silicon crystals.

References

[1] Moriya K and Ogawa T 1978 *J. Cryst. Growth* **44**, 53
[2] Moriya K and Ogawa T 1983 *Jpn. J. Appl. Phys.* **22**, L207
[3] Nango N, Ogawa T and Irisawa T 2005 *Jpn. J. of Appl. Phys.* **44**, 5898
[4] Moriya K 1991 *Phil. Mag. B*, **64**, 425
[5] Nango N, Iida S and Ogawa T 1999 *J. Appl. Phys.* **86**, 6000
[6] Moriya K and Ogawa T 1980 *Phil. Mag. A* **41**, 191
[7] Kissinger G, Vanhellemont J, Claeys C and Richter H 1996 *J. Cryst. Growth* **158**, 191
[8] Landau L and Lifshitz E and Piteavskii L P 1984 *Electrodynamics of Continuous Media* vol.8 (Oxford: Pergamon) p415
[9] Authier A 2003 Intern. tables for crystallography,D Physical properties of crystal, 175
[10] Hirth J and Lothe J 1968 *Theory of dislocations*, (New-York: Wiley)
[11] Kononchuk O, Monier V, Capello L and Pichaud B 2009 *Phys. Stat. Sol. c* **6** 1935

[12] Monier V, Capello L, Kononchuk O and Pichaud B 2009 *ECS Trans.* **25**, 91
[13] Hounsome L S, Jones R, Shaw M J and Briddon P R 2006 *Phys. Stat. Sol. (a)* **203** 12
[14] Biegelsen D K 1974 *Phys. Rev. Lett.* **32**, 1196

International Conference on Extended Defects in Semiconductors (EDS 2010) IOP Publishing
Journal of Physics: Conference Series **281** (2011) 012006 doi:10.1088/1742-6596/281/1/012006

Effects of speckle in Makyoh topography for the studies of extended defects

Ferenc Riesz

Hungarian Academy of Sciences, Research Institute for Technical Physics and Materials Science, P.O. Box 49, H–1525 Budapest, Hungary

E-mail: riesz@mfa.kfki.hu

Abstract. The effects of the speckle phenomenon on Makyoh-topography imaging is analysed. The speckle manifests itself as a quasiperiodic pattern in the Makyoh image which is easily mistaken for an image of a periodic surface morphology. The characteristic signature of speckles in Makyoh images is determined, thus allowing for its recognition. The effects of speckle on the Makyoh imaging is analysed as a function of surface roughness, illumination coherency and wavelength, light source size and instrumental parameters. It is shown that speckle effects are present even for incoherent illumination because of the coherency enhancement due to the limited source size. The findings are illustrated with experimental images of various semiconductor samples.

1. Introduction

Makyoh (or magic-mirror) topography (MT) is a simple but powerful optical method for the study of macroscopic surface defects in semiconductor wafers and structures [1-3]. The principle of the method is the following: any topographical irregularity of the sample surface acts as a (local) convex or concave mirror therefore a parallel or spherical light beam impinging on the surface produces an image on a screen at certain distance that reflects the sample morphology because of the light/bright contrast due to focusing or defocusing effects (figure 1). Practically realised set-ups usually contain additional optics and a CCD camera as an imaging tool. The advantages of the technique, as compared to other methods, are the following: simple, inexpensive, robust set-up, relative immunity to vibrations, real-time operation and high dynamic range.

MT emerged as a tool applied mainly for the qualitative assessment of the polishing quality of semiconductor wafers: saw and lapping marks, dimples and other defects are clearly visualised by MT [1-2]. However, it has been demonstrated that the method is sensitive enough to detect the surface fingerprints of extended defects as well, such as stresses of grain boundaries in GaSb crystals [4], swirl defects in *p*-Si [5] and bunches of dislocation slip lines in GaAs wafers [6]. Figure 2 shows an example of a concentric rippled morphology of a *p*-Si wafer associated with a swirl defect [5].

Makyoh is usually considered a geometrical optical technique. Coherency effects are observed only as diffraction fringes at sample edges and around isolated surface defects. However, another coherency effect, the speckle phenomenon, has been overlooked by researchers dealing with MT. The speckle pattern is caused by self-interference of a reflected beam because of the sample's surface roughness. In this paper, the effect of the speckle phenomenon on MT imaging is analysed. First, the background on Makyoh topography and the speckle phenomenon is summarised, then, the relevance to Makyoh topography is analysed. The findings are illustrated with experimental data with emphasis on problems related to the imaging of surface fingerprints of (point-like and periodic) extended defects. Since Makyoh topography is a chiefly qualitative technique, we restrict ourselves to a semi-quantitative analysis stressing the basic characteristics.

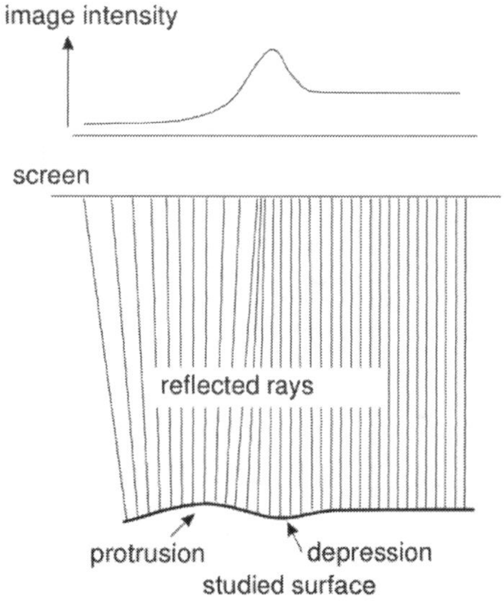

Figure 1. Schematic representation of the image formation in Makyoh topography.

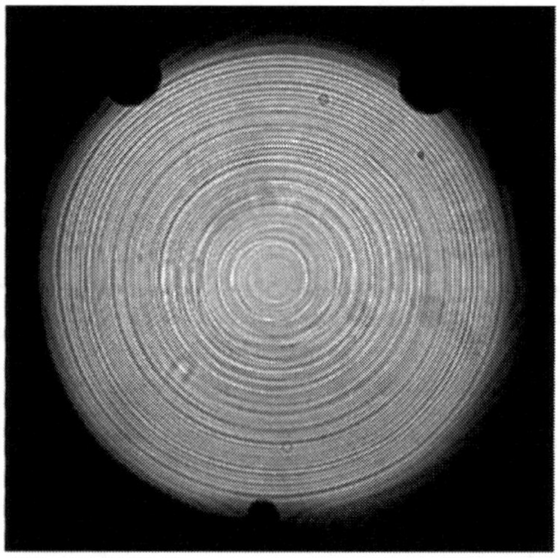

Figure 2. The Makyoh-topography image of a 3-inch *p*-Si wafer showing swirl defect.

2. Background on Makyoh topography

A geometrical optical model of image formation has been presented in [7]. The two basic imaging equations are:

$$\mathbf{f}(\mathbf{r}) = \mathbf{r} - 2L\,\mathrm{grad}\,h(\mathbf{r}) \tag{1}$$

and

$$I(\mathbf{f}) = \left|(1-2LC_{\min})(1-2LC_{\max})\right|^{-1}. \tag{2}$$

Equation (1) gives the position $\mathbf{f}(\mathbf{r})$ of the image of a sample point $\mathbf{r}$, while equation (2) gives the image point's intensity $I(\mathbf{f})$ normalised to that of a flat surface, for a given surface height profile $h(\mathbf{r})$.

We assume unity surface reflectivity. The key parameter of Makyoh imaging *is* L the screen-to-sample distance. C_{min} and C_{max} are the two principal curvatures at point **r**, that is, the minimum and maximum of the second derivatives of $h(\mathbf{r})$. Increasing L results in increasing contrast, thus higher sensitivity, but also the distortion of the imaging topology.

3. Basics of speckle phenomena

If a fully or partially coherent light beam is reflected from a rough surface, a speckle pattern is formed on a distant screen or sensor by the self-interference of the reflected beam [8-9] (this is the so-called objective speckle). The speckle is a quasi-periodic pattern with two main characteristic parameters: the speckle size and the contrast. The average speckle size s is given by

$$s \approx \lambda L/d, \tag{3}$$

where d is the diameter of the area from which the reflected rays can interfere at a given screen point, and λ is the illumination wavelength [8]. The intensity distribution follows an exponential decay. The associated speckle contrast C_S is defined as the rms intensity distribution divided by the average intensity. The speckle contrast is determined by the roughness [9]. Its exact nature depends on the actual surface statistics in a complex way even for the simplest models. Generally, the contrast monotonically increases with roughness and saturates at unity at the rms roughness of about $\lambda/4$. Note that the contrast does not depend on L for a given surface and wavelength.

4. Implications for Makyoh topography

The Makyoh arrangement is naturally prone to the speckle phenomenon. The lasers and LEDs used for illumination have coherence lengths of several meters and ~10 µm, respectively. As a result speckle effects are expected to occur. Although the coherence lengths of incandescent lamps (often used in Makyoh set-ups) is a few microns at most, an increase of the spatial coherency occurs because of the small source size [9], giving rise to speckle. This effect will be discussed later.

To account for the observed speckle contrast, we have to consider the surface roughness. Today's high-quality Si wafers have rms roughness in the sub-nm range, compound semiconductors have that in nm's, giving contrast of a few percent at most which is just noticeable. However, chemically etched semiconductor surfaces and deposited metal layers may have higher roughness values up to tens of nm, yielding C_S of maximum ~20%.

Since speckle is a quasiperiodic pattern, it can be easily mistaken for the Makyoh image of a periodic surface relief. However, two features make them distinguishable: (1) the speckle size linearly increases with L, while the feature size of a Makyoh image is independent of L (for a closely flat sample), and (2) the Makyoh contrast increases with increasing L, while the speckle contrast has no L dependence. Varying L thus facilitates recognizing the speckle. Independent knowledge on the surface roughness may also help in separating the speckle and the Makyoh contrasts.

How does an existing speckle pattern influence the "readibility" of the relief-related (Makyoh) contrast is a complex question. For surfaces that are subjects of MT studies, a dominant specular reflection exists and the speckle will only modulate the reflected beam's intensity. Consequently, the image contrast is still determined chiefly by the surface curvature variations. Speckle will pose a limit to the observation of the image only if the speckle pattern and the Makyoh patterns are matched in shape and periodicity. To quantify this effect, we consider the Makyoh contrast. It has been shown in [10] that the Makyoh contrast C_M of a sine-like surface having a peak-to-peak amplitude A and periodicity k is

$$C_M = 8p\pi^2 AL/k^2, \tag{4}$$

where p is the dimensionality of the surface (one for a surface of translational symmetry and two for a general surface). The contrast in equation (4) is defined in the same way as for the speckle, however, the intensity distribution of the Makyoh image is not exponential. Nevertheless, as a crude limiting approximation, it can be stated that the speckle contrast C_S sets an ultimate limit on the Makyoh sensitivity, that is, $C_M > C_S$ must be fulfilled to observe the Makyoh image. In the more probable case, where the Makyoh image has a characteristic spatial pattern different from the speckle pattern (isolated point or line defects, paralel lines etc.), the statistically uniform spotty speckle appears as a noise on the Makyoh image, usually degrading visibility, but still allowing recognition. This problem leads to the complex field of pattern recognition in gray-level images [11]. For a clearer observation of the Makyoh pattern, small speckle sizes are preferred.

5. Experimental results and discussion

Experiments were carried out using a home-made Makyoh-topography set-up [12] that employs a pigtailed LED for illumination (λ = 670 nm, estimated coherency length is about 10 μm, source diameter $d_S \approx 50$ μm). Our set-up uses a converging lens placed close to the sample for collimating the illumination and providing a "magnifyer" for the CCD camera used for imaging (focal length $f = 500$ mm). The set-up is optically equivalent to the original "parallel-beam and screen" arrangement [7]. Negative L values are also possible with this kind of set-up.

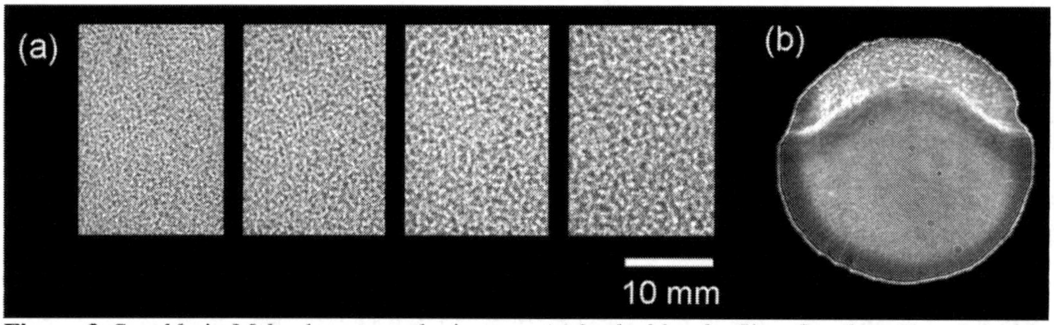

Figure 3. Speckle in Makyoh-topography images: (a) back side of a Si wafer, (b) a Si wafer with two regions of differing roughness. The L values for part (a) are –65 mm, –200 mm, –330 mm and –440 mm from left to right; for (b), $L = -200$ mm

Figure 3 shows some examples of Makyoh images exhibiting speckles: part (a) shows the image of the half-polished backside of a Si wafer, exhibiting a textbook example of a fully developed speckle, and (b) shows a Si wafer having two regions with different amounts of speckle contrast. The images have been corrected for the magnification of the optical system. The effect of changing L is clearly visible as a change of the speckle size. The results with samples having various degree of roughness indicated that equation (3) approximately holds but the resulting d is significantly lower than the sample size: we found roughly $s \approx |L|/1000$, leading to $d \approx 1.5$ mm. This confirms the reduction of the surface's coherency area due to finite source size. A semi-quantitative account of this effect can be outlined as follows. Since the speckle is formed by interference of the rays reflected by the sample surface, we have to consider the transverse coherence area at the sample plane (similarly to the Young double-slit experiment).The theoretical size (diameter) d_C of the coherency area is given by [9]

$$d_C = \lambda/\omega, \tag{5}$$

where ω is the angle subtended by the light source at a surface point. In our case,

$$\omega = d_S/f. \tag{6}$$

(For an arrangement without collimator lens, the source-to-sample distance replaces f.) For our set-up, we get $d_C = 6.7$ mm. Simple considerations dictate that d_C should be equal to d; our evaluation of $d \approx 1.5$ mm is contradictory. However, this discrepancy may be due to the simplified formulae used and to the fact that the actual intensity distribution was not taken into account.

Assuming that d_C equals d, combining equations (3), (5) and (6), we obtain for the speckle size

$$s = d_S L/f, \tag{7}$$

that is, the speckle size is independent of the wavelength. The factor L/f is an instrumental geometrical parameter depending only on the actual arrangement of the set-up which usually takes a value on the order of unity.

Finally, figure 4 illustrates the effect of the speckle pattern on real Makyoh images. The sample is a semiconductor wafer showing saw marks, a global deformation and some surface defects. The images taken at different L settings show the dominant presence of a fully developed speckle pattern, but the Makyoh contrast is also well visible.

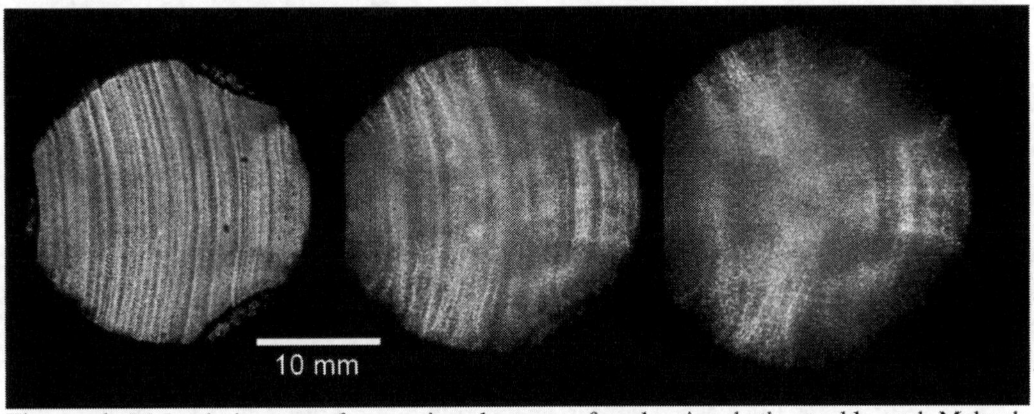

Figure 4. Makyoh images of a semiconductor wafer showing both speckle and Makyoh contrast. L values from left to right: ≈10 mm, 55 mm and 110 mm.

6. Conclusions

Coherency effects, including speckle, are present in Makyoh topography even for incoherent illumination because of the small light source size. Speckle effects as quasiperiodic patterns are encountered with contrast determined by the surface roughness and average size on the order of light source size. The speckle phenomenon appears as a disturbing effect on Makyoh topography.

Acknowledgements

This work was supported, in part, by the (Hungarian) National Scientific Research Fund (OTKA) through Grant K 68534 and by the Agency for Research Fund Management and Research Exploitation (KPI) under contract GVOP-3.2.1.-2004-04-0337/3.0.

References

[1] Blaustein P and Hahn S 1989 *Solid State Technol.* **32** 27
[2] Pei Z J, Fisher G R, Bhagavat M and Kassir S 2005 *Int. J. Machine Tools Manufact.* **45** 1140
[3] Riesz F 2004 *Proc. SPIE* **5458** 86
[4] Minchev G M, Pramatarova L D, Podor B and Szabo J 1994 *Cryst. Res. Technol.* **29** 1131
[5] Riesz F, Pap A E, Ádám M and Lukács I E 2008 *Thin Solid Films* **516** 8087
[6] Möck P 2001 *J. Crystal Growth* **224** 11
[7] Riesz F 2000 *J. Phys. D: Appl. Phys.* **33** 3033
[8] Sirohi R S and Chau F S 1999 *Optical methods of measurement – wholefield techniques*, Marcel Dekker, New York
[9] Goodman J W 1984 *Statistical properties of laser speckle patterns*, in: Dainty J C (Ed.) *Laser speckle and related phenomena*, Springer, New York, pp. 9-75.
[10] Riesz F 2011 *Optik* in press
[11] Pizer S M and ter Haar Romeny B M 1991 *J. Digital Imaging* **4** 1
[12] Riesz F 2001 *Mater. Sci. Eng. B* **80** 220

Journal of Physics: Conference Series **281** (2011) 012007 doi:10.1088/1742-6596/281/1/012007

Indentation-induced dislocations and cracks in (0001) freestanding and epitaxial GaN

I Ratschinski [1], H S Leipner[1], F Heyroth[1], W Fränzel[2], O Moutanabbir[3], R Hammer[4], M Jurisch[4]

[1]Interdisziplinäres Zentrum für Materialwissenschaften, Martin-Luther-Universität Halle-Wittenberg, Heinrich-Damerow-Straße 4, D-06120 Halle, Germany

[2]Institut für Physik, Martin-Luther-Universität Halle-Wittenberg, Von-Danckelmann-Platz 3, D-06120 Halle, Germany

[3]Max-Planck-Institut für Mikrostrukturphysik, Weinberg 2, D-06120 Halle, Germany

[4]Freiberger Compound Materials GmbH, Am Junger-Löwe-Schacht 5, D-09599 Freiberg, Germany

E-Mail: ingmar.ratschinski@cmat.uni-halle.de

Abstract. The (0001) surface of freestanding GaN and an (0001) oriented epitaxial GaN layer of 3 μm thickness on sapphire have been deformed at room temperature using a Vickers indenter. The samples were indented with two different orientations of the indenter with loads in the range from 0.02 to 4.90 N and 0.10 to 4.90 N, respectively. Dislocations and cracks at the indentations were observed by means of scanning electron microscopy, cathodo-luminescence, optical microscopy and transmission electron microscopy. Dislocations occur at all indentations for the loads used in the investigations. In both materials, the dislocation arrangement corresponds to the symmetry of the indented surface and the orientation of the indenter has only a marginal influence. Higher loads lead to radial cracks at the corners of the indentations and lateral cracks beneath the surface. The crack system is predominantly determined by the symmetry and orientation of the indenter. The dislocation arrangement and the crack system in freestanding GaN and an epitaxial layer of GaN on sapphire are compared.

1. Introduction

GaN is a III-V compound semiconductor crystallizing predominantly in the wurtzite structure ($\boldsymbol{a}$ = 0.319 nm and $\boldsymbol{c}$ = 0.518 nm [1]). It is an important material for optoelectronic and high-frequency devices on account of its direct band gap of 3.39 eV [2]. There are several studies of the mechanical properties of GaN single crystals, but there are just a few results of crack formation and propagation [3-6]. One model for the formation of crack seeds is the interaction of dislocations on different glide planes and the pile-up of dislocations at Lomer-Cottrell barriers [7]. If a crack seed is formed, subsequent crack propagation can take place and lead to a macroscopic crack. In addition to a deeper understanding of fundamental processes, a better knowledge of the mechanical behavior would be beneficial to optimize processing steps like sawing and grinding, as has been demonstrated for (001) GaAs [8].

The deformation of monocrystalline solids under increasing indenter load can be divided into three stages. In the elastic stage, the deformation is complete reversible. The initiation of the elastic-plastic stage is characterized by the pop-in event. It is associated with the onset of plasticity and the generation of dislocations [9]. In the elastic-plastic stage, the material is deformed by the generation and propagation of dislocations, but the deformation is partly elastic. The third stage is characterized by the formation and propagation of cracks. This can lead to chipping of the material and finally to a total destruction of the sample.

A detailed analysis of dislocations in the wurtzite structure has been given by Osipiyan et al. [10]. At indentations in (0001) GaN, dislocations are arranged in a star-shaped rosette with dislocation segments oriented in the $\pm\langle 2\bar{1}\bar{1}0\rangle$ directions [11]. Transmission electron microscopy investigations indicate ***a***-type screw dislocations in the rosettes [12]. The rosette consists of dislocations perpendicular and inclined to the surface [13]. Slip lines occur only in the area within the indentation [14]. GaN has $\{1\bar{1}00\}$ cleavage planes [15]. Drory et al. [3] indented a GaN single crystal using a Vickers indenter. In that work, radial cracks occurred at the corners of the indentations. Most of the investigations deal with GaN epitaxial layers instead of freestanding GaN. The dislocation arrangement or the crack formation has only been investigated separately. In the present study, both dislocations and cracks are investigated in freestanding GaN and in GaN on sapphire. Freestanding GaN and GaN on sapphire are compared. Between dislocations and cracks, geometrical correlations could be found.

In order to investigate the crack formation, the relative crack frequency $n(F)$ is defined as [16]:

$$n(F) = N(F)/N_0, \tag{1}$$

where N_0 is the total number of corners of the indentation and N the number of corners exhibiting a radial crack at a certain load F. The crack resistance F_c is defined as the load F where the relative crack frequency equals 0.5 [16]:

$$n(F_c) = 0.5. \tag{2}$$

2. Experimental details

In this study, three types of (0001) oriented, not intentionally doped GaN samples have been investigated: as-grown GaN single crystals, a GaN wafer and an epitaxial layer of GaN on sapphire (table 1). As-grown GaN crystals of 0.3 and 1.2 mm thickness have been produced using hydride vapor phase epitaxy on a sapphire substrate and exfoliation. The crystal directions were distinguished by X-ray Laue patterns. The samples were indented at room temperature (RT) using a Vickers indenter. A Hanemann MHP 100 micro hardness tester was used for loads from 0.02 to 1.00 N. In the range from 0.10 to 4.90 N, indentation experiments were performed with a Zwick 3202 hardness tester. For each material, indentations were performed with two different orientations of the indenter (0° series and 45° series). This was made to separate the influence of the indenter geometry from the symmetry of the crystal.

Table 1. Summary of the investigated (0001) oriented GaN samples.

Material	Density of in-grown dislocations	Thickness	Investigation
GaN wafer	10^9 cm^{-2}	0.3 mm	TEM preparation
		1.2 mm	Relative crack frequency
	1.7×10^7 cm^{-2}	0.17 mm	Cross section preparation
GaN on sapphire	6×10^8 cm^{-2}	3 μm	Relative crack frequency

In the 0° series, one indentation diagonal was oriented in $[2\bar{1}\bar{1}0]$ direction and the other diagonal pointed into the $[01\bar{1}0]$ direction (figure 1a). In the 45° series, the sample was rotated in an angle of 45 degrees with respect to the indenter (figure 2b). For each load and orientation, 10 indentations were evaluated.

The indentations and cracks were examined by scanning electron microscopy using a Philips ESEM XL 30 FEG. Bulk GaN was investigated in secondary electron contrast (SE) at an acceleration voltage of 5 kV. GaN on sapphire was examined in the gas mode of the environmental scanning electron microscope (ESEM) at 20 kV in order to reduce the electric charging. The dislocations were observed by cathodoluminescence imaging (CL) in a JSM 6400 scanning electron microscope at an acceleration voltage of 8 or 10 kV in panchromatic mode. Defects beneath the surface in GaN were observed by optical microscopy (OM) in reflection or transmission using a Leica DMR microscope. Selected indentations were inspected by atomic force microscopy using a Nano-R of Pacific Nanotechnology. For TEM investigations in diffraction contrast, a JEM 4000 FX transmission electron microscope was used at an acceleration voltage of 400 kV. In order to fabricate electron-transparent samples, the indented crystal was drilled out with an ultrasonic driller and cleaned in acetone and ethanol. The back side of the sample was ground and dimpled mechanically. Finally, the sample was thinned by 3 keV argon ions to a final thickness of 0.5 to 1 μm in the region of the indentations using a Gatan precision ion polishing system model 691.

3. Freestanding (0001) GaN

Dislocations occur in the surrounding of all indentations for the loads used in the present investigations. In-grown dislocation can be clearly distinguished from indentation-induced dislocations. In-grown dislocations form an irregular network with no preferential orientation (figure 1). Indentation-induced dislocations are arranged in a rosette with six dislocation arms running in the $\pm\langle 2\bar{1}\bar{1}0\rangle$ directions. Each arm is made up of two branches. Neighboring dislocation arms cut each other. As a consequence, the centre of the rosette appears star-shaped. In the arms, the dislocations are arranged in bands. In TEM micrographs, individual dislocations are visible in the bands. Long bands of straight dislocations indicate surface parallel dislocations. At a diffraction vector $\boldsymbol{g}$ of $2\bar{1}\bar{1}0$, six dislocation arms are visible. At $\boldsymbol{g} = 10\bar{1}0$, the contrast of dislocations in the $\pm$ $[\bar{1}2\bar{1}0]$ directions is at minimum (arrows in figure 1). This indicates ***a***-type screw dislocations in agreement with the results of Albrecht et al. [12].

Figure 1. Dislocations at a Vickers indentation in N-polar $(000\bar{1})$ GaN at a load of 0.25 N from the 0° series. TEM bright field images of the same position with a diffraction vector $\boldsymbol{g}$ of (a) $2\bar{1}\bar{1}0$ and (b) $10\bar{1}0$.

In GaN, dislocations act mainly as centers of non-radiative recombination. Hence, they appear dark in CL images (figure 2a, 2b). The speckled background in the CL images is caused by in-grown dislocations. The area of the indentation appears dark. No individual dislocations can be distinguished there. Near the indentation, line contrasts dominate in the dislocation arms, which indicate dislocation bands parallel to the surface. The line contrasts in the CL images correspond to dislocation bands found in TEM. In the outer regions, intersection points of dislocations with the surface dominate. The orientation of the indenter has got only a minor influence on the arrangement of the dislocations. However, the density of dislocation bands may vary, as can be seen in the CL images. The extension of the dislocation rosette is increased by electron beam irradiation (recombination-enhanced dislocation glide effect [17]) during the CL measurements. The size of the rosettes rises with increasing load, whereas the characteristic shape remains unchanged.

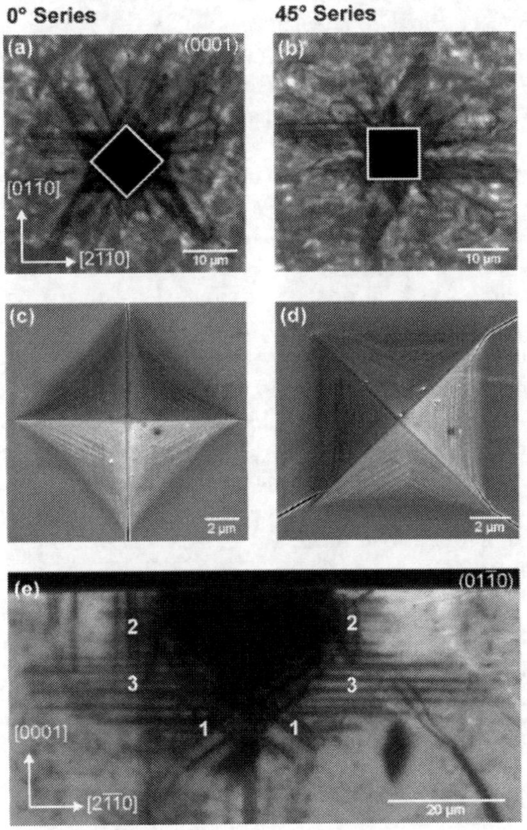

Figure 2. Dislocations and slip lines in the surrounding of Vickers indentations. Investigation of two indentations in Ga-polar (0001) GaN at a load of 0.98 N from the 0° and 45° series, respectively. (a) and (b): CL images at a temperature of 83 K. (c) and (d): SE images. (e) CL image at RT of an indentation from the 0° series at 4.90 N in (01$\bar{1}$0) cross section with three types of dislocation bands (1 – 3).

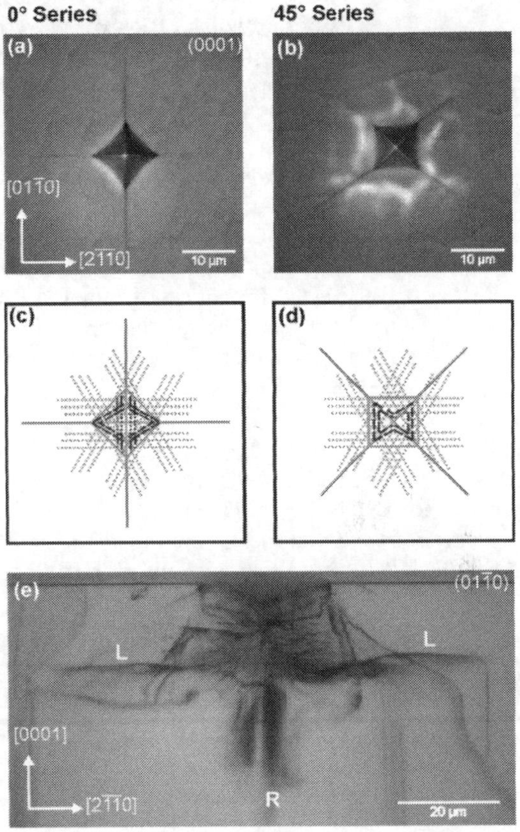

Figure 3. Cracks at Vickers indentations in (0001) GaN. (a) and (b): OM images in reflection corresponding to figure 2a and 2b, (c) and (d): Scheme of dislocations (dotted), slip lines (dashed) and radial cracks (solid) at an indentation. (e) Cross section transmission OM image of a Vickers indentation from the 0° series at 4.90 N, R radial crack, L lateral crack.

Slip lines occur in the area within the indentations (figure 2c, 2d). There are three sets of slip lines lying in the $\pm\langle\bar{1}100\rangle$ directions. In the 0° series, one set of slip lines is parallel to an indentation diagonal. The other sets are oriented at an angle of 60° to this diagonal and intersect at the other indentation diagonal. In the 45° series, one set of slip lines is parallel to the edges of the indentation in the $[01\bar{1}0]$ direction. The other sets of slip lines cut each other at a face of the indentation.

For cross section investigations, a set of indentations in the 0° orientation was aligned in $[2\bar{1}\bar{1}0]$ direction. Subsequently, the sample was cleaved though the indentations along the $(01\bar{1}0)$ plane. In the cross section CL images, three types of dislocation bands are visible (figure 2e). The area beneath the indentation appears dark. Near the indentation, no individual dislocations are visible, but in the deeper regions, dislocations inclined to the surface occur (1). Next to the indentation, dislocations perpendicular to the surface are visible (2). Beneath the indentation, dislocations parallel to the surface occur (3). The area of these dislocations reaches from the surface into deeper regions.

Higher loads lead to radial cracks at the corners of the indentations and lateral cracks beneath the surface (figure 3). The brightening at indentations in the OM images is caused by specular reflections at lateral cracks (figure 3a, 3b). The system of radial cracks is predominantly determined by the symmetry of the indenter. It changes with the rotation of the indenter. The crack patterns at Vickers indentations in (0001) GaN can be characterized as follows: Radial cracks emanate from the corners of the indentation. At higher loads, lateral cracks occur beneath the surface. At the highest loads used, radial cracks can emanate from the edges of the indentation in some cases. The characteristic features of the dislocation arrangement and the radial crack system are illustrated in figure 3c and 3d.

In the 0° series, the formation of radial cracks starts at a load of 0.05 N (figure 4a). In the 45° series, it begins at 0.1 N. The relative crack frequency n (equation 1) rises with the load. The crack resistance F_c for radial cracks (equation 2) obtained for measurements with the MHP 100 micro hardness tester amounts to (0.28 ± 0.12) N in the 45° series and (0.36 ± 0.14) N in the 0° series. The differences in the crack resistance between the 0° series and the 45° series are not significant. In addition to radial cracks, the formation of circular lateral cracks beneath the surface begins at 0.50 N in the 45° series and at 0.98 N in the 0° series (figure 4b). The crack resistance F_c for lateral cracks amounts to 0.81 N in the 45° series and 1.39 N in the 0° series. For a load of 4.90 N, radial cracks emanate from every corner of the indentations. Additionally, at two indentations from the 0° series, a radial crack emanates from one edge of the indentation.

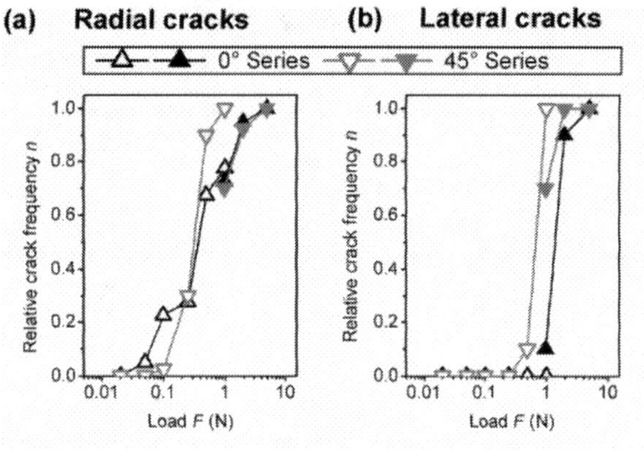

Figure 4. Relative crack frequency at Vickers indentations in Ga-polar (0001) GaN (a) of radial cracks and (b) of lateral cracks with a MHP 100 micro hardness tester (empty symbols) and a Zwick 3202 hardness tester (filled symbols).

Journal of Physics: Conference Series **281** (2011) 012007 doi:10.1088/1742-6596/281/1/012007

4. Investigation of (0001) GaN on sapphire and comparison with freestanding (0001) GaN

Similar to freestanding GaN, the speckled background in the CL images is caused by in-grown dislocations (figure 5a, 5b). Indentation-induced dislocations are arranged in a rosette with six dislocation arms running in $\pm\langle 2\bar{1}\bar{1}0\rangle$ directions. Each arm is made up of two branches. The orientation of the indenter has got only a minor influence on the dislocation arrangement. In contrast to freestanding GaN, the dislocation arms are broader. Line contrasts dominate and no intersection points of dislocations with the surface are visible. This indicates dislocations parallel to the surface. Slip lines in the different $\langle\bar{1}100\rangle$ directions occur in the area within the indentations (figure 5c, 5d), similar to freestanding GaN. Higher loads lead to lateral cracks beneath the surface (figure 5e, 5f). The frequency of lateral cracks is zero for loads up to 0.49 N (figure 5g). At 0.98 N, lateral cracks occur. For higher loads, the relative crack frequency equals 1. The crack resistance for lateral crack can be estimated to about 1.5 N. The differences between the 0° series and the 45° series are not significant. In contrast to freestanding GaN, radial cracks occur only marginally even at the highest load used.

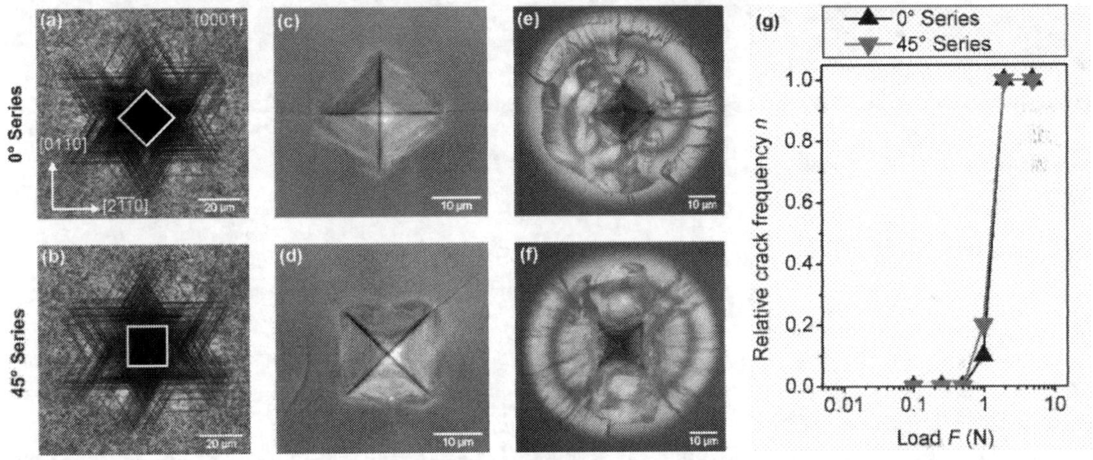

Figure 5. Dislocations and cracks at Vickers indentations in (0001) GaN on sapphire at a load of 4.90 N from the 0° and 45° series, respectively. (a) and (b): CL images at RT, (c) and (d): ESEM images. (e) and (f) OM images in reflection. (g) Relative crack frequency of lateral cracks at indentation in (0001) GaN with a Zwick 3202 hardness tester.

5. Discussion

In the wurtzite structure, there are two types of smallest translation vectors: The ***a***-type ($a/3\langle 2\bar{1}\bar{1}0\rangle$) and the ***c***-type ($c\langle 0001\rangle$). This leads to simple dislocations with a Burgers vector of $a/3\langle 2\bar{1}\bar{1}0\rangle$ and $c\langle 0001\rangle$. These Burgers vectors lead to five types of perfect dislocations: ***a***-type edge dislocations, ***a***-type screw dislocations, ***a***-type 60° dislocations, ***c***-type edge dislocations and ***c***-type screw dislocations [10]. The most densely packed planes in the wurtzite structure are the $\{0001\}$ basal planes and the $\{\bar{1}100\}$ and $\{2\bar{1}\bar{1}0\}$ prismatic planes. These planes are assumed to be the glide planes in the wurtzite structure [10].

For the comparison of the dislocation rosette (figure 1) with glide prism models, the most important results of the investigations are summarized as follows. The dislocation rosette at indentations in (0001) GaN consists of six arms running in the $\pm\langle 2\bar{1}\bar{1}0\rangle$ directions. TEM investigations showed long bands of straight dislocations with an ***a***-type Burgers vector in direction of the dislocation arms. In the outer regions of the rosette, CL images showed intersection points of dislocations with the surface. These factors indicate a dislocation arrangement for one rosette arm as sketched in figure 6. This model explains the experimental data for surface parallel dislocation arms.

The glide planes of these dislocations are perpendicular to the surface. As a consequence, the screw dislocation segments occur as line contrasts both in TEM and CL images. In the outer regions, intersection points of dislocations at the surface are visible in CL images. The slip lines in $<\bar{1}100>$ direction correspond to traces of the $\{2\bar{1}\bar{1}0\}$ planes.

Dislocations can interact with each other during their propagation. At the intersection of neighboring surface parallel dislocation arms, three different types of interactions of perfect edge and screw dislocations are conceivable (figure 7): 1. The intersection of two ***a***-type screw dislocations. 2. The intersection of an ***a***-type edge and an ***a***-type screw dislocation. 3. The interaction of two ***a***-type edge dislocations [18].

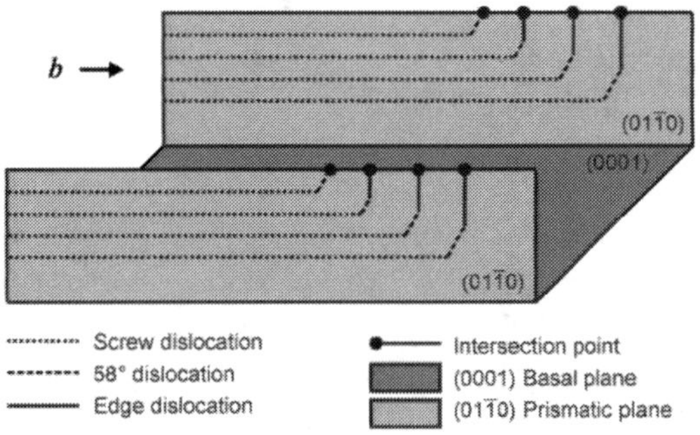

Figure 6. Glide prism configuration of dislocations in the rosette arms.

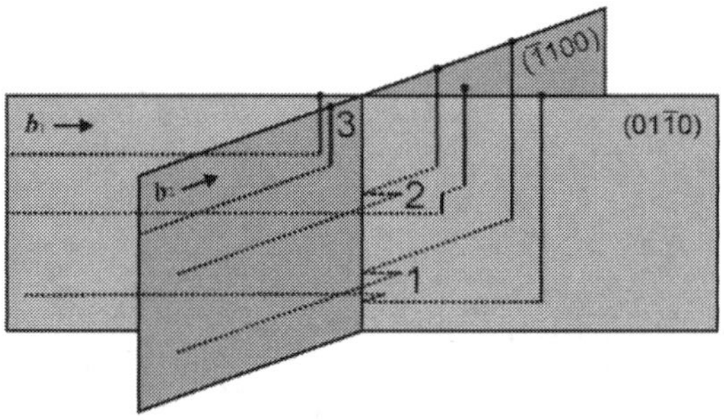

Figure 7. Possible interactions of dislocations on neighbouring dislocation arms.

1. The intersection of two ***a***-type screw dislocations leads to jogs on both dislocations, which have a 60° character (figure 7-1). The intersection of individual dislocations leads to elementary jogs, whereas the intersection of dislocation bands can lead to jogs with the height of several Burgers vectors. These jogs can easily move along the screw dislocations, but for a movement in the ± [0001] directions they have to climb. If the screw dislocations move in the ± [0001] directions, the 60° segments will lag behind. Subsequently, elementary jogs can lead to point defects and higher jogs to prismatic loops. The edge dislocation segments are not affected by the cutting process, and they can propagate unhindered in the $\pm<2\bar{1}\bar{1}0>$ directions on $\{1\bar{1}00\}$ slip planes. 2. The intersection of an ***a***-type edge and an ***a***-type screw dislocation leads to jogs on both dislocations, which have a 60° character (figure 7-2). The behavior of a jog on a screw dislocation was described above. The jog on the edge dislocation can move easily in the $\pm<2\bar{1}\bar{1}0>$ directions, but in the ± [0001] directions it can only climb. Hence, the edge dislocation can propagate unhindered in the $\pm<2\bar{1}\bar{1}0>$ direction of their slip plane. 3. Two ***a***-type edge dislocations ($\boldsymbol{b}_1 = \boldsymbol{a}$, $\boldsymbol{b}_2 = \boldsymbol{a}$) of neighboring dislocation arms include an angle of 60° (figure 7-3). The reaction of these dislocations would result in a dislocation with a Burgers vector $\boldsymbol{b}_3 = a <\bar{1}100>$. According to the $E \propto \boldsymbol{b}^2$ criterion (E is the elastic energy of the dislocation), this dislocation is expected to be unstable, since $\boldsymbol{b}_1^2 + \boldsymbol{b}_2^2 = 2\boldsymbol{a}^2$ but $\boldsymbol{b}_3^2$ equals $3\boldsymbol{a}^2$. This leads to repulsive forces between the edge dislocations and can result in a pile-up of dislocations.

The crack system is influenced by various factors. These are the cleavage planes of the crystal, the geometry of the indenter together with the strain field of the indentation and the interactions of dislocations. In the 0° series of freestanding GaN, radial cracks propagate in $\pm[2\bar{1}\bar{1}0]$ and $\pm[01\bar{1}0]$ directions (figure 3a). Cracks in $\pm[2\bar{1}\bar{1}0]$ direction propagate in the $(01\bar{1}0)$ cleavage plane. In this direction, the cutting of surface parallel dislocations arms is low. Cracks in $\pm[01\bar{1}0]$ direction do not propagate in a cleavage plane. In this direction, two surface parallel dislocations arms intersect. This may lead to the various dislocation cuttings or the pile-up of dislocations, as described in figure 7. For the formation of crack seeds, different models have been proposed. Two prominent examples are the models of Stroh and of Cottrell. The model of Stroh predicts the formation of a crack seed at the pile-up of dislocations on one glide plane [19] whereas the model of Cottrell describes the formation of a crack seed due to the pile-up of dislocations on different glide planes [20]. Hence, cracks in $\pm[01\bar{1}0]$ direction could be initiated by the interaction of ***a***-type dislocations on neighboring dislocation arms. In the 45° series, radial cracks are crystallographically not strictly oriented (figure 3b). They emanate from the corners of the indentation in a region with a high interaction of neighboring dislocation arms (figure 2b). These cracks could be initiated by cutting processes or pile-up of ***a***-type dislocations.

In GaN on sapphire, no intersection points of dislocations with the surface are visible (figure 5a, 5b). This indicates dislocations parallel and near to the surface. Another possibility is that the edge dislocation segments do not intersect the surface but they are bended to the interface GaN-sapphire. (0001) GaN on a (0001) sapphire substrate has a lattice mismatch of about 15 % [3]. The lattice relaxation is realized mainly by misfit dislocations with a Burgers vector in the (0001) plane [21]. However, there is only a partial relaxation, and a residual strain remains. Hence, the dislocations could be concentrated in the GaN layer. This would lead to an increased interaction between dislocations. The consequence is a modified dislocation rosette compared to freestanding GaN.

In (0001) GaN on a (0001) sapphire substrate, radial cracks occur only marginally at the highest loads used. In this orientation, the cleavage planes of GaN and sapphire do not coincide [15]. In addition, mechanical properties like hardness and fracture toughness are different [3]. Hence, the propagation of radial cracks due to cleavage may be suppressed. In GaN on sapphire, the formation of lateral cracks is the primary cracking mechanism. The formation of lateral cracks is independent from the orientation of the indenter.

In freestanding GaN, radial cracks are formed at lower loads than lateral cracks. The differences in the formation of radial cracks between the 0° series and the 45° series are not significant. The load of crack initiation and the crack resistance for lateral cracks are lower in the 45° series than in the 0° series. For lateral cracks in epitaxial GaN on sapphire, the differences between the 0° series and the 45° series are not significant. Their crack resistance is similar to the value of freestanding GaN from the 45° series. Radial cracks in GaN on sapphire occur only at the highest loads used.

6. Conclusions

In the study presented, the formation of dislocations and cracks at Vickers indentations in (0001) oriented freestanding GaN and 3 μm epitaxial (0001) GaN on sapphire have been investigated. Dislocation rosettes are found at all indentations. TEM investigations indicate dislocations with ***a***-type Burgers vectors parallel to the direction of the dislocation arms. From TEM and CL images it can be concluded that the dislocations propagate as quarter loops. They emanate as ***a***-type screw dislocations from the indentation and intersect the surface as ***a***-type edge dislocations or 58° dislocations. Dislocations of neighboring rosette arms can interact with each other and hinder their further propagation. However, the dislocation arrangement beneath the indentation requires further investigation. For a more detailed analysis, the splitting of perfect dislocations into partial dislocations and their reactions should be considered.

Cracks occur at higher loads at the indentations. In freestanding GaN, radial cracks emanate from the corners of the indentations. The investigations indicate that the formation of crack seeds has to be clearly separated from crack propagation. Crack seeds are attributed to the interaction of dislocations such as intersection or pile-up of dislocations, whereas the crack propagation is associated with the strain field of the indentation and the mechanical and crystallographical properties of the material. In contrast to $(\bar{1}\bar{1}\bar{1})$ GaAs, radial cracks in GaN are not aligned in cleavage planes in many cases. In future experiments, the formation as well as the length of radial cracks in different directions will be investigated with higher statistics in order to prove differences in the cracking behavior inside and outside a cleavage plane. The crack frequency in freestanding GaN with lower dislocation density will be investigated in more detail in order to determine the influence of in-grown dislocations on indentation-induced dislocations and the crack formation. In addition, this could allow a closer analysis of indentations-induced dislocations and their interactions.

References

[1] Sakai A, Sunakawa H and Usui A 1997 *Appl. Phys. Lett.* **71** 2259
[2] Chow T P, Khemka V, Fedison J, Ramungul N, Matocha K, Tang Y and Gutmann R J 2000 *Solid-State Electron* **44** 277
[3] Drory M D, Ager III J W, Suski T, Grzegory I and Porowski S 1996 *Appl. Phys. Lett.* **69** 4044
[4] Yonenaga I, Hoshi T and Usui A 2000 *J. Phys.: Condens. Matter* **12** 10319
[5] Jahn U, Trampert A, Wagner Th, Brandt O and Ploog K H 2002 *Phys. Status Solidi* A **192** 79
[6] Kavouras P, Komninou Ph and Karakostas Th 2007 *Thin Solid Films* **515** 3011
[7] Hirsch P B, Pirouz P, Roberts S G and Warren P D 1985 *Phil. Mag.* B **52** 759
[8] Hammer R, Bergner F, Flade T, Jurisch M, Kleinwächter A and Schaper M 2005 *Z. Metallkd.* **96** 787
[9] Bradby J E, Williams J S and Swain M V 2004 *J. Mater. Res.* **19** 380
[10] Osipiyan Yu A, and Smirnova I S 1968 *Phys. Status Solidi* **30** 19
[11] Kucheyev S O, Bradby J E , Willliams J S , Jagadish C , Toth M , Philips M R and Swain M V 2000 *Appl. Phys. Lett* **77** 3373
[12] Albrecht M, Strunk H P, Weyher J L, Grzegory I, Porowski S and Wosinski T 2002 *J. Appl. Phys.* **92** 2000
[13] Weyher J L, Albrecht M, Wosinski T, Nowak G, Strunk H P and Porowski S 2001 *Mat. Sci. Eng.* B **80**, 318
[14] Bradby J E, Williams J S, Wong-Leung J, Kucheyev S O, Swain M V and Munroe P 2002 *Phil. Mag* A **82** 1931
[15] Kuramata A, Horino K and Domen K 1998 *Fujitsu Sci. Tech. J.* **192** 34
[16] Wada M, Furukawa H and Fujita K 1974 *Proceedings of the 10th International Congress on Glass* Vol. **11** Ed. Kunugi M, Tashiro M and Soga N, (Kyoto Ceramic Society of Japan) 39
[17] Maeda K, Suzuki K, Ichihara M, Nishiguchi S, Ono K, Mera Y and Takeuchi S 1999 *Physica B* **273-274** 134
[18] Hull D Bacon D 1984 *Introduction to dislocations* (Oxford Pergamon Press) 141
[19] Stroh A N 1957 *Adv.Phys* **6** 418
[20] Cottrell A H 1958 *Trans. AIME* **212** 192
[21] Ambacher O 1998 *J. Phys. D: Appl. Phys.* **31** 2653

Journal of Physics: Conference Series **281** (2011) 012008 doi:10.1088/1742-6596/281/1/012008

Identification of dislocation-related luminescence participating levels in silicon by DLTS and Pulsed-CL profiling

Anton Bondarenko[1*], Oleg Vyvenko[1], Ivan Isakov[1]

[1] V.A. Fok Institute of Physics, St.Petersburg State University, Ulyanovskaya 1, 198504 St.Petersburg, Russia

[*] Corresponding author. E-mail: bondarenko@nano.spbu.ru

Abstract. We present a study of the dislocation network that occurs in the space charge region of a Schottky diode, by means of DLTS and our recently developed cathodoluminescent (CL) technique called Pulsed-CL. The details of the Pulsed-CL technique are provided. We establish a correspondence between the CL spectra of dislocation-related luminescence in silicon in the vicinity of the so-called D1 band and levels determined from DLTS measurements. The centres responsible for the 815 meV CL component are related to dislocations cores while the centres responsible for the 795 meV CL component are related to some defects outside of the dislocation cores.

Introduction

Dislocation networks (DN) in silicon are nowadays highly reproducible and controllable structures with a number of promising applications [1]. Among them is the perspective to use DN as a base of a light emitting diode (LED) fully compatible with silicon technology, which addresses a need of the modern microelectronics industry. Efficient and reliable LEDs that implement dislocation related luminescence (DRL) in silicon are impossible unless the intrinsic mechanisms of DRL as well as the electronic levels they involve are determined. One of the key problems that remain open so far is how to find among the electrical levels detected with space charge region (SCR) spectroscopy those that are responsible for DRL.

Previously, a new approach to solve this problem was suggested based on control of the electrical level occupancy in the SCR of the Schottky-diode. This was achieved by varying the bias voltage, and a correlation between the cathodoluminescent (CL) and electrical properties of DN was demonstrated [2]. Recently that approach was further developed into a new technique referred to as Pulsed-CL, which incorporates the use of a filling pulse instead of DC bias. It is therefore complimentary to conventional DLTS/MCTS profiling but is sensitive to optical transitions only. Details of this new technique are described in section 2.

The main purpose of the present work is to establish direct correspondence between DRL CL spectroscopy components and the levels determined from DLTS measurements, and the role of deep and shallow levels in DRL processes.

1. Experimental Method

We used boron doped Cz-Si 300 mm diameter wafers with initial doping concentration of about 10^{15} cm^{-3}. They have a (100) surface orientation and were hydrophilically bonded with a small misorientation tilt angle of about 0.5° and twist angle of about 3°. These were thinned so that the DN was 150 nm from the surface. Details of the bonding procedure can be found elsewhere [3]. The piece cleaved from the central part of the bonded wafers was etched in HF:10·H_2O solution for 2 minutes to remove the native oxide layer and then semitransparent titanium contacts of a thickness of 50-100 nm were deposited from the thinned side of bonded wafers to prepare a Schottky diode. The diode was examined in a planar geometry inside the chamber of a field-emission scanning electron microscope (FESEM) Zeiss Supra 40 VP with Gatan LHe cryostage and Gatan MonoCL3 system.

The structural properties of the DN were STEM examined using the same FESEM in planar geometry. The space between twist dislocations in the DN was 7 nm, in good agreement with theoretical prediction. The STEM results are reported in another proceeding paper of this conference.

CL was measured with sample excitation through a metal electrode in a spot mode to enable us to measure the electric field effects. The CL signal was acquired in the vicinity of the D1 band of DRL in the range 0.76 – 0.88 eV with LN2 cooled InGaAs 1024 elements array CCD detector Princeton Instruments OMA-V. The stable beam current of the FESEM (change of current is less than 0.2% per hour) coupled with the CCD acquisition system allowed us to perform CL measurements for sufficiently long timescales (typically tens of minutes) to reach desirable accuracy. The electron beam acceleration voltage was 30 kV and electron beam current was 10 nA. The presence of electroluminescence was monitored at each bias and was below the system sensitivity at every applied bias voltages used in our experiments.

DLTS measurements were performed using a SULA DLTS spectrometer and Janis closed cycle helium cryostat with Lakeshore LS-331 temperature controller.

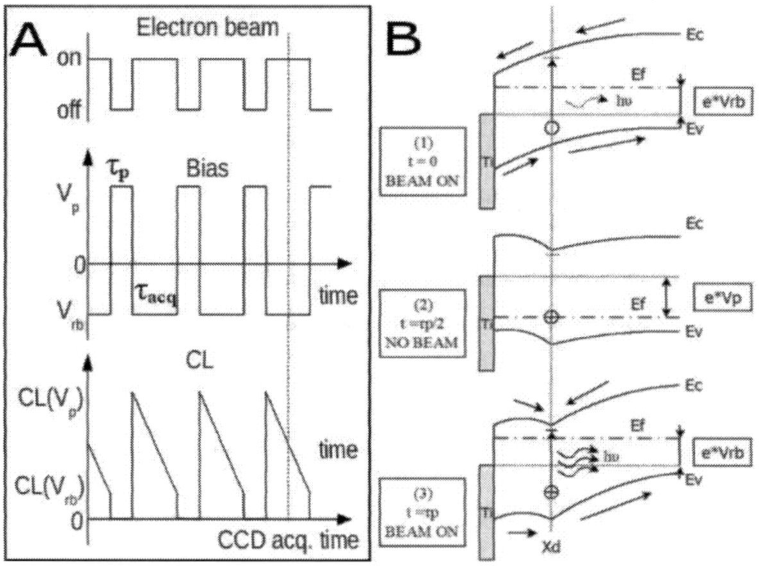

Figure 1. Experimental sequence of Pulsed-CL technique (A) and energy band bending diagrams (B) illustrating sample in three main states occurring during such measurements:

(1) Relaxed at reverse bias;

(2) During the filling pulse;

(3) Immediately after the filling pulse end.

The new technique named Pulsed-CL that we have recently developed was used to profile the levels participating in the DRL process. The temporal sequence of the Pulsed-CL measurements is shown in figure 1A and is similar to that of a DLTS experiment. The bias voltage pulses of value V_p and width τ_p are applied periodically to fill the traps with the majority carriers within the SCR of the Schottky diode. For the rest of time it is kept under a sufficiently large reverse bias V_{rb}.

Simultaneously with the refilling voltage pulses the electron beam is removed from the sample by the fast beam blanker. The CL signal is collected during the time interval τ_{acq} which should be short enough to neglect carrier thermo-emission at the measurement temperature.

The band bending diagrams in figure 1B show the electronic processes which occur during the three main stages of Pulsed-CL measurements.

Stage (1) corresponds to the situation when reverse bias V_{rb} was applied to the diode for a long time, allowing the DN levels, initially filled with majority carriers, to completely empty. In this stage, the CL signal arises only due to recombination of excess carriers generated by the electron beam. The minority carriers come to DN mostly from the sample bulk provided that the electron beam penetrates the sample much deeper than the SCR width. Only majority carriers that are generated in the close vicinity of the DN can participate in the recombination. Due to a high electrical field value at the DN the excess carriers are swept away from the SCR and only a weak CL signal is expected.

Stage (2) takes place during the application of the filling pulse V_p to the diode when the electron beam is switched off. The DN levels fill with the equilibrium majority carriers to the occupancy degree that is controlled by the V_p value in accordance with Fermi-Dirac statistics.

Stage (3) corresponds to the moment just after the restoration of the reverse bias V_{rb} and of the electron beam on the sample. Now, the minority excess carriers can recombine with majority carriers trapped by DN levels during the filling pulse. For a short period of time until the DN levels reaches the stationary occupancy defined by the carriers generated by the electron beam, the CL signal exceeds the value detected in the stage (1).

Since the occupancy degree of the DN levels is determined by the refilling pulse voltage in both DLTS and Pulsed-CL experiments the comparison of the dependencies of CL spectral line intensities and of DLTS-peak magnitudes on V_p can be used to establish one-to-one correspondence between the optical transitions and electric levels.

2. DLTS and Pulsed-CL investigations of DRL participating interfacial electronic states

DLTS spectrum of our Schottky-diode with built-in DN is shown in figure 2. The spectrum consists of four peaks, in agreement with previously obtained data on similar silicon samples with DN [3]. Analyzing the Arrhenius plots for DLTS spectral peaks we have derived energy levels for majority carriers in the lower half of the band gap which might be responsible for DRL. These are: shallow level Ev + 0.1 eV (narrow peak at about 60 K), and deep levels Ev + 0.25 eV, Ev + 0.3 eV, Ev + 0.35 eV (wide local maxima near 120 K, 150 K, and 200 K respectively). Note that the activation energies for the deep levels could not be obtained precisely due to the overlapping of neighbouring peaks.

The dependence of DLTS peak magnitudes on the refilling voltage pulse level is plotted in figure 3. The curve measured at T = 147 K corresponds to the deep levels near Ev + 0.3 eV. It exhibits a rapid increase starting from the pulse level of 0.25 V and the saturation at about 0.6 V. The second curve shown in figure 3 was measured at T = 60 K and is associated with shallow levels of about Ev + 0.1 eV. The DLTS-signal exhibits a rapid increase starting at V_p = 0.6 V followed by saturation at V_p = 1.3 V.

Both curves show an additional slow increase after plateau regions that occur starting from the filling pulse levels V_p = 1 V and V_p = 1.7 V for DLTS peaks at 147K and 60K respectively. The origin of the slow increase is not known precisely. They might be simply an artefact due to overloading of the capacitance bridge since the starting pulse voltages coincide well with the voltages responsible for the rapid increase of the diode forward current.

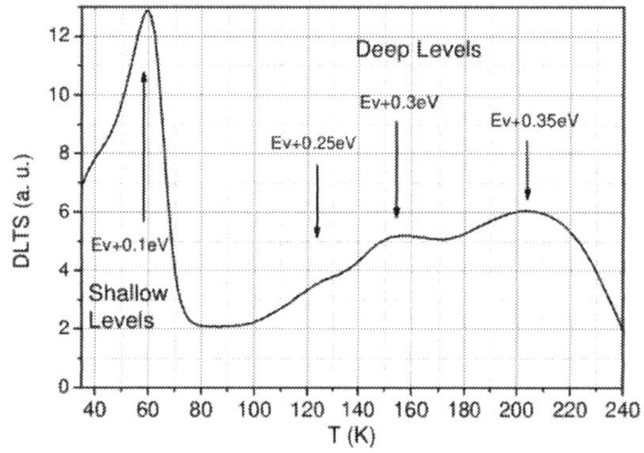

Figure 2. DLTS spectrum of p-Si sample with DN measured at reverse bias V_{rb} = -2 V, filling pulse level V_p = 1.1 V, and emission rate window of 20 ms.

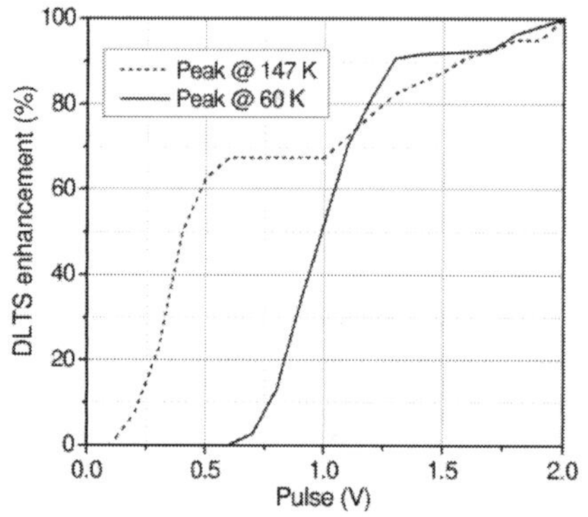

Figure 3. DLTS enhancement profiles measured at V_{rb} = -2 V versus filling pulse magnitude at temperatures corresponding to different peaks of DLTS spectra. DLTS signals at 60 K or 147 K correspond to the levels Ev + 0.1 eV or Ev + 0.3 eV respectively.

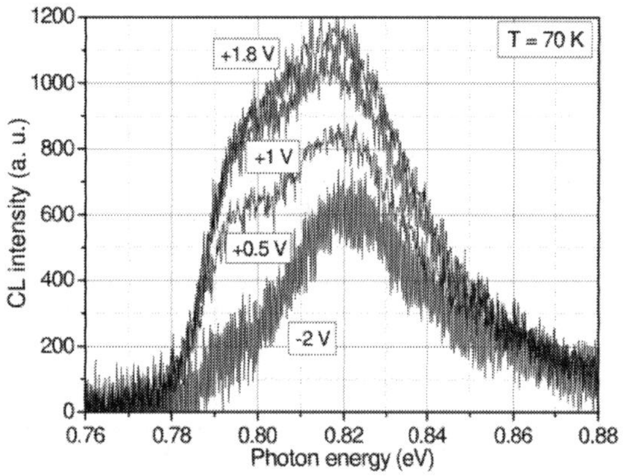

Figure 4. Pulsed-CL spectra acquired at the reverse bias of -2V after the traps filling with pulses of voltage level varying from -2 V to + 1.8 V. T=70 K.

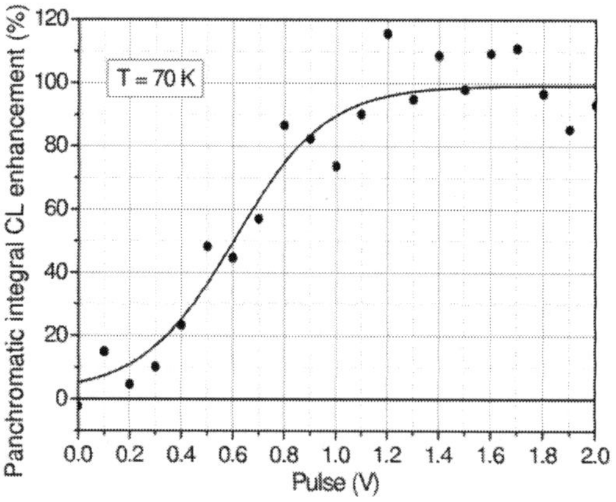

Figure 5. Panchromatic Pulsed-CL enhancement profile plotted as integral intensity of CL spectra enhancement in the vicinity of D1 band versus filling pulse magnitude measured at 70 K.

The CL measurements were carried out using filling pulse width τ_p = 1 μs that enabled complete filling of the trap. The acquisition time was τ_{acq} = 5 μs, significantly less than the thermoemission time constant for both deep and shallow levels but longer than the time constant of CL recombination.

Examples of Pulsed-CL spectra measured at different values of filling pulse are presented in figure 4. CL spectra exhibit the main maximum at about 815 meV conveniently called the D1-band of DRL in silicon, as well as a low energy shoulder. The relative enhancement of the panchromatic integral intensity of CL for different filling pulse magnitudes retrieved from experimental data of figure 4 is plotted in figure 5. The Pulsed-CL profile in figure 5 increases from the zero level of the filling pulse value up to saturation at about V_p = 1.3 V. Comparing the panchromatic Pulsed-CL voltage profile with DLTS profiles in figure 3 we conclude that the D1 luminescence goes via the shallow levels responsible for the DLTS peak at 60 K but not via the deeper levels responsible for the DLTS signal at 147 K since the latter saturates at significantly lower filling pulse values of about 0.6 V. On the other side, the minimum pulse voltage for CL-enhancement is significantly less than that for appearance of the DLTS-peak at 60K, indicating an influence of deep level occupancy on D1 DRL. The possible reasons for that will be discussed in the next section.

figure 6 shows DLTS spectra measured at different filling pulse values in the range from 0.8 V to 1.7 V. It is seen that the peak at 60 K has a low temperature shoulder which exhibits a faster increase with the increase of filling pulse magnitude than the peak at 60 K. This lower temperature shoulder reflects the presence of even shallower levels than Ev + 0.1 eV. figure 7 represents CL enhancement spectra that were obtained by subtraction of the CL spectrum registered without filling pulses and by smoothing. It is seen that the shape of the CL spectra changes upon applied voltage pulse magnitude and can be decomposed into two components: one at 785 meV and another at 815 meV.

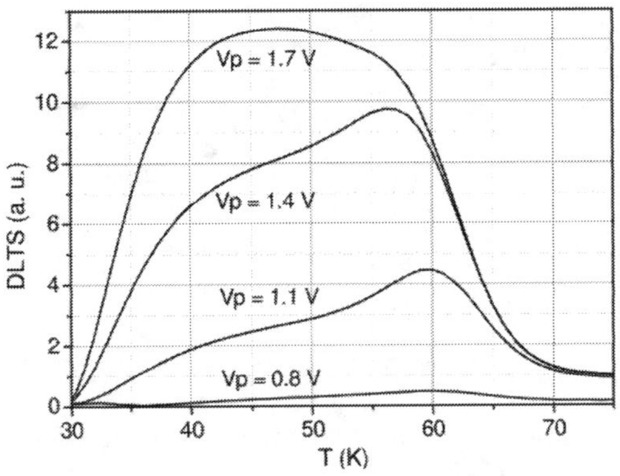

Figure 6. DLTS spectra measured with filling pulse values V_p varying from 0.8 V to 1.7 V (V_{rb} = -2 V) in temperature range of 30 – 75 K demonstrating different behavior of shallow levels. Emission rate window was 20 ms.

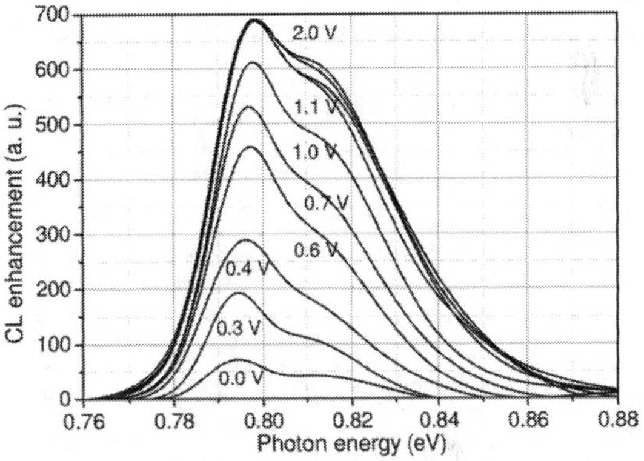

Figure 7. Pulsed-CL enhancement spectra retrieved from experimental Pulsed-CL spectra for filling pulse magnitudes varying from 0 V to + 1.8 V at temperature of 70 K. Calculation includes subtraction of CL spectrum registered without filling pulses and smoothing.

The relative DLTS signal enhancement versus filling pulse value plots measured at T = 44 K and at T = 60 K are shown in figure 8. As expected, the voltage profile for the shallower level with the peak at T = 44 K is shifted to the larger filling pulse values at about 0.15 V from the profile measured at 60 K. The voltage profiles for relative Pulsed-CL enhancement measured at photon energies of 795 meV and of 815 meV are plotted in figure 9. The profile for the 815 meV component is shifted to the higher values of filling pulse from the profile measured at a photon energy of 795 meV by a value of 0.15 V that coincides well with the pulse voltage difference for the above described DLTS peaks. Thus, from comparison of Pulsed-CL and DLTS pulse voltage profiling we can identify the CL components of 795 meV and of 815 meV with the levels responsible for DLTS peak at 60 K and at 44 K respectively.

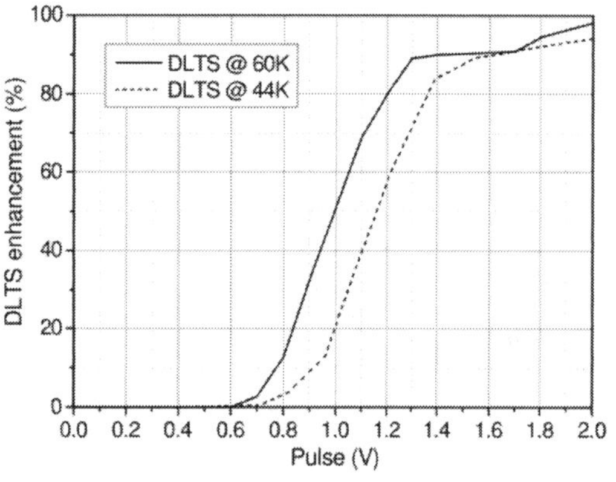

Figure 8. DLTS profiles measured at V_{rb} = -2 V versus filling pulse magnitude at temperatures corresponding to different peaks of DLTS spectra.

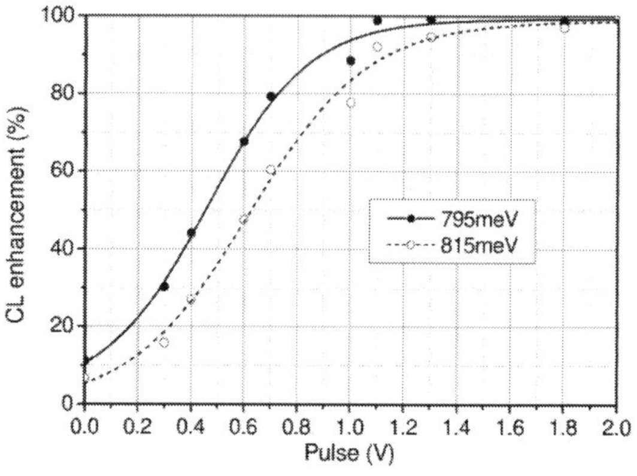

Figure 9. Pulsed-CL enhancement profiles of integral intensity of luminescence bands at 815 meV and at 795 meV versus filling pulse magnitude measured at 70 K.

3. Discussion

CL spectra measured in the vicinity of the so called D1 band of DRL in silicon corresponded in our sample with DN containing at least two components: 815 meV and 795 meV. The original Cz-Si wafers themselves do not show any luminescence in the region 0.7 eV – 0.9 eV, as confirmed by CL-measurements with backside excitation of our bonded wafers. Thus, both components of the CL spectra must be related to the presence of the DN.

From the comparison of DLTS and Pulsed-CL refilling voltage profiles it was concluded that both 815 meV and 795 meV DRL components could be identified with shallow levels near Ev + 0.1 eV. Moreover, the 815 meV and 795 meV components were found to have well distinguishable pulse voltage profiles. Lastly, the CL component at 795 meV vanishes at small reverse bias voltages (without application of refilling pulses) while the 815 meV component of the CL spectrum persists and does not change significantly within the reverse voltage ranges of several volts. That indicates a large difference in the carrier cross section for the defects responsible for these CL signals. There is a difference between the larger majority carriers capture cross-section of centres responsible for the 815 meV component, and those responsible for the 795 meV CL component. This difference can be explained by the existence of an attractive electrical potential which could be the deformation potential of dislocations. A confirmation of the existence of the attractive electrical potential is the Pool-Frenkel effect registered in [3] for the shallow level on the same sample. Removal of the 795 meV CL signal by an electrical field means that these centres do not have such an attractive electrical potential.

These findings lead us to the conclusion that the luminescence centres of the 815 meV CL component are related to dislocations cores but the centres responsible for the 795 meV CL component are related to defects outside the dislocations cores, and their deformation potential band bending. The last CL component often accompanies D1 and is commonly thought to be associated with some oxygen complexes or clusters in the silicon matrix [2, 4] that are segregated at the DN. These oxygen complexes might be either artefacts of hydrophilic process in which initial wafers are oxidized before bonding or products of segregation of oxygen dissolved in Cz-Si crystals.

Detailed comparison of refilling pulse voltage profiles of the Pulsed-CL and DLTS reveals that Pulsed-CL profiles are more expanded towards lower pulse voltages than DLTS profiles for the shallow levels (compare figures 8 and 9). The threshold pulse voltages for CL-enhancement are close to those that are observed for the DLTS signal from the deep levels (compare figures 9 and 3). The initial bias voltage for the voltage region where the Fermi-level is pinned by the shallow level appears as a plateau on capacitance-voltage curves (not shown for brevity). Thus it appears we can explain the smoothness of the Pulsed-CL voltage profile by an indirect effect of the emptying or filling of the deep levels with the majority carriers on CL-intensity as follows.

At the end of the period of the excitation of the sample with the electron beam, deep level occupancy decreases. This gives rise to a corresponding decrease of the potential at DN with respect to its equilibrium value without any excitation, and to a lower voltage pulse level to start the filling of the shallow states responsible for DRL at the beginning of the refilling period. In addition, just after the end of the refilling period and the recovery of the reverse bias voltage, the filled deep DN levels cause the reduction of the electrical field direction in the region between the DN and metal electrode, or even its inversion as was reported in [2]. Such an effect could be illustrated by the energy band bending diagrams plotted for different bias values in figure 1B. When the deep levels are empty the electrical field is large and the majority carriers generated by the electron beam are swept out from the SCR without recombination. When the deep levels are being filled with majority carriers during the filling voltage pulse the electrical field decreases and the after-pulse capture rate for majority carriers effectively increases.

4. Summary

Two components of CL spectra related to the presence of DN in the silicon wafer bonding interface were registered in the vicinity of the D1 band, at 795 meV and 815 meV. A new technique called Pulsed-CL capable of registering CL spectra depending on the initial occupancy conditions of carriers traps controlled by the filling pulse was described. The correspondence between DLTS and DRL peaks in the photon energy range from 0.75 eV – 0.9 eV was established by comparison of DLTS profiles and Pulsed-CL profiles. It was found that both 795 meV and 815 meV CL components are caused by recombination through shallow levels with energy of about Ev + 0.1eV. From the analysis of the data it is concluded that the centres responsible for the 815 meV CL component are related to dislocations cores but the centres responsible for the 795 meV CL component are related to defects outside the dislocations cores.

Acknowledgements

Experimental data presented in this work were obtained using the equipment of the Interdisciplinary Resource Center for Nanotechnology at St. Petersburg State University, Russia (http://nano.spbu.ru).

This work was partially supported by the RFBR (Grants 08-02-00893, 10-02-01010).

References

[1] M. Kittler, X. Yu, T. Mchedlidze, T. Arguirov, O. F. Vyvenko, W. Seifert, M. Reiche, T. Wilhelm, M. Seibt, O. Voß, A. Wolff and W. Fritzsche 2007 *Small* **3** 964

[2] A. Bondarenko, O. Vyvenko, N. Bazlov and O. Kononchuk 2009 *Phys. B: Cond. Matter* **404** 4608

[3] M. Trushin, O. Vyvenko, T. Mchedlidze, O. Kononchuk and M. Kittler 2009 *Solid State Phen.* **156-158** 283

[4] S. Pizzini, M. Guzzi, E. Grilli and G. Borionetti 2000 *J. Phys.: Condens. Matter* **12** 10131

Journal of Physics: Conference Series **281** (2011) 012009 doi:10.1088/1742-6596/281/1/012009

Giant Poole-Frenkel effect for the shallow dislocation-related hole traps in silicon

M Trushin[1], O Vyvenko[2], V Vdovin[2] and M Kittler[1,3]
[1] Joint Lab IHP/BTU, Konrad-Wachsmann-Allee 1, 03046 Cottbus, Germany
[2] V.A. Fok Institute of Physics, St. Petersburg State University, Ulyanovskaya 1, 198504 St. Petersburg, Russia
[3] IHP microelectronics, Im Technologiepark 25, 15236 Frankfurt (Oder), Germany

E-mail: vyvenko@gmail.com

Abstract. The results of a theoretical calculation of the Pool-Frenkel effect due to the strain field of screw and 60° dislocations upon the valence band in silicon, and of a detailed DLTS study of the electrical field impact on carrier emission from the dislocation-related states of two types of bonded samples are presented and discussed. A good agreement between the theory and experiment was established. It is concluded that the large Pool-Frenkel coefficient value that significantly exceeds the value for a Coulomb-like potential is a new distinguishing feature of the hole thermo-emission from dislocation-related levels in silicon.

1. Introduction

During the last few decades, dislocation-related electronic states in the band gap of silicon were intensively studied by means of deep level transient spectroscopy (DLTS) and numerous data were collected (for review see [1]). The main well-known features of the dislocation-related DLTS-signal are the broadening of the DLTS peaks and the logarithmic dependence of their magnitude on the refilling pulse duration. Very recently it was reported [2] that the main hole traps in p-type silicon with a dislocation network (DN) exhibited a strong field-enhanced thermal emission. A linear dependence of the activation energies for the hole emission on the square root of the electric field (the Pool-Frenkel effect) was established. In addition, the Pool-Frenkel coefficient value was found to be different in samples with different dislocation structures, and to exceed the theoretical value for the Coulomb potential. That led to the assumption [2] of a dislocation-strain-related origin of the phenomenon observed.

In this work we performed theoretical calculations of the Pool-Frenkel effect due to the strain field of screw and 60° dislocations for the valence band in silicon, and more detailed experimental investigations of the impact of the electrical field on the carrier emission from the dislocation-related states of similar types of samples produced by silicon direct wafer bonding by means of diverse techniques of capacitance-transient analysis and of transmission electron microscopy (TEM). A good agreement between the theory and experiment was established. From this fact we conclude that giant Pool-Frenkel coefficient value is a new distinguishing feature of 60° dislocations in silicon.

2. Theory of Pool-Frenkel effect due to elastic strains of dislocations

2.1. Isolated dislocations

The Pool-Frenkel effect is a lowering of the potential barrier for carrier thermoemission from a deep level as the result of the composition of its attractive, Coulomb-like potential that is inversely proportional to the distance, r, and of the potential of a uniform electrical field, F (see figure 1).

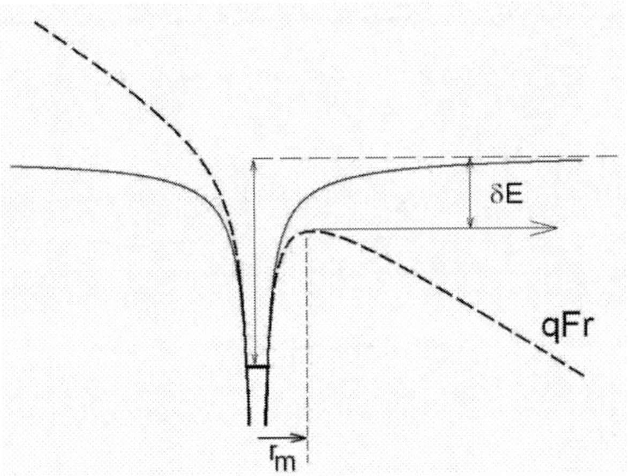

Figure 1. Coulomb attractive potential (solid line) and its changes in the presence of a uniform electric field (dashed line).

The resulting potential is three dimensional (3D) and depends on the angle θ between F and r as:

$$V(r) = -q(\frac{A}{r} + Fr\cos\theta) \tag{1}$$

It exhibits a maximum at the distance (see figure 1):

$$r_{\mathrm{m}} = \sqrt{\frac{A}{F\cos\theta}} = \frac{\beta}{2\sqrt{F}}, \tag{2}$$

giving rise to the reduction of the emission activation energy by

$$\delta E_T = -q\beta\sqrt{F} \tag{3}$$

where

$$\beta = 2\sqrt{A\cos\theta} \tag{4}$$

is the Pool-Frenkel constant and q is the elemental charge.

For a Coulomb potential $A=q/4\pi\varepsilon$ that gives the maximum β-value of $2.22\times10^{-4}\mathrm{V}^{-1/2}\mathrm{cm}^{1/2}$ for $\cos(\theta)=1$ corresponding to the one-dimensional (1D) case. The calculation of the field-enhanced emission in the 3D-case needs the integration over spatial angles [3] that can be performed analytically and gives rise to a lower β-value of about $2\times10^{-4}\mathrm{V}^{-1/2}\mathrm{cm}^{1/2}$.

The deformation potential of dislocations is also inversely proportional to the distance [4]. Thus, a qualitatively similar field effect on the thermal emission of the carriers from electronic states located close to the dislocation cores can be expected. In the following we perform calculations of the Pool-Frenkel effect for the hole emission from the dislocation-related level to the valance band (Γ-point) in silicon. For this zone point, the theory [5] predicts that both uniaxial and shear components of dislocation deformation tensor can contribute to the deformation potential.

The components of the deformation tensor are usually expressed in the dislocation coordinate system where x coincides with the dislocation line which is [110] for both screw and 60° dislocations in silicon. Two other coordinates directions are arbitrary and we will take z to be along [001] and y to be along [1-10] directions.

In this coordinate system, the deformation tensor for screw dislocations consists of two non-diagonal components ε_{xy}, ε_{zx}:

$$\varepsilon_{xz} = -\frac{B_x}{4\pi r}\sin\theta \qquad \varepsilon_{xy} = \frac{B_x}{4\pi r}\cos\theta \tag{5}$$

For edge dislocations, the tensor consists of three components that also contain hydrostatic deformation ε:

$$\begin{aligned}
\varepsilon_{yy} &= -\frac{B_y}{4\pi r(1-\nu)}\left(\cos\theta\{(1-2\nu)+2\sin^2\theta\}\right)+\frac{B_z}{4\pi r(1-\nu)}\left(\sin\theta\{(1-2\nu)-2\cos^2\theta\}\right)\\
\varepsilon_{zz} &= -\frac{B_y}{4\pi r(1-\nu)}\left(\cos\theta\{(1-2\nu)-2\sin^2\theta\}\right)+\frac{B_z}{4\pi r(1-\nu)}\left(\sin\theta\{(1-2\nu)+2\cos^2\theta\}\right)\\
\varepsilon_{zy} &= \frac{B_y}{4\pi r(1-\nu)}\left(\sin\theta\{1-2\cos^2\theta\}\right)-\frac{B_z}{4\pi r(1-\nu)}\left(\cos\theta\{1-2\sin^2\theta\}\right)\\
\varepsilon &= Sp(\varepsilon_{ii}) = \frac{(1-2\nu)}{2\pi r(1-\nu)}\left(B_z\sin\theta - B_y\cos\theta\right)
\end{aligned} \tag{6}$$

where B_x, B_y and B_z are the projections of the Burgers vectors along corresponding directions, r is the distance from dislocation, the angle θ is counting out from z axis and ν is the Poisson ratio (ν=0.3 for Si).

The calculations of the dislocation deformation potential have to be performed in the natural crystal coordinate system that for cubic lattices is x'=[100], y'=[010], z'=[001] [5]. In this system the deformation tensor for the dislocation with an arbitrary Burgers vector can obtained by a 45° rotation of the dislocation coordinates around the z-axis and has the general form as follows:

$$\begin{pmatrix}
\varepsilon_{yy}/2+\varepsilon_{xy} & \varepsilon_{yy}/2 & \frac{\sqrt{2}}{2}(\varepsilon_{zy}+\varepsilon_{zx})\\
\varepsilon_{yy}/2 & \varepsilon_{yy}/2-\varepsilon_{xy} & \frac{\sqrt{2}}{2}(\varepsilon_{zy}-\varepsilon_{zx})\\
\frac{\sqrt{2}}{2}(\varepsilon_{zy}+\varepsilon_{zx}) & \frac{\sqrt{2}}{2}(\varepsilon_{zy}-\varepsilon_{zx}) & \varepsilon_{zz}
\end{pmatrix} \tag{7}$$

Following the deformation potential theory [5] the valence band in silicon at the Γ-point exhibits a shift due to the hydrostatic component, and a splitting into two sub-bands due to shear components of deformation tensor. Taking into account the explicit form of deformation tensor (7), the resulting deformation potential of the dislocation can be written as:

$$\begin{aligned}
V_{1,2} &= aSp(\varepsilon_{ii}) \pm\\
&\pm\left(\frac{b^2}{2}[(2\varepsilon_{xy})^2+(\varepsilon_{zz}-\frac{\varepsilon_{yy}}{2}+\varepsilon_{xy})^2+(\varepsilon_{zz}-\frac{\varepsilon_{yy}}{2}-\varepsilon_{xy})^2]+\frac{d^2}{2}((\varepsilon_{zy}+\varepsilon_{zx})^2+(\varepsilon_{zy}-\varepsilon_{zx})^2+\frac{\varepsilon_{yy}^2}{2})\right)^{1/2}
\end{aligned} \tag{8}$$

where numerical values of the deformation potential constants for the valence band in silicon are a=5.1eV, b=2.2eV and d=5.1eV.

Using equation (8) together with the explicit form of the deformation components (5) and (6), we calculated the parameters $A = Vr/qB$ and β that quantify the Pool-Frenkel effect. In the calculations the electrical field was directed along z. For a pure screw dislocations we took $B_x = B = 0.384$ nm. For the 60° dislocation, $B_x = B/2$, $B_z = 0.707B$, whereas B_y was either $+B/2$ or $-B/2$ corresponding to those two possible orientations of the 60° dislocation with respect to the (001) plane marked as b1 and b2 in figure 2A, respectively.

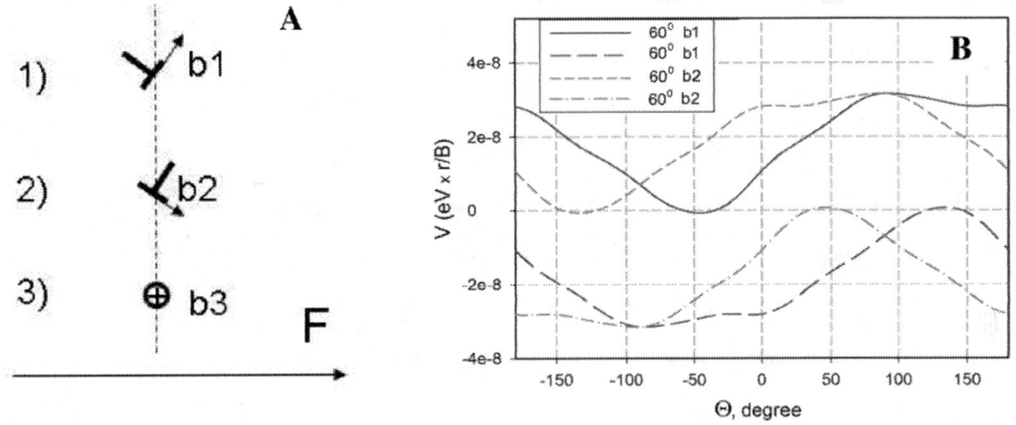

Figure 2. (A) Orientations of 60° (b1, b2) and screw (b3) dislocations in the projection on (110) plane. Dashed line shows the projection of the (001) plane with the dislocation network. Electric field F is directed to [001]. (B) shows the theoretical angle dependence of deformation potential (A-factor) for 60° dislocations with the orientations b1 and b2 shown in figure 1A. The angle is counted from the direction [001] that coincides with the direction of electric field.

Figure 2B represents the angle dependences of A-factor for 60° dislocations with two possible orientations in the DN plane depicted in figure 2A. The main features of the dependences are: 1) significant splitting of the valence band around the dislocations caused by shear deformations; 2) the lower sub-band forms an attractive potential for most directions that can cause a Pool-Frenkel effect; 3) the shapes of the corresponding potential for the two dislocation orientations are close to each other but shifted by the angle of about 70°.

Figure 3A shows $\beta(\theta)$ calculated for the lower valence band V_2 at the screw, at both types of 60° dislocations as well as for a 3D Coulomb potential. One can note that the $\beta(\theta)$ for screw dislocation is rather similar to the one for the Coulomb potential, while for 60° dislocations it has a larger maximum value especially for the b1 orientation.

The calculation of the field enhanced emission rate for dislocations was performed making use the integration according to:

$$\frac{e_p(F)}{e_{p0}} = \frac{1}{2\pi}\left(\alpha \int_{-\pi/2}^{\pi/2} d\theta \exp[\frac{\delta E_t(\theta)}{kT}] + 1 - \alpha\right) \quad (9)$$

where α is the ratio of the angles where enhanced emission takes place (α is 0.5 for screw and 60° dislocation b1, but is 0.325 for 60° dislocation b2).

The results of numerical integration of equation (9) for all dislocations as well as the previously reported [3] expression for the 3D Coulomb are presented in figure 3B. At sufficiently high electric fields, all of the curves can be well fitted with the Pool-Frenkel law: $e_p(F)/e_{p0} = \exp(\beta_{\text{eff}}\sqrt{F})$. The curves for the Coulomb potential and for the screw dislocation deformation potential for the valence band coincide practically with each other giving $\beta_{\text{eff}} \approx 2\times10^{-4}\text{V}^{1/2}\text{cm}^{-1/2}$. The values for 60° dislocations are $\beta_{\text{eff}}(\text{b1}) \approx 3.2\times10^{-4}\text{V}^{1/2}\text{cm}^{-1/2}$ and $\beta_{\text{eff}}(\text{b2}) \approx 2.4\times10^{-4}\text{V}^{1/2}\text{cm}^{-1/2}$.

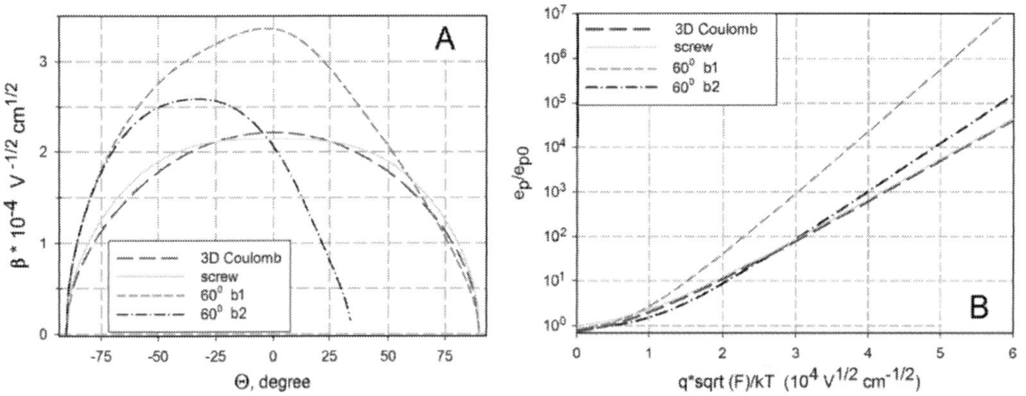

Figure 3. Angle dependence of Pool-Frenkel coefficient (A) and relative enhancement of the hole thermoemission under applied electric field (B) for 3D Coulomb (blue dashed line) and for deformation potential of screw (green solid line), 60 ° dislocation with two orientations depicted in figure 2A: b1 –red short-dashed line, b2- black dashed-dotted line).

2.2. Dislocation network

For the samples under investigation, the interaction of the elastic strains between the dislocations in a network must be taken into consideration that generally results in strain field localization.

Detailed examination of the contribution of different components of deformation tensor to the Pool-Frenkel constant for screw dislocations showed that the most important component is ε_{xy}. Using the explicit expression for the elastic strains of a dislocation row given in [4], we plotted the dependence for this component as a function of the reciprocal distance from the plane of dislocation row (figure 4A) and DN (figure 4B) and compared them with that for the isolated dislocation. The results are presented in figure 4C.

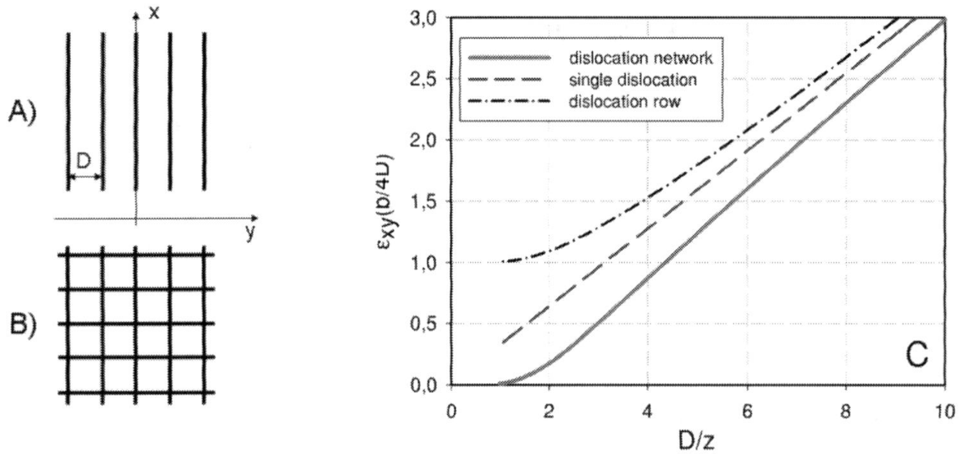

Figure 4. (A) Dislocation row and (B) square dislocation network. (C) shows the deformation of isolated dislocation (long dashed), dislocations in a row A (dashed-dot) and in a square network B (solid). For the DN, the z-coordinate goes through the centre of dislocation segments between the nodes.

One can see that the significant difference of the strains ε_{xy} from $1/r$ law starts at the distances larger than z_{crit} being of about $D/5$ and $D/2$ for dislocation row and for DN respectively. Obviously, the Pool-Frenkel law can be observed under application of the electric field above the value F_{min} that corresponds to Pool-Frenkel radius (r_m , equation 2) of about z_{crit}.

The latter value allows one to estimate F_{min} as a function of D by using relation:

$$F_{\min} > \left(\frac{\beta}{2z_{crit}}\right)^2 \tag{10}$$

or estimate D_{min} for the given electric field ranges:

$$D_{\min} > (2 \div 5)\frac{\beta}{2\sqrt{F}} \tag{11}$$

where the numbers 2 or 5 relate to DN or to the row, respectively. By using these expressions [4], it can be shown that similar estimates are valid for the row of edge type dislocations as well.

3. Samples and experimental technique

Samples were fabricated by direct bonding of 200 mm, p-type Si wafers with (001) surface orientations in a hydrophilic process, i.e. the surfaces of the wafers were covered with native oxide layer before the bonding process [6]. The doping level of the wafers was ~10^{15} atoms/cm^3 of boron. A detailed description of the bonding procedure can be found elsewhere [6],[7]. After the bonding and a thinning procedure of the top layer, a dislocation network was located at the depth of 160 nm from the sample surface. In order to dissolve the interfacial native oxide layer, samples were annealed at 1200°C in high purity conditions [6]. After the annealing neither secondary ion mass spectrometry nor TEM investigations revealed any variations of oxygen concentration near to the bonding interface. Wafers were bonded with two different twist angles of about 1 and 3 ° and are denoted as Gr-1 and Gr-3 respectively. Both samples have the same small tilt angle of about 0.5° that was a result of wafer polishing.

To produce Schottky contacts for DLTS measurements, 100 nm-thick titanium dots with 1.5 mm diameter were evaporated on the oxide free front surface of all samples. Ohmic contacts were prepared by rubbing of InGa on the rear side of the samples. DLTS measurements were performed by means of transient Fourier spectroscopy system DL-8000 (Accent).

4. Experimental results

DLTS spectra of both samples investigated possess similar features: the main, low-temperature peak (labelled as ST1 in Gr-1 and as ST3 in Gr-3) dominates the spectrum and a set of strongly overlapped DLTS peaks appear at temperatures of 100-250K (for details see [2]). It was also found, that the temperature positions of the ST1 and ST3 peaks were shifted toward lower temperatures with the increase of applied reverse bias, i.e. with the increase of the electric field in the depletion region, as shown in figure 5. Note that the sets of DLTS spectra for Gr-1 and Gr-3 samples presented in figures 5A and 5B were measured with the same refilling pulse voltage, U_p= -1V.

Detailed analysis of the shape of the ST1 and ST3 DLTS peaks revealed that their shapes differed from expected for point-like defects. The broadening of DLTS peaks can generally affect the precision of the E_a definition and can be caused either by the widening of the electronic spectrum or by spatial or transient variation of the electric field. That is why we performed another type of the experiment to evaluate the Pool-Frenkel coefficient value, that is isothermal relaxation Spectroscopy (ITS).

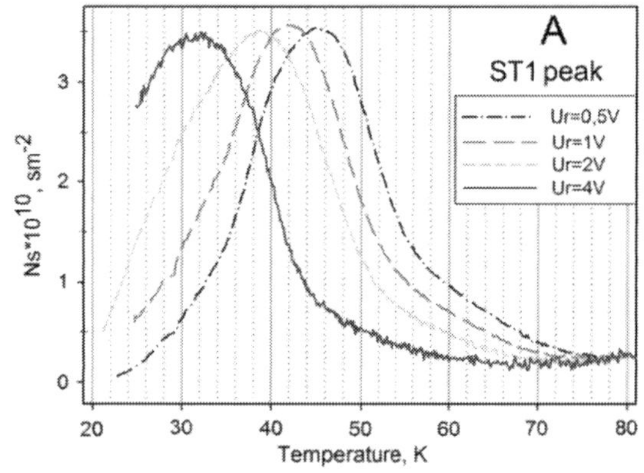

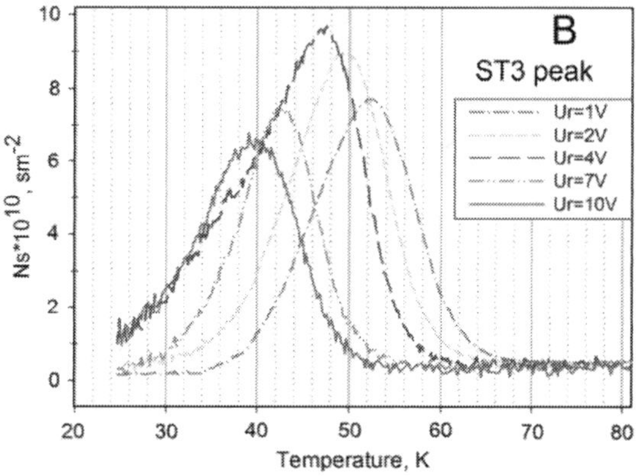

Figure 5. The low temperature part of the DLTS spectra of Gr-1 (5A, left) and Gr-3 (5B, right) samples, measured with different reverse bias voltages. DLTS-scan parameters: filling pulse duration t_p=100μs, filling pulse voltage U_p= −1V, and rate window T_w=20ms. Reverse bias voltages, U_r, are shown in the legends.

ITS measurements were performed at constant temperature under the gradually increasing rate window period using the same correlation procedure as DLTS. Since the response of the DLTS correlator is a function of the product of the emission rate, e_p, and the correlator period, τ, the dependence of τ on the electrical field obtained at a specific level of ITS signal (i.e. keeping the product $e_p\tau$ = constant) will be inversely proportional to the dependence on e_p. Thus, when the broadening of a DLTS line is caused by a widening of the electronic spectrum, the dependence of $\ln(e_p)$ on the square root of electric field must have the same slope, β, for all the ITS signal level. Alternatively, when the broadening is caused by electric field variations β might vary dependent on the ITS signal level.

Examples of ITS spectra of the Gr-1 and Gr-3 samples measured with different reverse bias voltages, U_r, and the same filling pulse voltage U_p =-1V are presented in figure 6. Measurement temperatures were chosen to be 40K for Gr-1 sample and to be 53K for Gr-3 sample that allowed one to use the largest interval U_r voltages for the ITS spectra acquisition inside the usable rate window ranges of our transient recorder.

The electric field at the position of the DN (d = 160nm below the front sample surface) for every reverse bias voltage was calculated using the standard relation $F = qN_a(w\text{-}d)/\varepsilon$, where N_a is the net acceptor concentration, and w is the SCR-width calculated from capacitance data at the DLTS/ITS measurement temperature. Generally, the charge contained within DN states should be taken into account for the electric field calculations as well. Unfortunately, the ground charge state of the detected traps is not known precisely. Nevertheless, we believe the charge collected at the DN will not significantly modify the results of the field definition since the total charge of the doping impurity inside the space charge region, eN_aw, is much larger than the total concentration of all traps detected by DLTS, i.e. $eN_aw > e\sum N_{\mathrm{DLTS}}$ for all used U_r voltages.

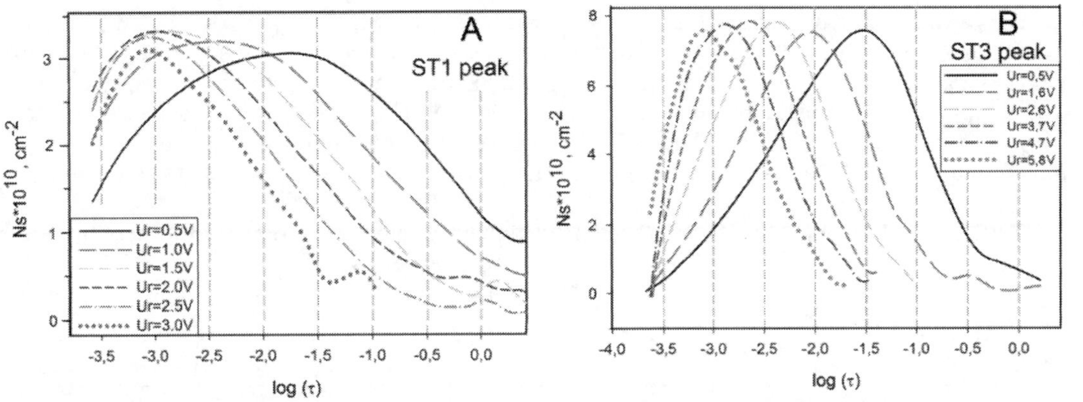

Figure 6. Sets of ITS spectra of (A) sample Gr-1 and (B) sample Gr-3, measured with different reverse bias voltages. Ordinate axes show the surface trap density, abscissa axes – the natural logarithm of the emission time constant τ. ITS-scan parameters: measurement temperatures are 40K for Gr-1 sample and 53K for Gr-3 sample, filling pulse duration t_p=100μs, and filling pulse voltage U_p= –1V. Reverse bias voltages, U_r, are shown in the legends.

Using the values of the electric field calculated in such a way, dependences of activation enthalpies E_a derived from the DLTS spectra, and $kT\ln(e_p)$ derived from the ITS measurements based upon the square root of electrical field $\sqrt{F}$ were plotted, as shown in figures 7A and 7B, respectively. One can see that all the dependences obtained for both ST1 and ST3 peaks could be well approximated by a straight line, and thus follow the Poole-Frenkel law.

The slopes of the linear dependences obtained, which correspond to Poole-Frenkel coefficients β, as well as the activation enthalpies, E_{a0}, at zero electric field approximated from DLTS data (figure 7A), are summarized in table 1. It should be noted also that β-values obtained from the ITS-data at different signal levels coincided within the measurement errors, indicating that the broadening of the DLTS spectra arises from the broadened spectra of the electronic states.

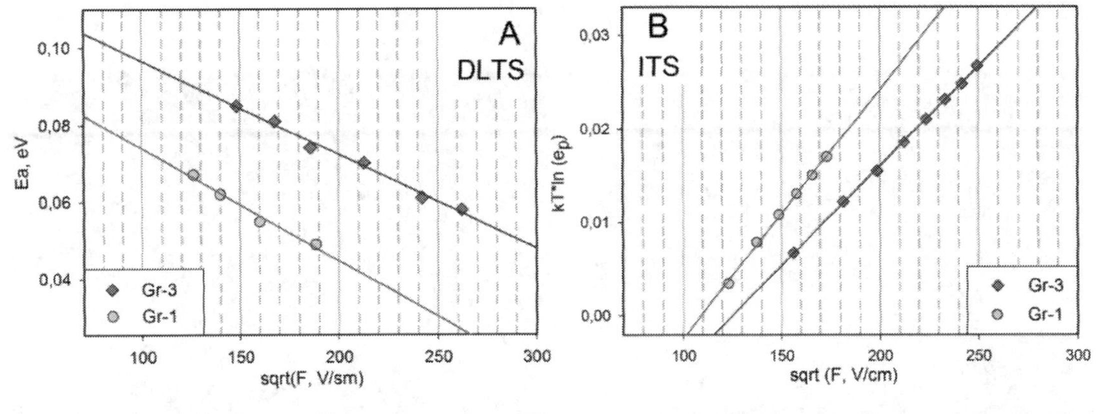

Figure 7. Dependences of the activation enthalpies, E_a, derived from (A) DLTS, and (B) the kT-corrected logarithm of emission rates derived from ITS dependence on the square root of electric field for the ST1 and ST3 peaks in Gr-1 and Gr-3 samples. Symbols show experimentally obtained values, and straight lines linear approximations.

The β–values are noticeably larger than those expected for a 3D Coulomb potential, especially for sample Gr-1. The values obtained from the DLTS measurements are slightly bigger than those from the ITS-data. The reason of this discrepancy is not known precisely and might be caused by a systematic error in temperature measurements affecting mostly the determination of the activation energies E_a from Arrhenius plots. The difference between the zero-field activation enthalpy, E_{a0}, for the ST1 trap level and that for the ST3 level at about 15 meV is close to the their definition error.

Table 1. Poole-Frenkel coefficient β and activation enthalpies E_{a0} at zero electric field for ST1 and ST3 peaks as derived from ITS and DLTS measurements.

		Gr-1 sample ST1 peak	Gr-3 sample ST3 peak
ITS	β, (10^{-4} cm$^{1/2}$V$^{-1/2}$)	2.8	2.1
DLTS	β, (10^{-4} cm$^{1/2}$V$^{-1/2}$)	3.0	2.4
	E_{a0}, (eV)	0.105	0.12

5. Discussion

Our theoretical calculations presented in section 2 showed that the β–coefficient value of the dislocation-related Pool-Frenkel effect depends strongly on the type, on the orientation and on the density of dislocations. Figure 8 shows TEM micrographs of the dislocation structure of the DN of both samples investigated.

Detailed analysis of the structure showed that most of the dislocations in sample Gr-1 (figure 8A) seemed to be 60° dislocations, though the exact direction of Burgers vector could not be defined precisely from TEM data. The structure in figure 8A looks like a row of parallel dislocations separated by distances of about 60 nm rather than a square DN.

The electric field values used in the experiment with this sample started from 14 kV/cm. The minimal distance when the theory of isolated dislocation is still valid can be estimated according to equation (10) for that field is of about 40 nm. Thus, the theoretical conditions are satisfied and the predicted Pool-Frenkel coefficient value is expected to be about 3×10^{-4}V$^{1/2}$cm$^{-1/2}$ for the case of the most favourable orientation b1 of 60^0 dislocations in the DN shown in figure 2. This is in good agreement with the β–value for this sample (see table 1).

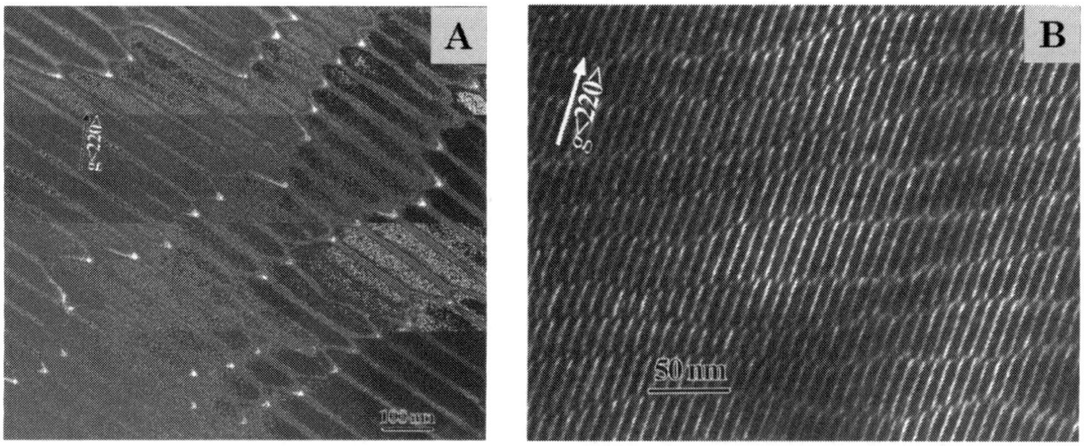

Figure 8. Dark field TEM images of the samples under investigation. (A) sample Gr-1 and (B) sample Gr-3.

The dislocation structure of sample Gr-3 consists of square mesh of perfectly straight screw dislocations (only one set of them is visible in figure 8B due to particular diffraction conditions) and the rows of curved 60° dislocations that are inclined to the screw dislocation lines on average at an angle of about 30°. The inter-dislocation distances for screw dislocations are of about 7 nm, while for the 60° dislocations they are varying from 30 to 50 nm. For the electric field ranges used in the experiments (see figure 7) the Pool-Frenkel radius is estimated to be of 8-14 nm for the mesh and 20-33 nm for the row. Hence, the validity of the theory for isolated dislocation is satisfied for 60° dislocations but not for screw dislocations. Thus, the screw dislocations have to be ruled out of the consideration as a possible origin for the observed Pool-Frenkel effect. The Pool-Frenkel coefficient value obtained from experiment $\beta \approx (2.1—2.4) \times 10^{-4} V^{1/2} cm^{-1/2}$ corresponds to the theoretically predicted value for 60° dislocation in the orientation b2 in figure 2.

The difference between the experimental β-value for the Gr-1 and Gr-3 samples can be then explained due to different crystallographic orientations of the handle and the top bonded wafers. The predominant 60° dislocation orientations b1 or b2 in figure 2 could be realized due to an asymmetry of the tilt misorientation between the handle and the top wafers, for example, when the handle wafer has exact (100) orientation and the tilt between the wafers is due to misorientaion of the top wafer only or vice versa. Since the tilt between the bonded wafers was not controlled during the bonding process and arose as the result of wafers polishing such an explanation seems to be reasonable. Nevertheless, one can not exclude that the lower β in the Gr-3 sample was due to the interaction of 60° dislocations with screw dislocations giving rise to a zigzag configuration of resulting dislocations [8]. Such a configuration, obviously, could have the strain field that differs significantly from that of a straight dislocation.

6. Conclusion

In conclusion we have demonstrated both theoretically and experimentally a new dislocation-related phenomenon that is the Pool-Frenkel effect caused be elastic strains of dislocations in silicon. We showed that Pool-Frenkel coefficient can reach the value that significantly exceeds the value for a Coulomb-like potential of a singly charged point defect. The presence of dislocations with different Burgers vectors and different orientations with respect to the electric field can cause a drastic broadening of DLTS-peaks.

Acknowledgements

The authors are grateful for helpful discussion to T. Mchedlidze and to O. Kononchuk.

References

[1] Schroeter W and Cerva H 2002 *Solid State Phenom.* **85-86** 67
[2] Trushin M, Vyvenko O, Mchedlidze T, Kononchuk O and Kittler M 2010 *Sol. St. Phenom.* **156-158** 283
[3] Hartke J L 1968 *J. Appl. Phys.* **39** 4871
[4] Hirth J P and Lothe J 1970 *Theory of dislocation.* (New York: McGraw-Hill)
[5] Bir G L and Pikus G E 1974 *Symmetry and deformation effects in semiconductors.* (Wiley, New York)
[6] Kononchuk O, Boedt F, and Allibert F 2008 *Sol. St. Phenom.* **131-133** 113
[7] Reiche M 2008 *Mater. Sci. Forum* **590** 57
[8] Föll H and Ast D 1979 *Philos. Mag. A* **40** 589

Electrical levels of dislocation networks in p- and n-type Si

I Isakov [a]*, A Bondarenko [a], O Vyvenko [a], V Vdovin [a], E Ubyivovk [a] and O Kononchuk [b]

[a] V.A. Fok Institute of Physics, St. Petersburg State University, 198504 St. Petersburg, Russia

[b] SOITEC, Parc des Fontaines, F-38560, Bernin, France

*Corresponding author. E-mail: isakov@nano.spbu.ru

Abstract. The results of deep level transient spectroscopy (DLTS) and minority carrier transient spectroscopy (MCTS) investigations on directly bonded n- and p-type silicon wafers with small twist misorientation angles ranging from 1 to 5 degrees are presented and discussed. Both shallow and deep levels in the upper half of a band gap are found and a good correspondence between the DLTS and MCTS data on n- and p-type samples was established. The dependence of DLTS-peak magnitude on twist and tilt misorientation angles (density of dislocations) was investigated and the origin of different levels is suggested.

1. Introduction

Dislocation networks (DN) obtained by the direct wafer bonding (WB) technique are perfect model objects for deeper understanding of the electronic properties of dislocations in silicon [1,2] since this technology allows one to create samples with DN of well controlled structure at depths well suited for space charge spectroscopy measurements and transmission electron microscopy (TEM) imaging. Such investigation might prompt development of models for dislocation-related phenomena that are attracting interest due to their importance for future application in dislocation related luminescence [2] and conductivity enhancement along DN in Si DN [3].

DLTS measurement of WB DNs were carried out mostly on p-type Si and only levels in the lower half of band gap were examined. In this work we present the result of a combined study of the dislocation-related electrical levels in the upper half of the band gap by means of DLTS measurements on n-type Si WB or by MCTS measurements on p-type Si and of the dislocation structure by TEM.

2. Experimental Method

A set of p-type (boron doped) and n-type (phosphorus doped) Czochralski grown Si wafers with initial doping concentration of 10^{15} and 10^{14} cm^{-3} respectively and (100) surface orientation were bonded hydrophilically with twist misorientation angles (TMA) of approximately 1° — 5°, and unavoidable tilt angles of about 0.5°. The wafers were thinned in order to have the DN depth of 200 nm from the surface. Details on the bonding procedure can be found elsewhere [2]. Semitransparent titanium (gold) contacts for p- (n-) type silicon with a thickness of about 20 nm were sputtered on top of the wafers to prepare Shottky diodes.

DLTS and MCTS measurements were conducted using a SULA Tech. Inc. instrument. Sample temperature was varied from 20 to 300 K by means of a Janis closed cycle helium cryostat. All measurements were carried out at the same initial conditions. The reverse bias voltage was kept –1 V. The filling pulse voltage for majority carrier excitation was +2 V with a duration of 200 μsec, time period between pulses was 50 ms, and the rate windows ranged from 0.2 ms to 20 ms. A light emitting diode with a wavelength of 890 nm was used to provide optical excitation of minority carriers. In order to examine the influence of the electrical field on the level position the applied reverse bias voltage was varying from 0 to 12 V. To retrieve the concentration of levels from data we used the following formula:

$$N_{ds} = \frac{(A\varepsilon\varepsilon_0)^2}{x_d} eN_d \frac{\Delta C}{C_3}, \quad (1)$$

where A is contact area, x_d is depth from the surface, N_d is doping concentration, C is capacitance, ΔC is the signal magnitude of DLTS (MCTS) peak.

TEM analysis was performed using a Zeiss Libra 200FE transmission electron microscope at different diffraction conditions. The electron beam acceleration voltage was 200 kV.

3. Results and discussion

3.1. Microstructure

TEM micrographs revealed regular screw dislocation networks compensating the twist, interrupted by dark lines related to tilt (figure 1a, b, c, right column). For low-angle samples edge dislocations are presented by zigzag lines of 60° dislocations (figure 1a right) described in the literature [4]. High-angle samples exhibit Moiré patterns which result in visually reducing the distance between dislocations by a factor of two (figure 1b right). We have extracted the twist and tilt angles between bonded wafers from measured distance between dislocations and using Frank's formula for fcc crystals

$$d \cong a / \sqrt{2} \sin\theta \quad (2)$$

(where d is a distance between dislocations, $a = 0.543$ nm is a lattice constant for Si, and θ is a misorientation angle). The summary of structural properties for all the samples is given in Table 1. It should be noted that the tilt in these samples was in the range 0.4° — 0.9°. Some pairs of p- and n-type samples, e.g. n-3.5°/0.6° and p-3.7°/0.7°, n-5.0°/0.45° and p-4.9°/0.9°, have close twist and tilt angle values and thus could be compared by charge spectroscopy techniques.

All the samples but one have 60°-dislocation families parallel to one of the component of screw dislocations square network (figure 1 right). Only the sample p-3.0°/0.7° has an edge dislocation line at about 30° (or 60°) to one component of the screw dislocation array.

Table 1. Summary data derived from TEM observations: *dtw* and *dti* — average distances between twist and tilt dislocations respectively. *Dsc* and *Ded* are the line densities of screw and edge dislocations. α_{TW} and α_{TI} are the twist and tilt misorientation angles calculated using Frank's formula (2). The labels are given with respect to the twist/tilt misorientation angles.

Sample label	α_{TW}, °	*dtw*, nm	Dsc, 10^6 cm^{-1}	α_{TI}, °	*dti*, nm	Ded, 10^6 cm^{-1}
"n-0.9°/0.55°"	0.9±0.1	23±3	0.43	0.55±0.15	39±9	0.26
"n-1.5°/0.55°"	1.7±0.1	12.9±0.7	0.78	0.55±0.1	38±7	0.26
"n-3.5°/0.6°"	3.5±0.2	6.3±0.3	1.6	0.6±0.1	35±7	0.29
"n-5.0°/0.45°"	5.0±0.4	4.5±0.4	2.2	0.45±0.1	46±6	0.22
"n-5.7°/0.5°"	5.7±0.2	3.8±0.2	2.6	0.5±0.1	40±4	0.25
"p-2.3°/0.7°"	2.3±0.1	9.6±0.3	1.0	0.7±0.1	32±5	0.31
"p-3.0°/0.7°"	3.0±0.1	7.4±0.2	1.4	0.7±0.05	30±4	0.33
"p-3.7°/0.7°"	3.7±0.4	6.0±0.6	1.7	0.7±0.05	32±3	0.31
"p-4.9°/0.9°"	4.9±0.5	4.6±0.4	2.2	0.9±0.05	25±2	0.4

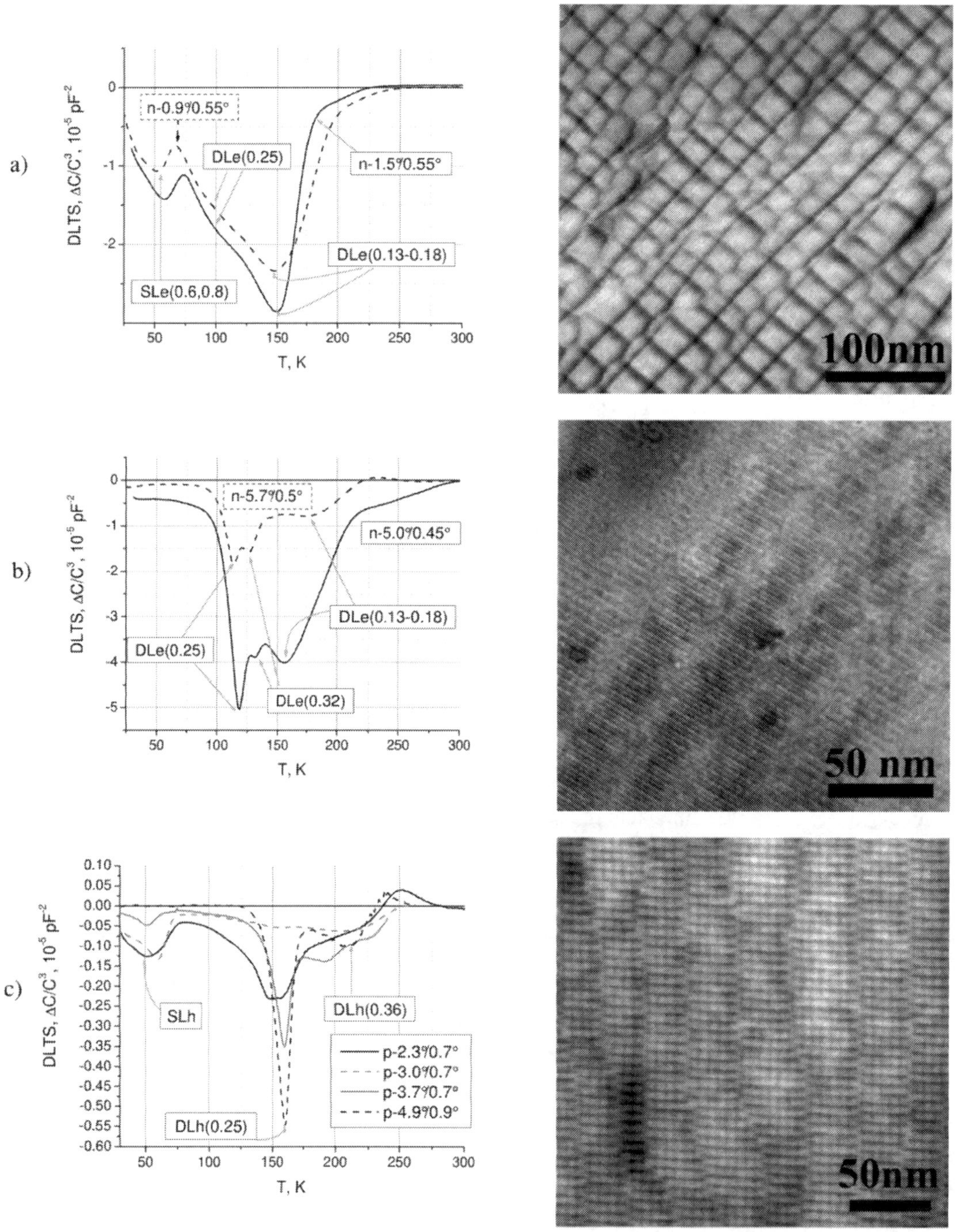

Figure 1. DLTS-spectra (left) for different subsets of samples. a) ""n-0.9°/0.55°"" ,"n-1.5°/0.55°" b) "n-5.0°/0.45°", "n-5.7°/0.5°" c) all the p-type samples. Corresponding TEM-images (right) of a) ""n-0.9°/0.55°"", b) "n-5.0°/0.45°", c) "p-3.7°/0.7°". DLTS conditions: $V_{rev} = -1V$; $V_{pulse} = +2V$; $t_{pulse} = 200\mu s$; $T_{period} = 50ms$, emission rate window $20ms$.

3.2. Space charge spectroscopy

Typical DLTS-spectra obtained on diverse samples are presented in figure 1 (left column). A +2 V pulse was used as it was determined to saturate all the levels. Summarized details of the DLTS and MCTS measurements are shown on the Arrhenius plots in figure 2 and in Table 2. In the present work for the first time we have observed levels in the upper-part of the band gap corresponding to dislocation network in WB Si. The overview of these levels is presented below.

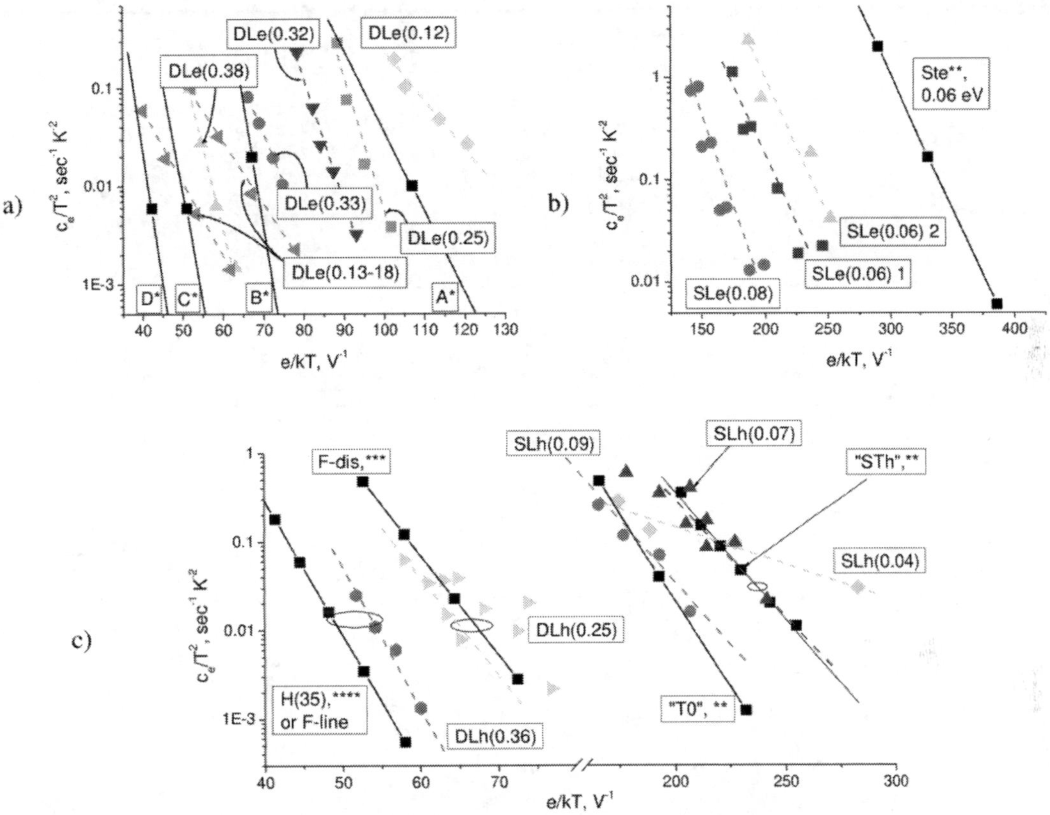

Figure 2. Arrhenius plots derived from DLTS and MCTS measurements for electron and hole traps. a) deep electron traps; b) shallow electron traps; c) hole traps. The dashed (colour) lines represent linear fitting of our data, solid black lines and black squares represent data reported in literature:
* – Omling [5]; ** – Castaldini et al. [6]; *** – Vyvenko et al. [7], **** – Kimerling and Patel [8]. Arrhenius plots attributed by us to the same level are joined by black ovals. See discussion in the text.

Table 2. The list of levels and samples with corresponding literature data (** – Castaldini et al. [6]; **** – Kimerling and Patel [8])

Level family	Level label (activation energy)	Cross-section, cm^2	Samples observed	Level family	Level label (activation energy)	Cross-section, cm^2	Samples observed
SLe	SLe(0.06) 1,	10^{-17}	n-0.9°/0.55°	SLh	SLh(0.07)/ STh** [6]	10^{-16}–10^{-17}	p-3.7°/0.7°, n-3.5°/0.6°, n-1.5°/0.55°
	SLe(0.06) 2	10^{-18}	p-3.0°/0.7°		SLh(0.09)/ T0** [6]	10^{-14}–10^{-15}	p-3.0°/0.7°
	SLe(0.08)	10^{-15}–10^{-16}	n-1.5°/0.55°, n-3.5°/0.6°, p-3.7°/0.7°		SLh(0.05)	10^{-17}–10^{-18}	p-2.3°/0.7°
DLe narrow	DLe(0.25)	10^{-12}–10^{-13}	n-0.9°/0.55°, n-1.5°/0.55°, n-3.5°/0.6°, p-3.7°/0.7°, n-5.0°/0.45°, n-5.7°/0.5°	DLh	DLh(0.25)	10^{-13}–10^{-14}	p-2.3°/0.7°, p-3.7°/0.7°, p-4.9°/0.9°, p-3.0°/0.7°
	DLe(0.33)	$5\cdot10^{-14}$	n-3.5°/0.6°, p-3.7°/0.7°, p-3.0°/0.7°		DLh(0.15)	10^{-18}	n-0.9°/0.55°, n-1.5°/0.55°, n-5.0°/0.45°, n-5.7°/0.5°
	DLe(0.38)	10^{-15}	n-3.5°/0.6°				
	DLe(0.32)		n-5.0°/0.45°, n-5.7°/0.5°		DLh(0.36)/ F-line,****[8]	10^{-13}	p-3.0°/0.7°, p-2.3°/0.7°, p-3.7°/0.7°, p-4.9°/0.9°, n-3.5°/0.6°
DLe wide	DLe(0.12)	$5\cdot10^{-12}$–$5\cdot10^{-13}$	p-3.0°/0.7°				
	DLe(0.13-0.18)	10^{-19}–10^{-21}	n-0.9°/0.55°, n-1.5°/0.55°, n-3.5°/0.6°, n-5.0°/0.45°, n-5.7°/0.5°				

3.2.1. Levels in the upper half of the band gap in Si

Shallow electron traps (SLe) were found in some n- and p-type samples. Shallow 1D bands have been predicted near both conduction and valence bands in the dislocated semiconductors [9]. Shallow electron traps were first observed in [6] in plastically deformed Si (figure 2b, "Ste**"). The activation energies of SLe(0.06) 1 and SLe(0.06) 2 coincide well with this shallow electron trap ($E_{Ste^{**}} = 0.06eV$), however, the cross-sections differ significantly from those reported previously (figure 2b). The activation energies measured in the experiment could represent a difference between certain trap energy and the bottom of 1D-band ($E_C - E_{Dle}$) as was proposed by Kveder *et al.* in [10]. This is an unlikely scenario since in that case the level corresponding to transition between trap and conduction band with $E_t = 0.06(0.08) + E_{Dle}$ should be present in the DLTS-spectrum, but it is absent.

Hence the transition SLe(0.08;0.06) should arise between the trap level (or 1D-band) and the conduction band directly.

Shallow electron trap concentration as a function of TMA is presented in figure 3a. The concentration of SLe-level grows with TMA from 1° to 3.5° and then rapidly decreases to zero. This dependence might be explained as a result of the changes of the dislocation deformation potential caused by increasing interaction of the strain field of screw dislocations.

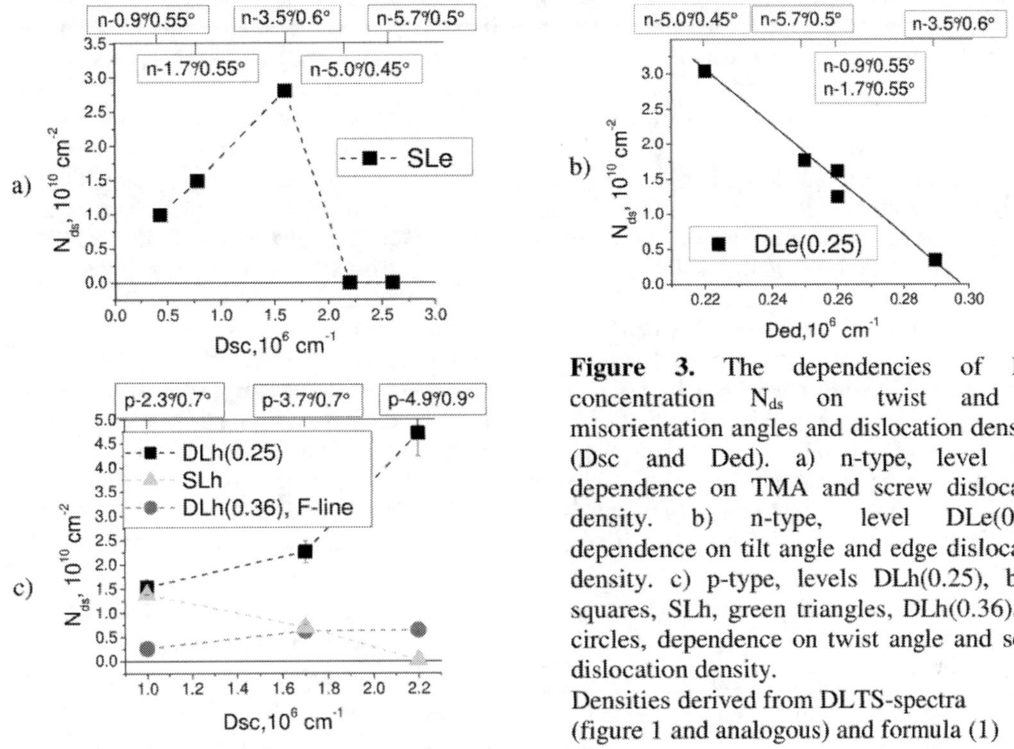

Figure 3. The dependencies of level concentration N_{ds} on twist and tilt misorientation angles and dislocation densities (Dsc and Ded). a) n-type, level SLe, dependence on TMA and screw dislocation density. b) n-type, level DLe(0.25), dependence on tilt angle and edge dislocation density. c) p-type, levels DLh(0.25), black squares, SLh, green triangles, DLh(0.36), red circles, dependence on twist angle and screw dislocation density.
Densities derived from DLTS-spectra (figure 1 and analogous) and formula (1)

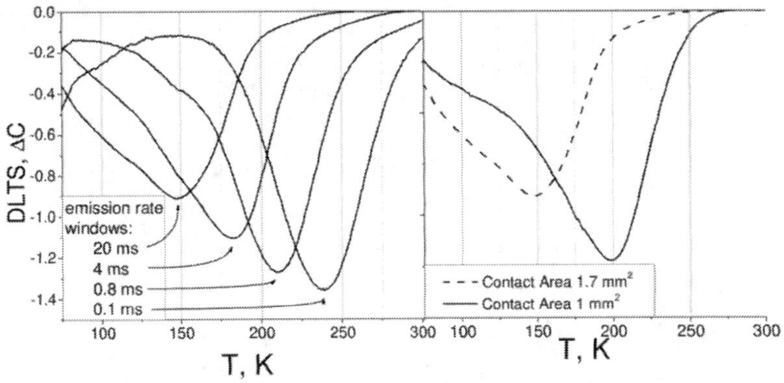

Figure 4. DLTS spectrum of n-2 sample. Typical behaviour of DLe(0.13-0.18) line. Left: peak intensity dependence on emission rate window. Right: peak position dependence on contact area (emission rate window $0.8ms$). DLTS conditions:

$V_{rev} = -1V$; $V_{pulse} = +2V$;
$t_{pulse} = 200\mu s$;
$T_{period} = 50ms$

Secondly, the level DLe(0.25) is observed in all n-type samples. Whereas its concentration has no correlation with TMA it increases with decrease of the tilt misorientation angle, or with the decrease of the edge-type dislocation density (see figure 3b). This proves the 60°-dislocation related origin of level DLe(0.25).

Thirdly, level DLe(0.13-0.18) also observed in all the n-type samples shows peculiar behaviour. This trap has an unexpectedly small cross-section of about 10^{-21} cm^2 with activation energy of 0.15 eV. The DLTS-peak is shifted towards higher temperatures after reducing the area of contact (figure 4 right). The magnitude of the DLTS-peak is increasing with smaller rate window (figure 4 left). Our tentative explanation for these phenomena is as follows: the DLe(0.13-0.18) peak represents a transition between a deep level and a 1D shallow band. The charge thermally emitted from the deep level is in some way trapped on the 1D-band and could be captured by the same deep level or diffuse away from the contact via the conducting channel of the 1D-band. These mechanisms could cause the small cross-section due to much smaller carrier concentration in the 1D-band in comparison with the conduction band, and the dependence on area due to stronger diffusion in the DN plane. The same behaviour of traps was reported in [11] in SiGe quantum wells and explained by lateral diffusion. No correlation between DLe(0.13-0.18) line intensity and twist or tilt misorientation angle was found.

For electron traps there are no direct coincidence with electron levels observed by both Omling [5] ("A","B","C","D" in figure 2a) deep in the band.

n- and p-type samples with similar twist and tilt misorientation angles of ~3.6° and 0.6° respectively were studied by means of DLTS and MCTS to get congruent and comparable data. Figure 5 shows nearly perfect correspondence between DLTS of n-type and MCTS of p-type samples. Levels SLe(0.08), DLe(0.25) and DLe(0.33) are present in both spectra. For DLe(0.25) and DLe(0.33) the concentration of traps are almost the same for both samples, however the low-temperature peak SLe(0.08) is three times larger in the n-type material. The coincidence of peaks allows us to suggest that these peaks are only DN-related and do not originate from contamination or interaction with doping material.

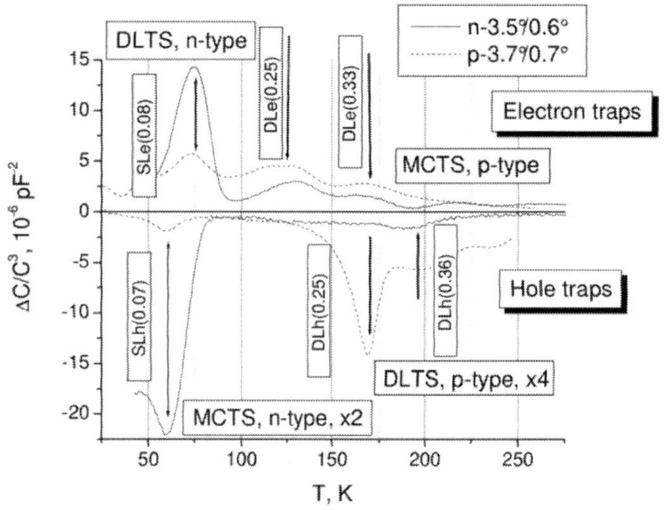

Figure 5. DLTS and MCTS spectra and observed levels for n- and p-type samples with similar misorientation angles of about 3.6°. The corresponding spectra (especially for electrons) coincide with each other well showing the same levels at the same temperatures.

DLTS conditions:
$V_{rev} = -1V$; $V_{pulse} = +2V$;
$t_{pulse} = 200\mu s$; $T_{period} = 50ms$,
emission rate window $0.8ms$.

Figure 1a,b shows DLTS-spectra and corresponding TEM-micrographs to illustrate the dependence of electrical properties of DN on microstructure. While the dislocation network has relatively big cells and pronounced zigzag lines due to interaction between screw and edge dislocations (figure 1a), the most important lines in the spectra are the shallow level SLe and deep level DLe(0.13-0.18). When the distance between dislocations becomes less than 5 nm and zigzag lines are not visible (figure 1b), the shallow levels disappear whereas the narrow deep levels DLe(0.25) and DLe(0.38) appear.

3.2.2. Levels in the lower half of the band gap in Si

Level SLh(0.07) in p-3.7°/0.7°, n-3.5°/0.6° and n-1.5°/0.55° coincides well with shallow traps for holes in Si observed in [6]. Level SLh(0.09) in p-3.0°/0.7° could be related due to their similar behaviour to SLh(0.07). However, it has intersecting points on the Arrhenius plot and some similar features with the T0-level [6], assigned by the authors to FeB pairs. The concentration of SLh levels in the samples p-2.3°/0.7°, p-3.7°/0.7° and p-4.9°/0.9° (figure 3c) is decreasing with increasing twist misorientation angle or screw dislocation density. This behaviour could be explained by the overlapping strain fields between screw dislocations for higher angle sample, which could in principle result in disappearance of higher energy levels.

The activation energy of level SLh(0.05) observed only in the p-2.3°/0.7° corresponds to the 1D-band energy calculated in [9] for dislocations in p-type Si.

Level DLh(0.25) is observed in all the p-type samples. It corresponds to level "F-dis" observed in [7] and "SD" observed in [12]. The dependence of peak intensity on TMA in samples p-3.7°/0.7°, n-3.5°/0.6°, n-1.5°/0.55° shows the monotonous increase with TMA, *i.e.* with screw dislocation density (figure 3c).

We attribute the DLh(0.36) line, which is also present in all the p-type samples to the F-line first observed by Kimerling and Patel [8]. Investigation of the peak intensities from deconvoluted DLTS-spectra shows that the trap concentration increases with the TMA.

These properties of DLh(0.25) and DLh(0.36) could be due to a number of things: proof of the levels' dependence on screw dislocation line length, intersection points between screw dislocations, or intersection points between screw and edge dislocations. The smaller DLh(0.25) and DLh(0.36) level density in low-angle samples could result in less non-radiative recombination via these levels. This behaviour could increase radiative recombination and thus enhance D1-line intensity.

Similarly to the study of the upper half of the band gap (sec. 3.2.1) we have studied the lower half by comparing corresponding DLTS-spectra of p-type samples and MCTS-spectra of n-type samples (figure 5, Hole traps). The temperature position of SLh(0.07) and DLh(0.36) peaks coincide well. However, the p-type sample exhibits a sharp peak near 170 K which is absent in the n-type sample. This can lead to a suggestion that level DLh(0.25), which is present only in p-type samples (see Table 2), can be a boron-related or processing-related level.

3.3. The Pool-Frenkel effect

The wafer bonding technique introduces dislocations at the interface situated parallel to surface. We assume that charge trapped at the interface doesn't determine the electrical field at the interface. Using Poisson's equation we can derive a simple relation between electrical field and voltage to quantitatively investigate the dependence of DLTS-lines on electrical field:

$$E = eN_d \cdot \left(\frac{S}{C(V)} - \frac{x_d}{\varepsilon\varepsilon_0} \right) - \frac{Q_{ss}}{2\varepsilon\varepsilon_0}, \tag{3}$$

where e is elementary charge, N_d is the doping concentration, S is the contact area, $C(V)$ is the capacitance dependence on applied voltage, x_d is the interface depth (200nm) and Q_{ss} is the charge trapped on all the deeper levels, calculated from DLTS-spectra using formula (1).

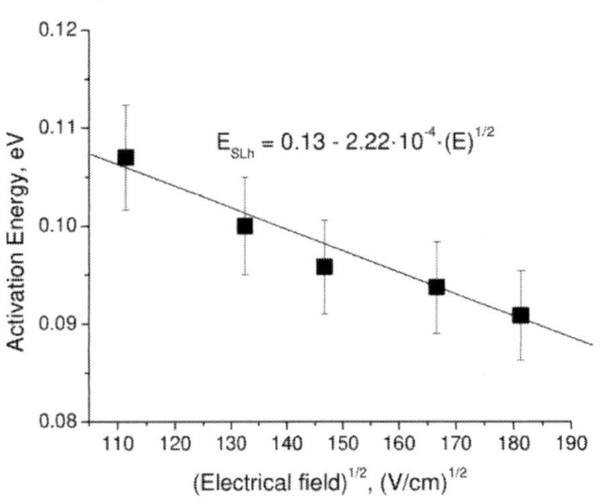

Figure 6. Dependence of level energies on the square root of electrical field. Data measured on sample p-3.0°/0.7°. The Pool-Frenkel effect was observed on level SLh (shallow hole trap, black squares, black solid line represents the linear fitting),

It was reported recently that the SLh-level in analogous samples with TMA = 3° exhibits a Pool-Frenkel shift [12]. The Pool-Frenkel effect was observed only on one of our samples, namely p-3.0°/0.7°. The SLh-level activation energy is plotted versus the square root of electrical field in figure 6. The linear approximation of this data gives the dependence pre-factor of $2.2\cdot10^{-4}$ e $V^{1/2}cm^{1/2}$, corresponding well with previously reported data [12]. The extrapolation results in an activation energy of 0.13 eV at zero electrical field. The Pool-Frenkel effect for the shallow electron trap SLe-level was not observed. The absence of the effect might be caused by different reasons: the electrical field was not high enough due to a lower doping level, or the origins of these two shallow levels differ from each other. The absence of Pool-Frenkel effect in similar samples p-3.7°/0.7° and p-2.3°/0.7° could be due to different dislocation microstructure. In fact, all the samples except the p-3.0°/0.7° have tilt axis exactly parallel to [110], or to the screw dislocation line direction (see figure 1). The angle between tilt axis and [110] in sample p-3.0°/0.7° is about 30°. The difference in this angle results in different types of dislocation reactions between 60°- and screw dislocations, which can in turn result in different strain fields around the dislocations. This feature of DN could also cause the difference between DLe(0.25) intensities in p-3.0°/0.7° and p-3.7°/0.7°.

4. Summary

A set of n- and p-type samples with different twist misorientation angles was investigated by means of DLTS, MCTS and TEM measurements. The whole energy band gap in Si was studied. Shallow and deep levels for electrons and holes were found and characterized in all the samples.

The maximum concentration of shallow traps for electrons was found to be in the 3-degrees misoriented sample. The dependencies of activation energy of shallow levels on electrical field were examined. Only one shallow hole trap in one sample (p-3.0°/0.7°) exhibited the Pool-Frenkel effect.

The levels that might have impact on the D1-luminescence are: SLe and SLh which could be a direct source for transition leading to D1-line luminescence, and levels DLh(0.25) and DLh(0.36) which could be active non-radiative centers reducing the amount of direct SLe to SLh transitions. These assumptions will be verified in forthcoming experiments.

We have found that microstructure and self-arrangement of dislocations can dramatically affect electrical properties of the material. The dislocation density derived from twist and tilt misorientation angles is not sufficient to know the exact properties of DN. The difference in the interaction type between 60° and screw dislocations should be examined. The zigzag lines of 60°-dislocations, the angle between rows of tilt dislocations, square meshes of screw dislocations, and probably other interactions between them can lead to some crucial changes in the electrical and optical properties of the DN.

Acknowledgements
Experimental data presented in this work were obtained using the equipment of Interdisciplinary Resource Center for Nanotechnology of St. Petersburg State University, Russia (http://nano.spbu.ru).
This work was partially supported by RFBR (Grants 08-02-00893, 10-02-01010).

References
[1] Mchedlidze T, Kononchuk O, Arguirov T, Trushin M, Reiche M and Kittler M 2010 *Sol. State. Phenomena* **156-158** 567–572
[2] Kveder V and Kittler M 2008 *Mat. Sc. Forum* **590** 29–56
[3] Ishikawa Y, Yamauchi K, Yamamoto C and Tabe M 2005 *Mater. Res. Soc. Symp. Proc.* **864**
[4] Föll H and Ast D 1979 *Phil. Mag. A* **40** No.5 589–610
[5] Omling P, Weber E R, Montelius L, Alexander H and Michel J 1985 *Phys. Rev. B* **32** 6571
[6] Castaldini A, Cavalcoli D, Cavallini A and Pizzini S 2005 *Phys. Rev. Letters* **95** 076401
[7] Vyvenko O F, Kittler M, Seifert W and Trushin M 2005 *Phys. Stat. Sol. C* **2** No.6 1852-1858
[8] Kimerling L C and Patel J R 1979 *Appl. Phys. Lett.* **34** 73
[9] Farvacque J-L and Franois P 2001 *Phys. Stat. Sol. B* **223** 635–648
[10] Kveder V, Badylevich M, Schröter W, Seibt M, Steinman E and Izotov A 2005 *Phys. Stat. Sol. A* **202** No. 5 901–910
[11] Schmalz K, Yassievich I N, Collart E J and Gravesteijn D J 1996 *Phys. Rev. B* **54** 23
[12] Trushin M, Vyvenko O, Mchedlidze T, Kononchuk O and Kittler M 2010 *Sol. State Phenomena* **156-158** 283-288

Behaviour of oxygen-related thermal donors in Ge crystals Czochralski-grown from the melt covered fully by B_2O_3

Toshinori Taishi[a, *], Yoshio Hashimoto[b], Hideaki Ise[c], Yu Murao[c], Takayuki Ohsawa[c], Yuki Tokumoto[c], Yutaka Ohno[c], Ichiro Yonenaga[c]

[a] Institute of Carbon Science and Technology, Shinshu University, Wakasato, Nagano 380-8553, Japan
[b] Faculty of Engineering, Shinshu University, Wakasato, Nagano 380-8553, Japan
[c] Institute for Materials Research, Tohoku University, Katahira, Aoba-ku, Sendai 980-8577, JAPAN

E-mail: taishi@shinshu-u.ac.jp

Abstract. Oxygen-related thermal donors (OTDs) in oxygen-enriched Czochralski Ge crystals grown from a melt fully covered by B_2O_3 liquid were investigated by infrared spectroscopy. Interstitially dissolved oxygen concentrations [O_i] and thermal donor concentrations N_{TD} in Ge specimens annealed at 350°C for 64h and at 550°C for 1h, followed by subsequent fast cooling to room temperature, were measured in comparison with those in as-grown Ge. By annealing at 350°C, an absorption peak developed at 780 cm^{-1} and the peak height at 855 cm^{-1}, related to [O_i], decreased. The absorption coefficient at 780 cm^{-1} showed the same correlation to the difference between the total concentration of oxygen atoms and the dissolved oxygen concentration in the annealed specimens. It was found that the number of oxygen atoms forming the OTD increases with increasing annealing time at 350°C.

1. Introduction

Recently, Ge has been revisited for its unique properties in the semiconductor industry [1]. Owing to its lattice matching to GaAs, Ge has been used as a substrate or a bottom cell of III-V-based triple-junction solar cells for space use. Indeed, a very high conversion efficiency of 41.1 % has been reported in a system with focused sunlight [2]. In addition, the application of Ge to high-speed MOSFET devices is expected due to faster carrier mobility in Ge than in Si [1]. In these circumstances, Ge crystals with higher quality will be demanded for production of such devices with higher performance.

In the case of Si, crystals are generally grown by the Czochralski (CZ) method using a silica crucible. During the growth, oxygen atoms dissolve from the silica crucible into the Si melt and incorporate inevitably into the grown crystal. Such interstitially dissolved oxygen atoms (O_i) contribute some benefits such as suppression of dislocation generation, leading to high strength stability [3]. Also, interstitially dissolved oxygen atoms O_i form oxygen-related thermal donors (OTD)s or act as nuclei for precipitates. Therefore, it is well recognized that for practical applications the oxygen concentration in Si crystals has to be strictly controlled. Recently, magnetic fields have been used for controlling melt convection during the crystal growth. Basic behavior of OTDs and oxygen precipitates in Si crystals have been widely studied [4-6], and suitable growth and post-anneal conditions have already been established for application in industry.

Oxygen can be expected to offer a similar advantage in Ge as in Si for the wide applications of Ge crystals [7,8]. Ge crystals are generally grown by the CZ method using a graphite crucible in a vacuum or inert gas (N_2 or Ar) ambient [9]. That is, the oxygen concentration in conventional Ge crystals is very low. Germanium-oxide-related particles coming from residual oxygen in the growth furnace are easily formed on the melt surface, and then they attach to the growing Ge crystal surface. At these locations, dislocations generate into the crystal. Dislocations seriously influence device performance. Oxygen-enriched Ge crystals have been grown under oxygen-containing ambient [7,8,10,11]. Indeed, maximum concentration of interstitially dissolved oxygen is reported using a later reported method to be $7 \times 10^{17} cm^{-3}$ in Ge grown in ambient [10], although the dislocation density is high. Behavior of oxygen in such Ge crystals was intensively studied using infrared spectroscopy. Oxygen atoms occupy interstitial sites in the Ge lattice similar to Si, and the concentration $[O_i]$ of interstitially dissolved oxygen can be determined from the peak height of an absorption at 855 cm^{-1} in the spectrum. The absorption relates to the asymmetric (ν_3) stretching mode of Ge-O-Ge quasi-molecules.

OTDs in such Ge crystals have also been investigated. First, a structure with four atoms of oxygen was proposed for OTD [10] and experimentally confirmed by a subsequent study [12]. According to the latter study, OTDs were saturated at donor concentrations of 9×10^{16} cm^{-3} by prolonged annealing at 350°C. Recently, double donor structures have been accepted for OTDs in Ge crystals with a relevant model [13]. Clauws et al. [14,15] reported behavior of OTDs by annealing in detail and showed three absorption bands at 600, 740 and 780 cm^{-1} in infrared spectra.

Recently, we grew Ge crystals with various concentrations of interstitially dissolved oxygen atoms up to 5.5×10^{17} cm^{-3} using B_2O_3 and a silica crucible [16]. Full coverage of the Ge melt by B_2O_3 liquid and also addition of GeO_2 powder are keys to enhance oxygen concentration in Ge crystals. In the crystals, B concentration was lower than the detection limit of SIMS analysis, and contamination of Si atoms and formation of GeO_2-related precipitates could not be detected from infrared spectra [17]. Behavior of OTDs and other oxygen-related defects in such crystals using B_2O_3 and silica crucible have not yet been well understood.

This paper reports some preliminary results of OTDs formed by post-annealing in an oxygen-enriched Ge crystal grown using B_2O_3 liquid. Formation and annihilation of OTDs were evaluated by Fourier-transform infrared spectroscopy (FTIR) and by the Hall effect. The interstitial oxygen concentrations and oxygen-related thermal donor concentration were compared in as-grown and post-annealed Ge specimens.

2. Experimental procedures

A Ge crystal was grown from a melt fully covered by B_2O_3 liquid without any doping of electrically active impurities. The interstitial oxygen concentration of the as-grown crystal was 5.5×10^{17} cm^{-3}. Detailed crystal growth conditions were as described previously [16,18]. The oxygen concentration in this crystal was also analyzed chemically by secondary ion mass spectroscopy (SIMS) to be 6.5×10^{17} cm^{-3}. The accuracy of the measurement was ±5%.

In order to determine interstitially dissolved oxygen concentrations, Ge blocks of 3 mm thickness were cut from the grown crystal. The blocks were then mechanically lapped, followed by chemical polishing. Infrared absorption spectroscopy was conducted at room temperature (RT) using a Fourier-transform spectrometer (JASCO FT-IR 610). The spectral resolution was 1 cm^{-1}. From the absorption peak height at 855cm^{-1}, oxygen concentration $[O_i]$ was determined using the calibration coefficient of 1.05×10^{17} cm^{-2} [8]. To evaluate carrier concentrations, specimens 5×5×1.5 mm in size were also prepared. The carrier concentration in the Ge specimens was determined by Hall-effect measurements using the Van der Pauw method at RT. As-grown specimens were annealed in a vacuum ampoule as follows: One was at 350°C for 64 h to increase the quantity of OTDs [15], the other at 550°C for 1 h in order to annihilate OTDs [8]. In both cases, subsequent fast cooling to RT was carried out after annealing.

Table 1. Anneal condition and characteristics related to oxygen of Ge specimens grown from melt fully covered by B_2O_3.

Anneal condition	Interstitial oxygen concentration $[O_i]$ [cm^{-3}]	Carrier concentration N_{TD} [cm^{-3}]	Absorption coefficient at 780 cm^{-1} α_{780} [cm^{-1}]
(a) as-grown	5.5×10^{17}	2.0×10^{16}	0.04
(b) 350°C × 64h	2.5×10^{17}	5.4×10^{16}	0.19
(c) 550°C × 1h	6.2×10^{17}	1.4×10^{14}	~ 0

3. Results and discussion

With regard to three Ge specimens of (a) as-grown, (b) annealed at 350°C for 64 h and (c) annealed at 550°C for 1h, interstitial oxygen concentration $[O_i]$, absorption coefficients around 780 cm^{-1} (α_{780}) obtained from infrared spectra and thermal donor concentrations (N_{TD}) determined from Hall-effect measurement are summarized in Table 1.

Infrared absorption spectra of three Ge specimens in the range 750–900 cm^{-1} are shown in Figure 1. Absorption peaks related to O_i were observed at 855 cm^{-1} in all the specimens, though the peak height was quite different, dependent on the post-annealing conditions. $[O_i]$ in specimens (a), (b) and (c) was determined to be 5.5×10^{17}, 2.5×10^{17} and 6.2×10^{17} cm^{-3}, respectively. It was found that $[O_i]$ in the as-grown state was reduced by 3×10^{17} cm^{-3} after annealing at 350°C for 64 h. On the other hand, an enhancement of $[O_i]$ by 0.8×10^{17} cm^{-3} was seen after annealing at 550°C for 1 h. An absorption peak was observed around 780 cm^{-1} due to OTDs, especially clear in specimen (b). The absorption coefficient (α_{780}) at 780 cm^{-1} of specimens (a), (b) and (c) was 0.04, 0.19 and ~ 0 cm^{-1}, respectively.

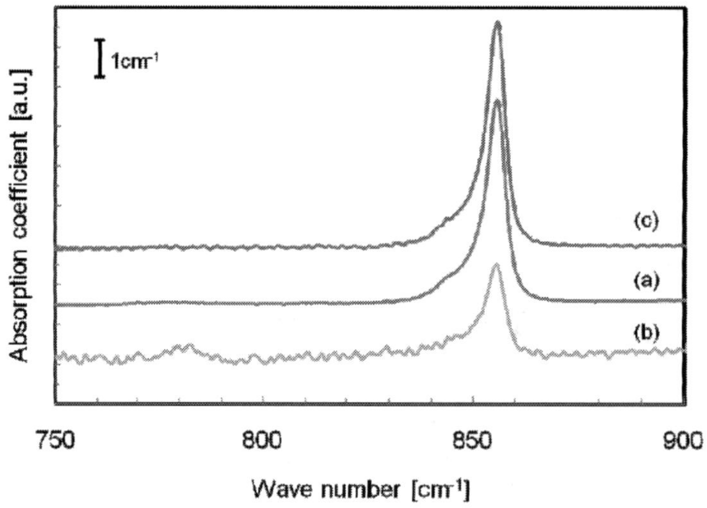

Figure 1. Infrared absorption spectra of three Ge specimens, (a) as-grown, (b) post-annealed at 350°C for 64 h and (c) post-annealed at 550°C, for 1 h in the range 750–900 cm^{-1} measured at RT. The vertical line in the figure shows an absorption intensity of 1 cm^{-1}.

In the Hall-effect measurement, all the three specimens were found to be *n*-type. N_{TD} in specimens (a), (b) and (c) were 2.0×10^{16}, 5.4×10^{16} and 1.4×10^{14} cm^{-3}, respectively. Specimen (b) was additionally annealed at 550 °C for 1 h. The specimen showed a reduction of N_{TD} to 1.5×10^{14} cm^{-3} which is in good agreement with the magnitude of N_{TD} in specimen (a). This means that OTDs in the as-grown specimen were all dissolved by annealing at 550 °C for 1h. In that case, [Oi] in Ge specimens was 6.2×10^{17} cm^{-3} free from OTDs.

The oxygen concentration in all the specimens was determined to be 6.5×10^{17} cm^{-3} by SIMS analysis. The difference between the values determined by infrared spectra and SIMS analysis was within the accuracy of the measurement, therefore we assumed that the former value of 6.2×10^{17} cm^{-3} was the total oxygen concentration [O_{total}]. Reduction of the interstitial oxygen concentration $\Delta[O_i]$ by annealing at 350°C from the as-grown specimen was calculated to be [O_{total}] – [O_i]. Figure 2 shows the relationship between $\Delta[O_i]$ or N_{TD} and α_{780}. Solid circles represent $\Delta[O_i]$. $\Delta[O_i]$ seems to show a linear increase with an increase of the intensity of α_{780}. It is reported that the intensity of absorption at 780cm^{-1} may relate to the concentration of oxygen atoms specifically at the ends of the TD structure, not of all oxygen atoms in the TD [19]. Probably some concentration proportions among the growing OTDs in annealed specimen remains constant, leading to the observed linear relationship. On the other hand, N_{TD} also increased with increasing α_{780} with saturation at 5.4×10^{16} cm^{-3}. This implies that the structure of OTDs formed by annealing at 350°C is not the same as that responsible for the absorption peak at 780 cm^{-1}. Clauws et al. [15] reported that another absorption band at 600 cm^{-1} developed earlier than that at 780 cm^{-1} during early stage of annealing at 350°C. Therefore, the structure of OTDs related to the absorption band at 780 and 600 cm^{-1} may be different, and the number of oxygen atoms contributing to the formation of OTDs related to the 780 cm^{-1} band may be more than that for 600 cm^{-1}.

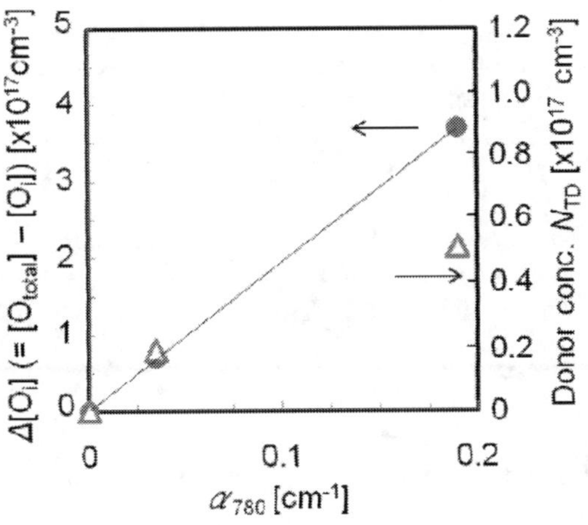

Figure 2. Relationships between interstitial oxygen concentration reduced by annealing, $\Delta[O_i]$ (=[O_{total}] – [O_i]) (circles) and donor concentration, N_{TD} (triangles) as a function of absorption coefficient at 780 cm^{-1}, α_{780}.

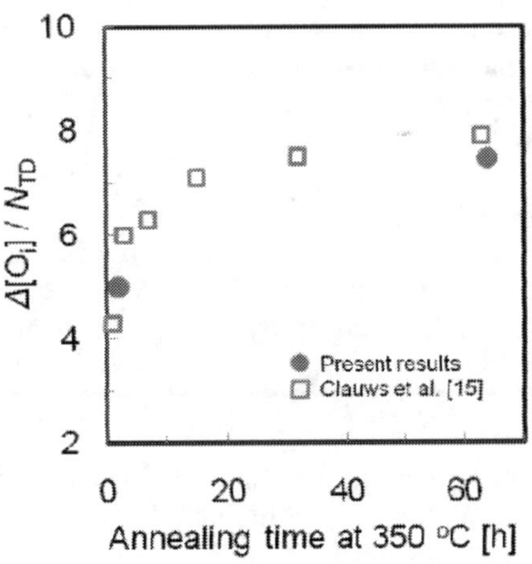

Figure 3. Dependence of the ratio of $\Delta[O_i]$ and N_{TD} on the annealing time at 350°C. Data reported by Clauws et al. [15] are also shown as squares.

Such speculation may be supported by Figure 3, which shows a dependence of the ratio of $\Delta[O_i]$ and N_{TD} on the annealing time at 350°C. The ratio means the number of oxygen atoms per OTD. Solid circles represent our results obtained using Ge crystal grown from a melt fully covered by B_2O_3. The symbol close to the left-side axis represents a result for the as-grown specimen, the annealing time at 350°C can be estimated to be 2 h or less considering cooling rate (500°C/h) after the crystal growth. Values of $\Delta[O_i]/N_{TD}$ are 5 and 7.5 for annealing times of < 2 h (as-grown) and 64 h, respectively. Results reported by Clauws et al. [15] are also superimposed in Figure 3 as square symbols. Their results of $\Delta[O_i]/N_{TD}$ logarithmically increase with increasing annealing time, a similar tendency is seen in the present results. According to Clauws et al. [15], the value finally approaches eight after prolonged annealing and it is in agreement with Fuller's model which identified them as GeO_4 assuming single donors [12].

In our preliminary results, we see that the observed behaviour of OTD formation and annihilation in Ge crystals grown from a melt fully covered by B_2O_3 is similar to those in Ge grown under oxygen-containing ambient. Effects of very long or short time annealing on the formation and annihilation of OTDs and detailed structures of OTDs in Ge crystals are now being investigated. Oxygen-enriched Ge crystals grown using B_2O_3 described above have the advantage that it is possible to reduce dislocation density. Therefore, new knowledge related to the formation of OTDs and oxygen precipitates in such Ge crystals would be useful in the future.

4. Conclusion

The behaviour of OTDs in oxygen-enriched CZ-Ge crystals grown from a melt fully covered by B_2O_3 liquid was investigated by infrared spectroscopy. A Ge crystal with $[O_i]$ of 5.5×10^{17} cm^{-3} in the as-grown state was prepared. The specimens were post-annealed at 350°C for 64h or at 550°C for 1 h, both with subsequent fast cooling to room temperature. $[O_i]$ and N_{TD} in the Ge specimens as-grown or annealed were compared. In the case of a specimen annealed at 350°C, an absorption peak grew at 780 cm^{-1}, while a peak at 855 cm^{-1} became small. The absorption intensity α_{780} shows a linear correlation to $[\Delta O_i]$ in the annealed specimen. On the other hand, N_{TD} increased with the intensity of α_{780}, showing similar tendency with the previously reported results. This implies that the structure of OTDs formed by annealing at 350°C is not same which relates to the absorption peak at 780 cm^{-1}. It was found that oxygen atoms contributing to the formation of OTDs increased with increasing annealing time at 350°C. Such behavior of OTDs is similar to that of Ge crystals grown in oxygen-containing ambient.

Acknowledgement

This work was supported in part by a Grant-in-Aid for Science Research (20760003 and 22686002) from the Ministry of Education, Science, Sports and Culture. This work was also performed under the Inter-university Cooperative Research Program of the Institute for Materials Research, Tohoku University.

References

[1] Vanhellemont J, Simoen E 2007 J. Electrochem. Soc. **154** H572
[2] Guter W, Schone J, Philipps S P, Steiner M, Siefer G, Wekkeli A, Welser E, Oliva E, Bett A W, Dimoroth F 2009 Appl. Phys. Lett. **94** 223504.
[3] Yonenaga I, Sumino K 1996 J. Appl. Phys. **80** 734.
[4] Fuller G, Ditzenberger J, Hannay N, Buehler E 1954 Phys. Rev. **96** 833.
[5] Kaiser W 1957 Phys. Rev. **105** 1751.
[6] Kanamori A, Kanamori M 1979 J. Appl. Phys. **50** 8095.
[7] De Gryse O, Vanmeerbeek P, Vanhellemont J, Clauws P 2006 Mater. Sci. Semicond. Process. **9** 246.
[8] Litvinov V V, Svensson B G, Murin L I, Lindström J L, Markevich V P, Peaker A R 2006 J. Appl. Phys. **100** 033525.
[9] Depuydt B, Jonghe M, Baets W 2007 Germanium-Based Technologies ed C Claeys and E Simoen (Oxford: Elsevier) p 17.
[10] Bloem J, Haas C, Penning P 1959 J. Phys. Chem. Solids **12** 22.

[11] Kaiser W, Thurmond C 1961 J. Appl. Phys. **32** 115.
[12] Fuller C, Kaiser W, Thurmond C 1961 J. Phys. Chem. Solids **17** 301.
[13] Jones R, Coutinho J, Öberg S, Briddon P 2001 Physica B **308-310** 8.
[14] Clauws P 1996 Mat. Sci. Eng. **B36** 213.
[15] Clauws P, Vanmeerbeek P 1999 Physica B **273-274** 557.
[16] Taishi T, Ise H, Murao Y, Ohsawa T, Suezawa M, Tokumoto Y, Ohno Y, Hoshikawa K, Yonenaga I 2010 J. Cryst. Growth **312** 2783.
[17] Taishi T, Ise H, Murao Y, Ohsawa T, Tokumoto Y, Ohno Y, Yonenaga I 2011 Microelectro. Eng. in press.
[18] Taishi T, Ohno Y, Yonenaga I 2009 J. Cryst. Growth **311** 4615.
[19] Coutinho J, Jones R, Murin L I, Markevich V P, Lindström J L, Öberg S, Briddon P R 2001 Phys. Rev. Lett. **87** 235501.

Journal of Physics: Conference Series **281** (2011) 012012 doi:10.1088/1742-6596/281/1/012012

Interfacial sources of extended defects in nonpolar and semipolar III-Nitride semiconductors

G P Dimitrakopulos

Physics Department, Aristotle University of Thessaloniki, GR 541 24, Thessaloniki, Greece

E-mail: gdim@auth.gr

Abstract. The principal characteristic of nonpolar and semipolar heteroepitaxial interfaces is their anisotropy that leads to distinct ways of misfit accommodation along two in-plane directions. One direction can require misfit dislocations with out-of-plane Burgers vectors that induce epilayer tilts and low-angle grain boundaries, and the other necessitates the introduction of multiple basal-plane stacking faults. The reduced bicrystalline symmetry also favours the coexistence of distinct epilayer orientation relationships. This may result in nanocrystalline interfacial zones comprising low-energy grain boundaries that act as defect sources for the matrix epilayer, generating threading dislocations and stacking faults.

1. Introduction

III-Nitride compound semiconductors with wurtzite structure are promising materials for extending the operational range of optoelectronic devices from UV to IR wavelengths. Towards this goal, a principal obstacle is the strain-induced piezoelectric polarization at quantum nanostructures (wells and dots) grown along the polar *c*-axis orientation. It leads to carrier separation and band structure distortion, reducing dramatically the internal quantum efficiency with increasing alloy content [1, 2]. Therefore III-Nitride heterostructures grown along nonpolar orientations are currently under intense investigation [2-4]. Semipolar orientations, whereby the *c*-axis is inclined relative to the heteroepitaxial interface, are another option, offering partial reduction of the polarization field. High quality heterostructure deposition on foreign substrates such as sapphire is required for effective technological advancement. To date such epilayers exhibit threading dislocation densities in excess of 10^{10} cm^{-2} and stacking fault densities of ~10^5 cm^{-2} .

The principal characteristic of these heterostructures compared to polar III-N epitaxy is that they exhibit distinct ways of misfit accommodation along different in-plane directions. The strain anisotropy and the reduced bicrystal symmetry lead to novel relaxation mechanisms, and we review and generalize such mechanisms for nonpolar and semipolar epilayers, based on recent results obtained for *a*-plane GaN grown on sapphire [5, 6]. For this purpose we employ transmission electron microscopy (TEM) observations and topological analysis in order to elucidate the role played by III-N/sapphire interfaces in the introduction of defects in nonpolar and semipolar epilayers.

For the purpose of demonstrating general conclusions, representative heterostructures are considered, namely (i) nonpolar *a*-plane $(11\bar{2}0)$ GaN grown on *r*-plane $(1\bar{1}02)$ sapphire, and (ii) semipolar $(11\bar{2}2)$ AlN grown on *m*-plane $(1\bar{1}00)$ sapphire [5-7]. The heteroepitaxial orientation relationships are (i) $[0001]_{III\text{-}N}//[1\bar{1}0\bar{1}]_{sapph.}$, $[1\bar{1}00]_{III\text{-}N}//[\bar{1}\bar{1}20]_{sapph.}$ for *a*-plane epilayers, and (ii)

$[\bar{1}\bar{1}23]_{III-N}//[0001]_{sapph.}$, $[1\bar{1}00]_{III-N}//[\bar{1}\bar{1}20]_{sapph.}$ for $(11\bar{2}2)$ growth. The layers have been grown by plasma-assisted molecular beam epitaxy (PAMBE) as described elsewhere [5-7].

2. Emanation of stacking faults

In both the nonpolar and semipolar orientations, two misfit dislocation arrays, with edge Burgers vector components vertical to one another, suffice for misfit accommodation. However, the two vertical in-plane directions exhibit distinct structures. Moreover, the misfit differs significantly along each direction. In *a*-plane GaN grown on *r*-plane sapphire, the misfit along $[0001]_{GaN}$ is only +1.2% and reaches +16% along $[1\bar{1}00]_{GaN}$. In $(11\bar{2}2)$ AlN grown on *m*-plane sapphire, the mismatch is tensile +13.4% along $[1\bar{1}00]_{AlN}$ and compressive -9.6% along $[\bar{1}\bar{1}23]_{AlN}$.

Accommodation of the misfit along the [0001] *c*-axis in the case of nonpolar heteroepitaxy, or along the in-plane projection of the *c*-axis in the case of semipolar, necessitates the introduction of misfit dislocations with edge Burgers vector components that comprise single (0002) extra half-planes. However, since the lattice dislocations with [0001] Burgers vector components require two extra half-planes (due to the ...*ABAB*... hexagonal stacking), such misfit dislocations can only be partials with Burgers vectors of type $\mathbf{b} = 1/6\langle 2\bar{2}03\rangle$. Each such partial must be associated with a stacking fault. There is then the possibility that the stacking fault is superimposed on the heteroepitaxial interface, or it emanates into the epilayer, as shown schematically in Figures 1(a) and (b) respectively. In the latter case, the basal stacking faults are of I_1 (0002) intrinsic type, since they are delimited by Frank partials [8]. Although the latter process necessitates dislocation climb, the TEM measurements show that the stacking fault density is consistent with this being a principal process of misfit relaxation [5]. The stacking fault emanation from the heteroepitaxial interface was verified by plan-view high resolution TEM (HRTEM) observations of *a*-plane GaN epilayers showing termination of Moiré fringes at the delimiting partial dislocations of edge-on I_1 stacking faults [5].

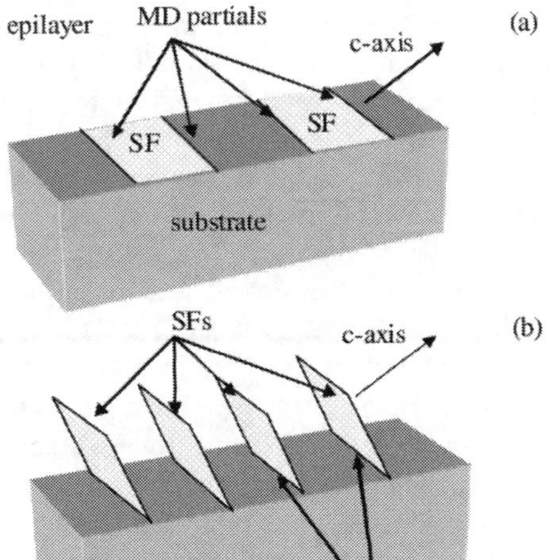

Figure 1. Schematic of the plausible mechanisms of misfit accommodation along the *c*-axis (nonpolar epilayer) or its in-plane projection (semipolar epilayer). The misfit dislocations (MDs) are $1/6\langle 2\bar{2}03\rangle$ partials. In (a) the associated stacking faults (SFs) are parallel to the heteroepitaxial interface thereby introducing alternating interfacial structure. In (b) the SFs are parallel to the (0002) basal plane and emanate into the epilayer.

3. Misfit dislocations with inclined Burgers vectors

A second relaxation mechanism concerns the misfit along the in-plane direction vertical to the *c*-axis. In this case the misfit dislocations have line directions along [0001] or its in-plane projection. The shortest lattice Burgers vectors are the $\mathbf{b} = 1/3\langle\bar{1}2\bar{1}0\rangle$ vectors. However, in both the $(11\bar{2}0)$ and $(11\bar{2}2)$ epitaxies, such Burgers vectors do not lie in the heteroepitaxial interface. In fact the

1/3<$\bar{1}2\bar{1}0$> Burgers vectors are in-plane only for prismatic *m*-plane {$1\bar{1}00$} and for pyramidal type {$1\bar{1}0w$} growth orientations. In all other nonpolar or semipolar epitaxial orientations the 1/3<$\bar{1}2\bar{1}0$> vectors are inclined relative to the substrate, and an example is shown in Figure 2 which illustrates a misfit dislocation array at the ($11\bar{2}2$)AlN/*m*-sapphire interface. A Burgers circuit has been drawn around one misfit dislocation showing the inclined Burgers vector. Such misfit dislocations comprise two edge components. The first component is in-plane accommodating the misfit, while the second is normal to the interface. In the case of Figure 2, these components have magnitudes 0.269 nm and 0.132 nm respectively.

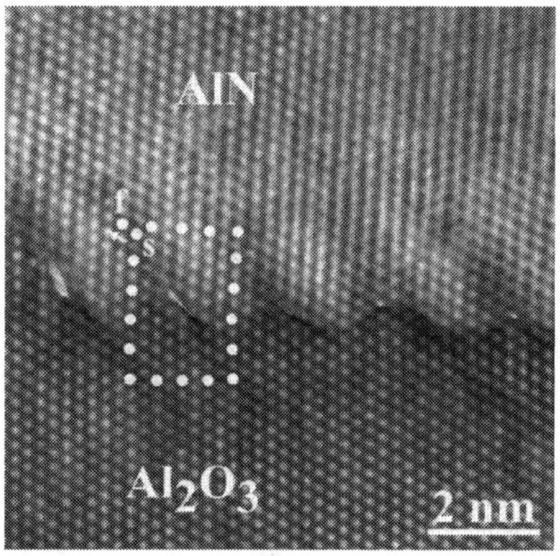

Figure 2. Cross sectional HRTEM image of ($11\bar{2}2$)AlN / *m*-plane sapphire viewed along [$\bar{1}\bar{1}23$]$_{AlN}$/[0001]$_{sapph.}$. The image shows a periodic array of misfit dislocations. The two-dimensional map of the lattice strain along the [$1\bar{1}00$]$_{AlN}$ in-plane direction is superimposed (this map has been obtained by geometrical phase analysis [9]). Circuit mapping identifies the Burgers vector of the misfit dislocations as an inclined vector **sf** of 1/3<$\bar{1}2\bar{1}0$> type.

As a result of the normal Burgers vector component, such misfit dislocation arrays induce lattice tilts by acting effectively as low-angle tilt boundaries [5]. Since the sapphire substrate is practically unaffected due to its large thickness and rigidity, the rotational effect of the misfit dislocation array is adopted by the epilayer. Figure 3 illustrates schematically the Volterra operations that introduce a misfit dislocation with inclined Burgers vector in *a*-plane GaN, in accordance with the topological theory of interfacial defects [10]. In this case the induced epilayer tilt is in opposite sense relative to the Burgers vector, since the extra half-plane is in the substrate. For a regular array of misfit dislocations with parallel Burgers vectors and spacing *d*, the maximum induced epilayer tilt is $\vartheta_{max} = b_n / d$ where b_n is the magnitude of the normal to the interface component of the Burgers vector. This tilt may be clockwise or anticlockwise depending on the upward or downward sense of the Burgers vector. However, if upward and downward-oriented Burgers vectors co-exist in the misfit dislocation array, then the rotational effect locally cancels out. Hence a range of misorientations of angles $\vartheta \leq \vartheta_{max}$ is anticipated. For *a*-plane GaN on *r*-plane sapphire, the misfit dislocations accommodating the 16% misfit introduce tilts by up to ±5.4° as verified by selected-area electron diffraction patterns obtained from epilayers grown by PAMBE under N-rich conditions [5]. For semipolar AlN the misfit is 13.4% corresponding to an average misfit dislocation spacing of 2 nm; hence $\vartheta_{max} = 3.8°$.

Since the tilt induced by misfit dislocations may be clockwise or anticlockwise, low-angle tilt grain boundaries (LAGBs) of tilt angle $\leq 2\vartheta_{max}$ are introduced between abutting islands as shown schematically in Figure 4. Such LAGBs consist of lattice dislocations with their line directions parallel to the heteroepitaxial interface and their occurrence has been confirmed by HRTEM in *a*-plane GaN [5]. The LAGB dislocations may later participate in dislocation interactions and bend upwards to become threading dislocations.

In the case of semipolar epitaxies, the 1/3<$\bar{1}2\bar{1}0$>-type misfit dislocations also comprise screw components since the basal planes on which these Burgers vectors lie are inclined relative to the substrate. The magnitude of the screw component increases with increasing inclination angle of the basal plane. Hence such arrays are capable of introducing twists in the orientation of the epilayer crystal planes. For example, in the representative case of $(11\bar{2}2)$ AlN on sapphire, the magnitude of this screw component is b_s = 0.082 nm, and the maximum twist angle normal to the $[\bar{1}\bar{1}23]$ dislocation lines is $\phi_{max} = b_s / d = 2.35^{\circ}$. This twist may be compensated by extra defects.

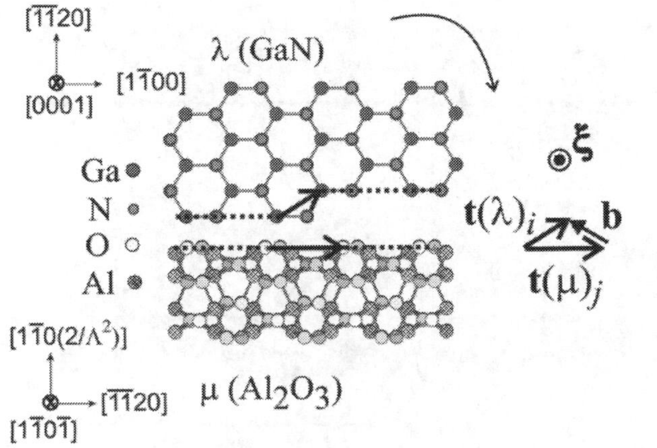

Figure 3. Volterra-like schematic illustration showing the translation operations $\mathbf{t}(\lambda)_i$ and $\mathbf{t}(\mu)_j$ that introduce a $\mathbf{b} = 1/3\langle\bar{1}2\bar{1}0\rangle$ misfit dislocation in *a*-plane GaN grown on *r*-plane sapphire. The curved arrow shows the sense of epilayer tilt. [Circle size and colouring denote distinct atomic species as indicated on the image, shading denotes distinct levels along the projection direction, vector ξ denotes the out-of-paper sense of the dislocation's line direction, and $\Lambda = (2/3)^{1/2}\,(c/a)_{sapph.}$].

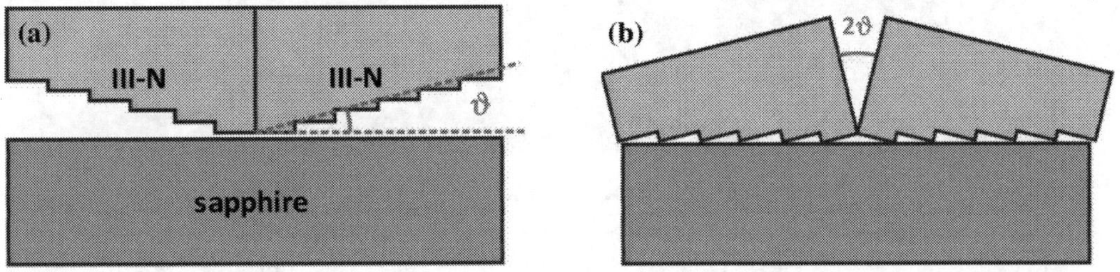

Figure 4. (a) Volterra-like schematic showing the clockwise and anticlockwise rotational senses induced by two misfit dislocation arrays with opposite Burgers vector senses in two abutting III-N islands. (b) Sequential accommodation of the two Volterra cuts and introduction of a relative tilt 2ϑ between the two islands. This 2ϑ wedge is filled by dislocated material comprising a LAGB.

4. Intergranular boundaries as defect sources

Anisotropy also favours the coexistence of distinct epilayer orientation relationships. In fact, just one low misfit in-plane direction can suffice for the introduction of a second epitaxial orientation. Depending on the growth conditions, one orientation may eventually dominate, but TEM observations in MBE-grown epilayers [5-7] have shown that the secondary orientation can still persist close to the substrate in the form of interfacial nanocrystals that are quickly overgrown. Such nanocrystals are surrounded by intergranular boundaries that are defect sources for the matrix epilayer, causing introduction of threading dislocations, stacking faults, nanotwins, and cubic pockets [5].

Figure 5(a) illustrates an HRTEM image of such a nanocrystal with an emanating threading dislocation inside PAMBE-grown *a*-plane GaN epilayer. The nanocrystal has a semipolar orientation described by a 90° $\langle 2\bar{1}\bar{1}0\rangle$ rotation relative to the $(11\bar{2}0)$ epilayer. Interestingly, the same orientation relationship was also observed between semipolar $(11\bar{2}2)$ and nonpolar *m*-plane $(1\bar{1}00)$ AlN nanocrystals grown on *m*-sapphire [7]. The 90° $\langle 2\bar{1}\bar{1}0\rangle$ relationship ensures a high order of coincident symmetry between the matrix epilayer and the nanocrystals, as illustrated in Figure 5(a). As a result, the (0002) planes in each crystal component are matched to the $\{10\bar{1}0\}$ planes in the other crystal with a 6.5% misfit for GaN and 8.2% for AlN. This misfit introduces dislocations at the intergranular boundaries as illustrated by geometrical phase analysis in Figure 5(b). Furthermore, the intergranular boundaries appear to promote the introduction of stacking faults. In fact the overlap of multiple stacking faults can lead to …*ABCABC*… cubic stacking resulting to 'pockets' of sphalerite structure as illustrated in Figures 5(a) and 5(c).

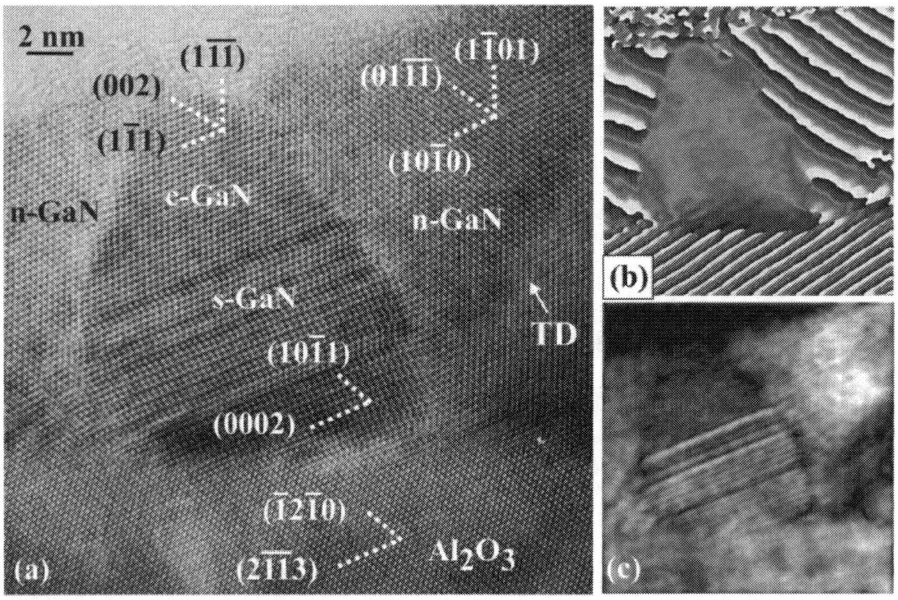

Figure 5. (a) HRTEM image of a semipolar interfacial nanocrystal (s-GaN) viewed along $[\bar{1}2\bar{1}0]$. Nonpolar $(11\bar{2}0)$ GaN (n-GaN) is viewed along $[\bar{1}2\bar{1}3]$ whereas $(1\bar{1}02)$ sapphire is viewed along $[10\bar{1}\bar{1}]$. The sphalerite-cubic part of the nanocrystal (c-GaN) is viewed along [110]. A threading dislocation (TD) emanates from the semipolar/nonpolar intergranular boundary (arrowed). (b) Phase map of (a) obtained using the $0002/10\bar{1}0$ GaN spatial frequencies. Fringe terminations at the interfaces indicate the locations of interfacial dislocation cores. [(a) and (b) are reprinted with permission from Smalc-Koziorowska *et al.*, J. Appl. Phys. **107**, 073525, 2010. Copyright 2010, American Institute of Physics]. (c) Amplitude image where the distinct structures inside the nanocrystal (wurtzite in lower part and sphalerite in upper part) can be discriminated.

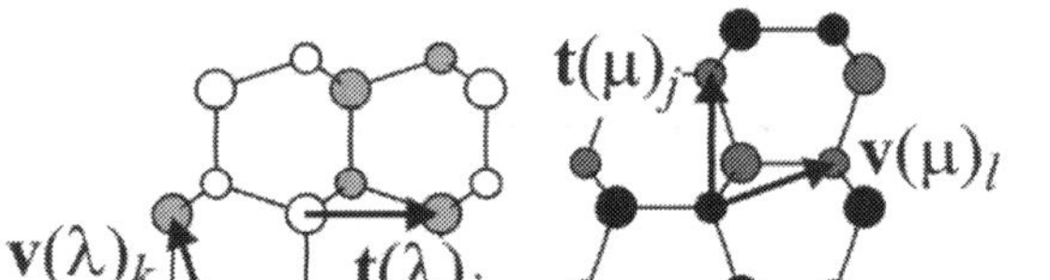

Figure 6. Schematic illustration showing the relative orientations of $\mathbf{t} = 1/3\langle\bar{1}2\bar{1}0\rangle$ lattice vectors and $\mathbf{v} = 1/6\langle 2\bar{2}03\rangle$ structural vectors in two wurtzite crystals that are rotated by 90° about $\langle 2\bar{1}\bar{1}0\rangle$.

An understanding of the influence of the intergranular boundaries on the introduction of threading dislocations and stacking faults can be attained by reference to the schematic illustration of Figure 6, whereby one crystal is arbitrarily designated as white (λ) and the other as black (μ). It is seen that, as a result of the 90° rotation, $1/3\langle\bar{1}2\bar{1}0\rangle$ lattice vectors in one crystal component are juxtaposed with $1/6\langle 2\bar{2}03\rangle$ structural vectors in the other crystal and vice versa. Hence either vectors types may comprise the Burgers vectors of the interfacial dislocations that relieve the $(0002)/\{10\bar{1}0\}$ misfit. The former vectors are the vectors of threading dislocations while the latter are responsible for the introduction of I_1 stacking faults as discussed earlier. Therefore, the intergranular boundaries can act as sources of such defects, without needing to satisfy complex nodal balance requirements. One possible mechanism for defect emanation is the wing coalescence due to lateral overgrowth of the nanocrystals by the matrix.

5. Conclusions

Based on the present review of nanoscale mechanisms we can deduce general conclusions and recommendations concerning the heteroepitaxial growth of nonpolar and semipolar III-Nitride epilayers. Regarding the large stacking fault density, this is an intrinsic property due to the misfit accommodation by partial dislocations comprising single (0002) extra half planes, and novel approaches involving interlayers and strain engineering are needed in order to overcome this problem. At the same time, the misfit dislocation arrays accommodating the mismatch along the in-plane direction vertical to the *c*-axis can induce tilting and twisting of epilayer planes due to their inclined Burgers vectors, and thus introduce extra extended defects, such as dislocations forming low-angle grain boundaries. In this regard the $\{1\bar{1}00\}$ and $\{1\bar{1}0w\}$ orientations are more promising heteroepitaxies. The third important issue concerns the control of the crystalline quality of the interfacial region with the substrate. Secondary epitaxial orientations appear easily due to the low common symmetry with the foreign substrate. Such an interfacial zone can even be difficult to detect since it may be confined to below 10 nm of epilayer thickness. However, the low energy of the intergranular boundaries, promoted by the 90° $<\bar{1}2\bar{1}0>$ relative rotation, and their associated intrinsic dislocations make them potent defect sources for the epilayer. It is recommended that growth conditions, such as substrate pre-treatment, temperature, and flux ratios, are correlated to detailed HRTEM observations of the interfacial region with the substrate in order to optimize them towards eliminating any parasitic nanocrystals.

Acknowledgements

The author is grateful to J. Smalc-Koziorowska, Th. Kehagias, Ph. Komninou, and Th. Karakostas for invaluable input and critical reading, as well as to A. Georgakilas and E. Monroy for providing nonpolar and semipolar specimens grown by PAMBE. This work was performed under the support of EU projects DOTSENSE (Grant No. STREP 224212) and PARSEM (MRTN-CT-2004-005583).

References

[1] Humphreys C J 2008 *MRS Bulletin* **33** 459
[2] Speck J S and Chichibu S F 2009 *MRS Bulletin* **34** 304
[3] Masui H, Nakamura S, DenBaars S P, and Mishra U K 2010 *IEEE Trans. Electron. Devices* **57** 88
[4] Paskova T (ed) 2008 *Nitrides with Nonpolar Surfaces: Growth, Properties, and Devices* (Weinheim: Wiley)
[5] Smalc-Koziorowska J, Tsiakatouras G, Lotsari A, Georgakilas A, and Dimitrakopulos G P 2010 *J. Appl. Phys.* **107** 073525
[6] Smalc-Koziorowska J, Dimitrakopulos G P, Sahonta S-L, Tsiakatouras G, Georgakilas A, and Komninou Ph 2008 *Appl. Phys. Lett.* **89** 021910
[7] Kehagias Th, Lahourcade L, Lotsari A, Monroy E, Dimitrakopulos G P, and Komninou Ph 2010 *Phys. Stat. Sol.* **247** 1637
[8] Hirth J P and Lothe J 1982 *Theory of Dislocations*, 2nd Edition (New York: Wiley Intersciences) p 354
[9] Hÿtch M J, Putaux J-L, and Pénisson J-M 2003 Nature **423** 270
[10] Dimitrakopulos G P, Komninou Ph, Karakostas Th, and Pond R C 2003 Topological Analysis of Defects in Nitride Semiconductors *Nitride Semiconductors: Handbook on Materials and Devices*, ed P Ruterana, M Albrecht, and J Neugebauer (Weinheim: Wiley)

EBIC investigation of InGaN/GaN multiple quantum well structures irradiated with low energy electrons

P S Vergeles, E B Yakimov[1]

Institute of Microelectronics Technology RAS, 142432, Chernogolovka, Russia

E-mail: yakimov@iptm.ru

Abstract. Light emitting structures with InGaN/GaN multiple quantum wells have been studied in the Electron Beam Induced Current (EBIC) mode of a scanning electron microscope (SEM). Special attention is paid to the channels of enhanced carrier transport across the active region demonstrating bright EBIC contrast. Defects with the smallest bright contrast are associated with threading dislocations while the large defects present in lower density could be associated with dislocation bunches. Low energy (10 keV) electron beam irradiation in the SEM was found to suppress the bright EBIC contrast of these defects. A comparison with the cathodoluminescence spectra changes due to similar exposure allows us to assume that recombination enhanced diffusion is responsible for the effects observed.

1. Introduction

Despite intensive industry development of blue light emitting diodes (LED) based on the Multiple Quantum Well (MQW) InGaN/GaN system, aimed at the production of effective solid-state lighting, some fundamental properties of this system are still not completely understood. One such problem is the effect of extended defects on LED parameters. In spite of very high dislocation density (10^8-$10^{10}cm^{-2}$) the quantum efficiency of LEDs based on GaN is rather high. It was believed that the high quantum efficiency was associated with the carrier localization inside the InGaN quantum wells. It was shown that dislocations could play a significant role in such localization [1]. Extended defects in semiconductor structures usually produce dark contrast in the Electron Beam Induced Current (EBIC) mode due to the enhancement of excess carrier recombination. However, in the MQW InGaN/GaN structures two types of defects demonstrating bright EBIC contrast were observed in [2-5]. It was shown that bright EBIC contrast of 1-2% could be associated with threading dislocations. The nature of other defects with a density of about 10^5cm^{-2} demonstrating bright contrast up to 50% was not clarified up to now. As shown in [4,5], these large defects enhance carrier transport across the active layer that determines the bright EBIC contrast formation. A further problem is associated with the effect of extended defects on LED degradation. A study of low energy electron beam irradiation (LEEBI) effect on the EBIC signal could provide the useful information concerning this problem and could also help to clarify the pronounced LEEBI effect on the cathodoluminescence in the InGaN/GaN MQW structures [6,7].

In the present paper the LEEBI effect on the electrical properties of defects with bright EBIC contrast in the MQW InGaN/GaN LED structures has been studied in the EBIC mode. The profile

[1] To whom any correspondence should be addressed.

width of EBIC contrast was studied for two types of threading defects with bright contrast. The kinetics of collected current changes both near the defects and far from them due to e-beam exposure have been investigated.

2. Experimental

The LED structures were grown by metal-organic chemical vapor deposition on (0001) sapphire substrates. They consist of a 3 µm thick n-GaN lower layer doped with Si (N_d~ $3\cdot10^{18}cm^{-3}$), a buffer superlattice InGaN/GaN with low In content (<10%), a MQW InGaN/GaN layer with three and five periods of 3 nm InGaN (In 20%) and 12 nm GaN layers and 0.1 µm thick p-GaN cap layer with Mg concentration of about $10^{20}cm^{-3}$.

For the EBIC measurements mesa structures with a diameter of 0.45 mm were etched by Ar ion sputtering and metal contacts with a diameter of 0.35 mm were evaporated in the middle of them. The EBIC measurements were carried out on uncovered parts of the mesa structures using a JEOL JSM 840A scanning electron microscope in the normal geometry, with e-beam perpendicular to the p-n junction plane. Irradiation by the focused e-beam was carried out at room temperature in the same microscope at beam energy of 10 keV and beam current smaller than 10^{-9}A.

As usual the collection efficiency β was obtained from the current collected in the EBIC mode I_c as

$$\beta = \frac{I_c E_i}{I_b E_b \eta}, \tag{1}$$

where E_i is the average energy necessary for the electron-hole pair creation, I_b and E_b are the beam current and energy respectively, and η is the average beam energy absorption coefficient. For GaN the η / E_i ratio was found to be close to $8\cdot10^{-2}$ [8]. The following expression was used to simulate the $\beta(E_b)$ dependence [9]

$$\beta = \int_0^\infty \psi(z)h(z)dz, \tag{2}$$

where $\psi(z)$ is the collection probability (the probability for a minority carrier created at a depth z to be collected in the EBIC mode) and $h(z)$ is the depth-dose dependence, which was assumed to be the same as for GaN [8]. For the LED structures studied $\psi(z)$ was calculated as $\psi(z) = \psi_1(z,W,z_0)\cdot\psi_{DR}$, where $\psi_1(z,W,z_0)$ is the probability for carriers to reach the depletion region, (obtained by a numerical solution of the diffusion equation for p- and n-regions), W is the depletion region width, z_0 is the p-n junction depth and ψ_{DR} is the probability for excess carriers to permeate through the depletion region, which was assumed to be independent of z. Usually this probability is close to 1 and the excess carrier recombination inside the depletion region could be neglected. It needs to be taken into account only for low-doped materials or for structures with very high recombination rates (very small diffusion lengths). However, in the MQW structures the excess carriers are captured by quantum wells inside the depletion region and could be collected only if they overcome the barriers between the wells by the thermal activation or by tunneling. As shown in [2,5] in this case the measured $\beta(E_b)$ dependence could be fitted by multiplying $\psi_1(z,W,z_0)$ by a coefficient ψ_{DR}, which depends on the position of quantum wells inside the depletion region and on the dopant concentration. This coefficient for the MQW structures could be rather small. For the structures in which at least a few quantum wells are located close to the depletion region edge, ψ_{DR} could be as low as 0.25, i.e. 75% of excess carriers reaching the depletion region recombine in the quantum wells. The ψ_{DR} value together with the diffusion length L in the lower n-GaN layer, the effective diffusion length in p-region and W could be estimated by fitting the $\beta(E_b)$ dependence obtained from the experimental data. That could be done because different parts of the $\beta(E_b)$ dependence are determined mainly by one of these parameters and ψ_{DR} determines the absolute collection efficient value only. Thus, at small E_b the $\beta(E_b)$ dependence is mainly determined by the effective diffusion length in the p-region, while the position of the maximum essentially depends on W. The $\beta(E_b)$ dependence decay at large E_b is mainly determined by the diffusion length L in the lower n-GaN layer.

3. Results and discussion

The dependences of collection efficiency β on beam energy E_b measured on the MQW structure with 3 quantum wells (curve 1) and on two structures with 5 quantum wells are presented in figure 1. The calculated dependences are shown in the same figure by solid lines. The collection efficiency decay at large E_b could be fitted rather well with the excess carrier diffusion length L in the n-GaN layer of 130 (1), 110 (2) and 180 nm (3). Thus, the difference in diffusion length values for the structures studied is not large and the observed difference in the collection efficiency for these three structures is mainly determined by the different recombination rate inside the depletion region characterized by the ψ_{DR} value. For the structure with 3 quantum wells (curve 1) ψ_{DR} value obtained from fitting is practically equal to 1, i.e. the $\beta(E_b)$ dependence is close to that calculated neglecting recombination inside the depletion region. But the collected efficiency measured in the LED structures with 5 quantum wells (curves 2 and 3) is significantly lower than that calculated under such an assumption (ψ_{DR} = 0.5 and 0.28, respectively). Thus for these structures recombination inside the depletion region is essential and should be taken into account.

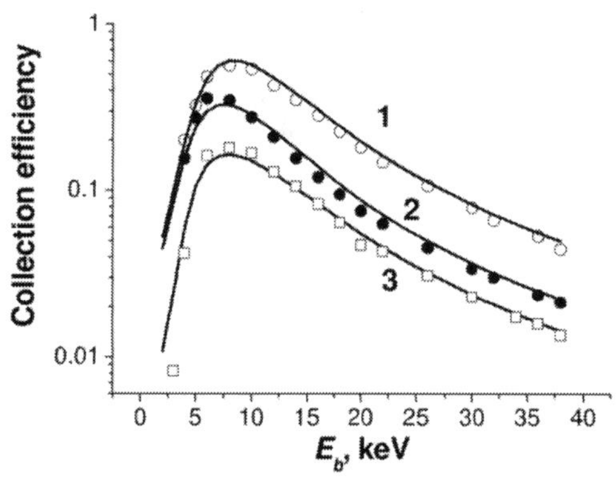

Figure 1. Measured (symbols) and simulated (lines) $\beta(E_b)$ dependence for the LED structures with three (1) and 5 quantum wells (2,3). ψ_{DR} = 0.97 for (1), 0.5 for (2) and 0.28 for (3).

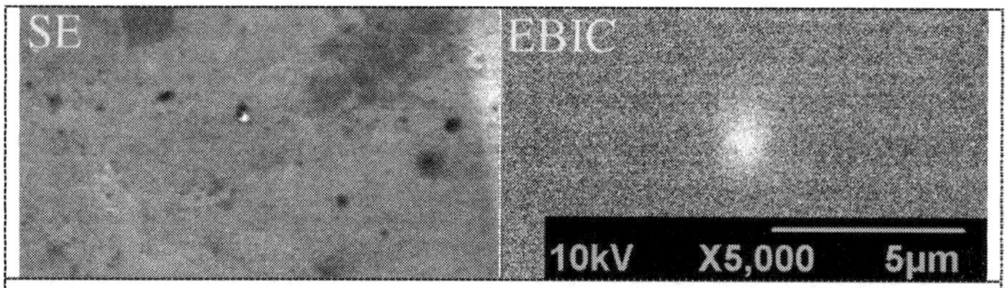

Figure 2. Images of region with a bright contrast defect in the secondary electron (left) and in the EBIC modes (right)

As shown in [4,5] by fitting the $\beta(E_b)$ dependence, near the defects with the large bright contrast the ψ_{DR} value increased. Thus, these defects provide channels of enhanced excess carrier transport across the InGaN/GaN MQW active layer by suppressing the recombination inside the InGaN wells or by increasing the probability to overcome the barriers between the wells. But the microscopic nature of these defects is not clear up to now. They could be micropipes, dislocation bunches or V-defects [10, 11]. As could be seen in figure 2, where the EBIC image of bright defect and the image of the same region in the secondary electron mode are presented, no pits or hillocks are revealed on the surface at the defect position. Thus, if the bright contrast is associated with V-defects or micropipes, it should be assumed that under the upper p-GaN layer growth the defects were filled and the surface was totally planerized, which seems unlikely.

The size of these defects could be estimated from the EBIC investigations. Indeed, the bright contrast is formed inside the active layer containing quantum wells, i.e. at a depth smaller than 200 nm. At such depth the lateral resolution in the EBIC mode for the structures with the small diffusion length is rather high, especially at large E_b [12]. This could be seen in figure 3 (curve 1), where the profile of the EBIC bright contrast normalized for its maximum value for an individual threading dislocation is presented. The contrast is calculated as $C = I_{def}/I_0 - 1$, where I_{def} and I_0 are the collected current values with e-beam located at the defect and far from it. In the same figure the normalized contrast profile for a large bright defect is shown by the curve 2. It is seen that while the Full Width at Half Maximum (FWHM) for the individual dislocation is about 70 nm, the FWHM for the large defect is 900 nm. The FWHM for the individual dislocation confirms that the generation region size in the active region is indeed small. Thus the larger FWHM for the large bright defect shows that the FWHM is determined by its size. The size of the large bright defect could be estimated as several hundred nm, which is significantly larger than the size of commonly observed V-defects. Thus, we believe that the defects with the large bright EBIC contrast could be associated with dislocation bunches, although we cannot totally exclude V-defects or micropipes as the origin for such contrast.

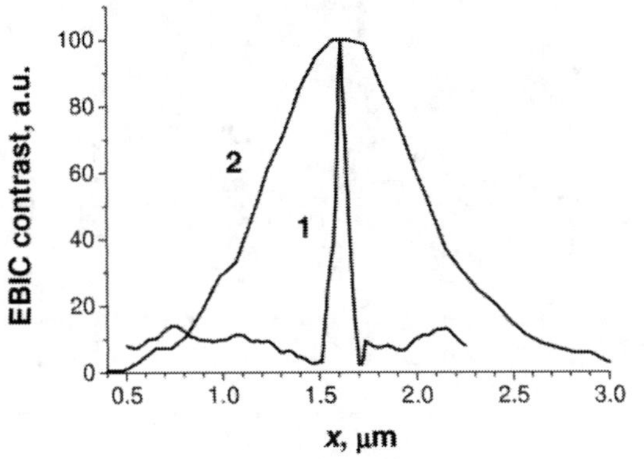

Figure 3. Profiles of normalized bright EBIC contrast for a threading dislocation (1) and for a large bright defect (2). $E_b = 35$ keV.

To study the effect of low-energy e-beam exposure on electrical behaviour, a 19 μm^2 area containing the defects with large bright EBIC contrast was irradiated with a beam current of about 10^{-10}A. In EBIC mode (figures 4b-4c) the irradiation region can be seen as a rectangular region in the center of images surrounded by bright boundaries. In figure 4 two defects with bright contrast are located inside the irradiated region and one is outside it. The evolution of EBIC signal I_{cirr} normalized on the signal from the non-irradiated region I_{cnir} for two bright defects in the irradiated region and for the irradiated region far from them with exposure time is shown in figure 5. As can be seen in figures 4 and 5, the bright contrast monotonically decreases with irradiation time and at irradiation doses larger than 1.5 $C{\cdot}cm^{-2}$ these defects could not be revealed in the EBIC image of the irradiated region. Thus, the difference between the EBIC signals from the defect region and far from it monotonically decreases under the LEEBI. This suggests that such treatment leads to healing of the defective regions. The EBIC signal from the irradiated region also changes due to irradiation. However, it increases for a small exposure time and then started to decrease. At doses larger than 1.3 $C{\cdot}cm^{-2}$ it becomes smaller than that from the non-irradiated region. This small decrease could be explained qualitatively by the build-up of hydrocarbons on the surface after a long e-beam exposure, which is revealed in the secondary electron mode.

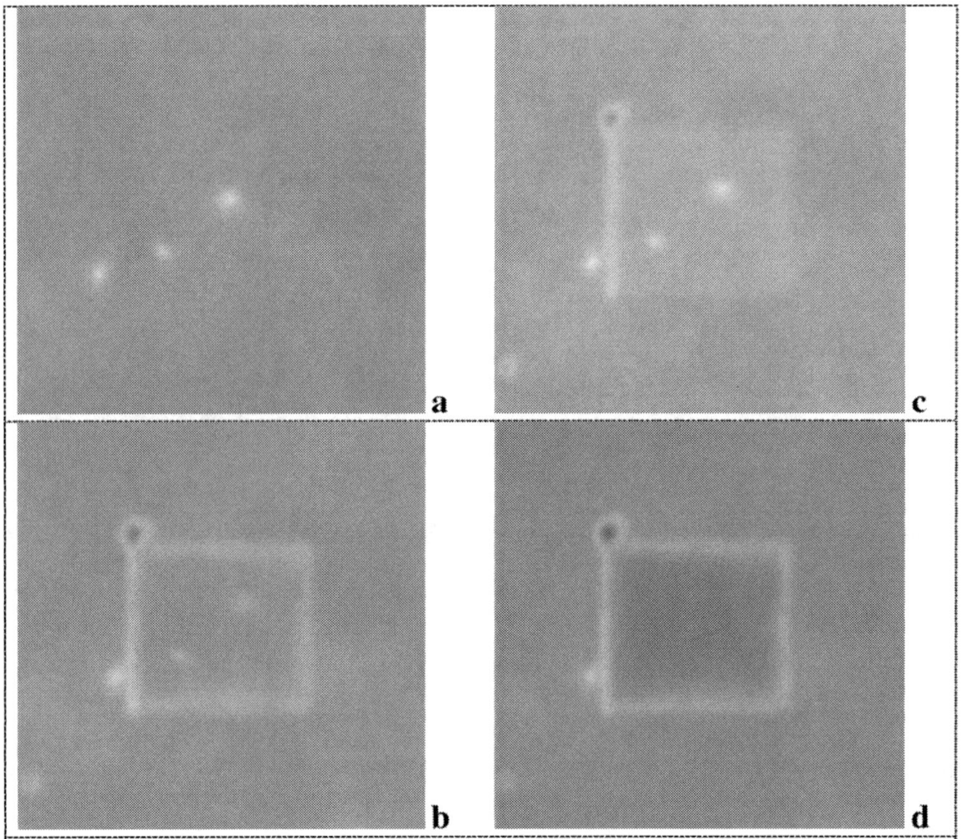

Figure 4. EBIC images of irradiated region (in the centre of figures) with two bright defects before LEEBI (a) and after LEEBI for 900 (b), 2100 (c) and 3300 s (d). The irradiation region size is 4.35×4.35 μm^2, E_b = 10 keV.

Our recent cathodoluminescence (CL) investigations have shown [7] that irradiation of the same structures by low energy electrons leads to the formation of a new InGaN related emission band blue shifted in comparison with the initial one. The total InGaN related luminescence intensity increases at the first stage of irradiation and then starts to decrease. Such behaviour of InGaN quantum well related luminescence was very similar to that shown for the EBIC signal in figure 5 for the irradiated region. In both cases the maximum was observed at similar e-beam exposure doses. That allows us to assume that the LEEBI effect on the EBIC signal and on the InGaN related emission was determined by similar mechanisms.

The analyses carried out in [7] showed that the LEEBI effect on the CL could be explained under the assumption that e-beam irradiation led to a local change of In content due to In or Ga recombination enhanced diffusion. As a result, some new quantum dot like regions could be formed due to the e-beam exposure and the new emission bands arise in addition to existing ones. The inhomogeneous potential distribution formed under this process could lead both to a suppression of excess carrier recombination inside the quantum wells and to a change of tunneling probability, that in turn could increase the EBIC signal. It is seen in figure 4 that the boundary of the irradiated region gives the bright contrast in the EBIC mode. If indeed irradiation changes the In content, the electric field arising at the boundary between irradiated and non-irradiated regions could suppress the excess carrier recombination and produce the bright contrast. Besides, it is well known [13] that LEEBI destroys Mg-H pairs in p-GaN. If such pairs are present in the structures studied, the released hydrogen could passivate the recombination centers and thus increasing the EBIC signal. The observed healing of defect regions could mean that the enhanced diffusion processes and/or their effect on the local well structure and composition near the bright contrast defects are more effective than in the defect-free region.

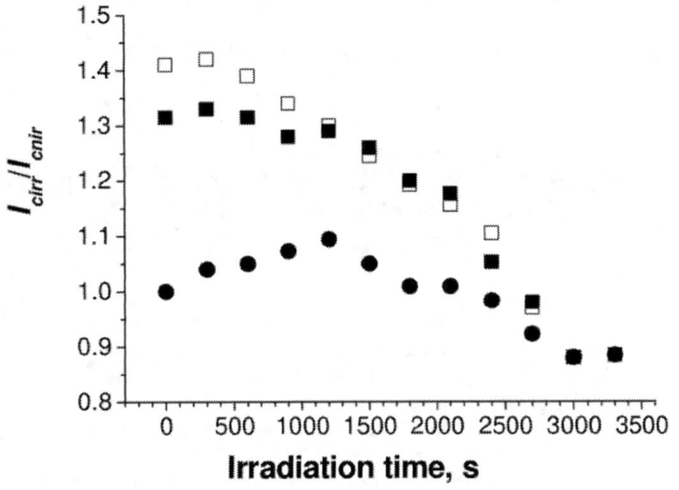

Figure 5. Dependence of I_{cirr}/I_{cnir} ratio on irradiation time for two bright defects shown in figure 4 (squares) and for the irradiated region (filled circles)

Acknowledgements
The work was partially supported by the Program of basic researches of RAS "The fundamentals of basic researches of nanotechnologies and nanomaterials".

References

[1] Oliver R A, Bennett S E, Zhu T, J Beesley D, Kappers M J, Saxey D W, Cerezo A and Humphreys C J 2010 *J. Phys. D: Appl. Phys.* **43** 354003
[2] Shmidt N M, Vergeles P S and Yakimov E B 2007 *Semiconductors* **41** 491
[3] Greshnov A A, Chernyakov A E, Ber B Y, Davydov D V, Kovarskyi A P, Shmidt N M, Snegov F M, Soltanovich O A, Vergeles P S and Yakimov E B 2007 *Phys. Stat. Sol.* (*c*) **4** 2981
[4] Bel'nik S A, Vergeles P S, Shmidt N M and Yakimov E B *J. Surf. Investig. X-ray, Synchrotr. Neutron Techniq.* 2007 **1** 394
[5] Yakimov E B 2008 *Springer Proceed. Phys.* **120** Eds. A. G. Cullis and P. A. Midgley 481
[6] Jahn U, Dhar S, Kostial H, Watson I M and Fujiwara K 2003 *Phys. Stat. Sol.* (c) **0** 2223
[7] Vergeles P S, Shmidt N M, Yakimov E E and Yakimov E B 2011 *Phys. Stat. sol.* (*c*) to be published
[8] Yakimov E B, Borisov S S and Zaitsev S I 2007 *Semiconductors* **41** 411
[9] Yakimov E B 1992 *Bul. Russian Acad. Sci.* **56**, 312
[10] Sharma N, Thomas P, Tricker D and Humphreys C 2000 *Appl. Phys. Lett.* **77** 1274
[11] Henley S J and Cherns D 2003 *J. Appl. Phys.* **93** 3934
[12] Shmidt N M, Sirotkin V V, Usikov A S, Yakimov E B and Zavarin E E 2003 *Inst. Phys. Conf. Ser.* No 180 ed. by Cullis A G and Midgley P A 597
[13] Amano H, Kito M, Hiramatsu K and Akasaki I 1989 *Jpn. J. Appl. Phys.* **28** L2112

Photoluminescence from $Al_xIn_{1-x}N$ layers doped with Tb^{3+} ions

Pascal Gehring, Ye Weng, Zhenyu Wu, Horst P Strunk
Institute of Materials Science, Chair of Material Physics, University of Stuttgart, Heisenbergstr. 3, 70569 Stuttgart, Germany

E-mail: strunk@mf.mpg.de

Abstract. $Al_xIn_{1-x}N$ films ($0 \leq x \leq 1$) have been sputter deposited and annealed, both without and with terbium co-doping, to obtain a series of matrices whose band gap energies span the range from around 2 eV to 6 eV. The terbium green luminescence spectra (excitation wavelength 230 nm, i.e. 5.4 eV) are measured at room temperature as a function of the aluminium content (band gap route) and of the terbium concentration (concentration route). The green luminescence assumes a maximum of the integrated intensity at a band gap energy of 4.1 eV ($x = 0.7$) which can be argued to result from a resonant energy transfer from the host matrix into the Tb luminescence centres. Furthermore, the dependence of this maximum integrated intensity as a function of the Tb concentration, i.e. of the average distance between the Tb centres, suggests the energy transfer from the host material into the Tb luminescent centres to be due to dipole-dipole interaction via an exciton bound to the centre.

1. Introduction

Thin semiconductor films in large areas, such as for solar cells or thin film transistor arrays in displays, are generally produced by rapid deposition methods. Apart from chemical vapor deposition, which generally requires higher temperatures, magnetron sputtering is frequently used. Sputtering is applicable at any temperature, preferentially at room or modestly elevated temperatures because of economical reasons. Common to these methods is the fact that single crystalline material cannot be obtained. In most cases one obtains amorphous or nanocrystalline films whose structures can subsequently be improved by annealing causing crystallization and grain coarsening.

Crystal defects can impair the electronic and optical properties of semiconductor materials. This disadvantage is especially prominent in the 'classical' semiconductors like Si, Ge or the families of the IV-IV, II-VI and III-V semiconductor compounds, e.g. [1] and references therein. Interestingly enough, the III-nitride family (more precisely IIIA-nitride), i.e. GaN, InN, and AlN and its ternary and quaternary compounds, are much more tolerant to crystal defects [2]. Therefore this III-N family is very interesting for being investigated with regard to optoelectronic properties especially as it also offers a broad range of band gaps from less than 2 eV (InN) to a little larger than 6 eV (AlN) with the same crystal structure.

In this paper we report on an investigation of the luminescent properties of a series of $Al_xIn_{1-x}N$ compound films ($0 \leq x \leq 1$) doped with Terbium in varied concentrations of one to a few at-%. Measurements of the intensity of the characteristic green Tb luminescent lines are discussed as a function of the annealing temperature, of the Tb concentration, and of the In content, i.e. the band gap of the host. The observed dependence of the appearance of a luminescence intensity maximum with increasing annealing temperature can be understood in terms of formation and decay of efficient Tb luminescence centres, e.g. [3,4], the formation of a luminescence maximum as a function of the band

Published under licence by IOP Publishing Ltd

gap energy can be understood in terms of a resonance model of the energy transfer from the excited host electronic system into the Tb luminescence centres [5,6].

2. Experimental procedures

2.1. Film deposition and annealing

All $Al_xIn_{1-x}N$:Tb-films in the composition range $0 \leq x \leq 1$ were deposited by reactive co-sputtering in an ATC 1500-F (AJA International) magnetron system. The Al-In composition and the Tb concentration were adjusted by respective deposition rates (controlled by the selected sputtering powers) from Al-, In-, and Tb targets. Only in the case of rather In-poor films, the In target was replaced by an Al target onto which some In wires were placed. The substrates were kept floating in temperature at room temperature, with the utmost temperature excursion during sputtering reaching around 40°C. In each deposition run three different substrates have been used (in order to match the different experimental requirements of the following measurements): sapphire (0001), Si (001) covered with 50 nm amorphous silicon oxide (SiO_x), and Si (001) covered with 50 nm a-SiO_x covered with 50 nm a-SiN_x. Two experimental strategies are followed. For the *concentration route,* AlN samples with different Tb concentrations between 1 and 5 at.%, and for the *band gap route* $Al_xIn_{1-x}N$:Tb samples ($0 \leq x \leq 1$) with practically constant Tb concentration (around 1at.%) have been prepared. The layers had a thickness of about 400-500 nm, occasionally also around 100 nm.

All samples were annealed after deposition. The AlN films (*concentration route*) are isochronically annealed in the range of 100 to 1100°C under 1 bar of nitrogen in steps of 100°C (from 800°C in 50°C), each annealing step took 20 minutes. Because of the low nitrogen pressure a loss of InN occurred in these samples, so that their band-gap was investigated again after annealing. The AlInN films (*band gap route*) are annealed at 660°C (melting temperature of Al) for 20 minutes at 8 bar of nitrogen. The samples on sapphire were annealed for 20 min at 1100°C under 1 bar nitrogen.

2.2. Film structure, composition and optical properties

The structure and the Al content, x, of the samples were investigated by X-ray diffraction (XRD, D5000, Siemens, Cu-K_α-radiation, Bragg-Brentano-geometry). The spectra showed the wurtzite structure throughout, and indications of cubic phase or other phases were absent. Specifically the region $2\Theta = 40 - 52°$ (Θ: diffraction, or Bragg angle) comprising the characteristic 102-peak of the hexagonal wurtzite phase was characterized to determine its shift with Al content, x. This shift is significant for the lattice parameter and can, by the use of Vegard's law, be converted into the Al concentration. As an independent method the band gap energy is also utilized for characterization of the chemical composition.

The microstructure of the films has been characterized by transmission electron microscopy in a Philips CM 200 microscope operated at 200 kV accelerating voltage. Plan view and cross-sectional samples prepared by grinding and ion milling have been investigated. The chemical composition has been determined in detail by X-ray photoelectron spectroscopy (Thetaprobe of Thermo-VG, Al-K_α-radiation). Oxygen distributed homogeneously across the thickness of the films is the main impurity.

The optical properties are evaluated from transmission spectra (samples deposited on sapphire) taken with a dual-beam spectrometer Cary 5000 (Varian) in the wavelength range 280 – 800 nm. The excitation light (Xenon-lamp) had a wavelength of 230 nm. A 280 nm-edge filter was used to avoid signals from the excitation source.

3. Results and Discussion

3.1. Terbium luminescence in dependence of the band gap energy

A number of research activities have shown that annealing of sputter deposited semiconductor films, Si or AlN, can lead to a considerable improvement of the luminescence intensity of rare earth ions, say Tb^{3+} as in the present case, contained in these films [7,8]. Figure 1 shows the integrated intensities of the green line family $^5D_4 \rightarrow {}^7F_J$ from all of the $Al_xIn_{1-x}N$:Tb films as a function of the band gap of the host as measured after annealing at 660°C. Examples of the spectra are displayed in figure 2. The three spectra relate to the large square data points in figure 1. It is obvious that the run of the intensity curve exhibits a pronounced peak at the band gap energy of 4.1 eV and an intensity increase again towards the high band gap energy end. In order to interpret the peak, the microstructure has been investigated (see the three micrographs in figure 3). One can see that the layers are made up by fibre like grains with an average diameter of around 50 nm, essentially standing upright on the substrate. There is no noticeable change of this structure with Al content, i.e. with band gap. We may therefore rule out a microstructural origin of the intensity peak in figure 1. In view of the rather sharp peak in the intensity, a resonant excitation process, i.e. a resonant energy transfer from the host's matrix to the luminescent Tb centre, leading to this peak can be envisaged. In fact, measurements by photo luminescence excitation spectroscopy have indicated that a direct optical excitation of the Tb centres occurs at a negligible rate only [6].

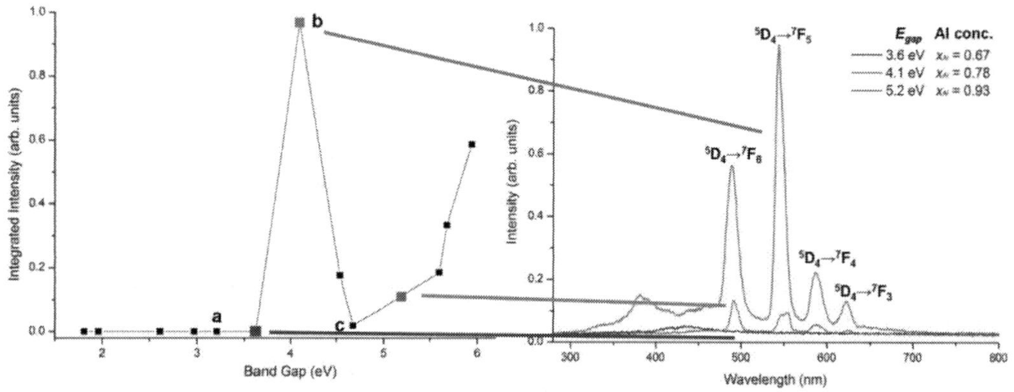

Figure 1. Integrated intensities of the Tb green luminescence lines in dependence of the host band gap energy. The intensity peaks at around the band gap energy of 4.1 eV. The coloured squares indicate the spectra pertaining to figure 2. The lettered data points indicate the respective film microstructures in figure 3.

Figure 2. Examples of Tb luminescence spectra measured from hosts having band gap energies given in the inset. Measurements after a 20 min anneal at 660 °C in atmosphere.

A rather transparent explanation for the occurrence of the intensity maximum relies on the work of Lozykowski and Jadwisienczak [9] who considered, for the case of AlN, isovalent rare earth induced centres (RESI) formed by RE^{3+}-ions to which an exciton is bound. They discriminate three types of centres formed by a single, two nearby and three Tb ions ($RESI_{I, II, III}$). The exciton binding mechanism results from the difference in electronegativity (and possibly also in size) by which a hole is attracted, which in turn binds an electron due to coulomb interaction to form a bound exciton. The associated defect level can be located within the band gap, its energy position depending upon band gap, exciton energy and energy by which the hole is bound to the RE^{3+}-isovalent ion.

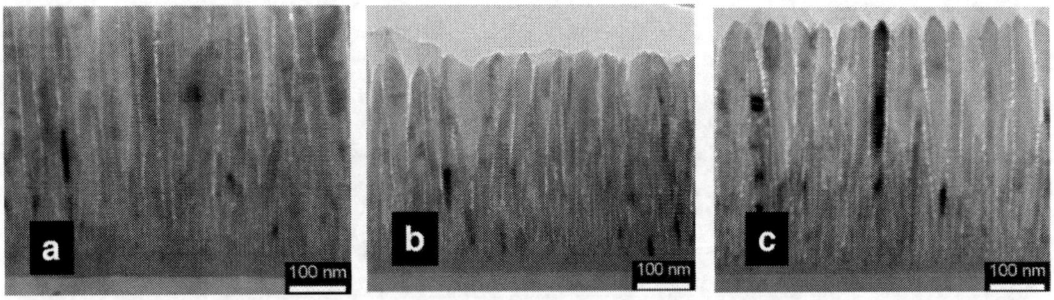

Figure 3. Microstructures of AlInN films with Al contents $x_{Al} = 0.61, 0.78$ and 0.82 in a, b and c, respectively. All samples have fibre structure, fibre grain diameter around 40 nm. The grain boundaries are decorated by white contrasts very probably due to preferential ion etching during sample preparation for the cross-sectional transmission electron microscopy.

Unfortunately these energy data are not known for the present material system (Al,In)N for whole the In concentration range since the necessary measurements could not yet be performed. Instead we proceed in modelling by analogy to the results of Nepal et al. who measured Tm induced RESI levels in the system $Al_xGa_{1-x}N$ by thermal quenching experiments [10]. The evaluated data support the conclusions that with smaller band gap energy, E_g, the RESI levels move nearer to the conduction band edge in agreement with theoretical results [11], and that parallel to it also the binding energy, E_b, of the exciton to the Tb centre decreases. Haynes [12] has estimated this binding energy to relate to the trap position, E_{trap}, with reference to the conduction band edge as $E_b = 0.1\ E_{trap}$. Based on these results the properties of the present system $Al_xIn_{1-x}N$ can now be discussed at least in principle.

Figure 4 shows the band gap energy as evaluated for this system [5] with reference to the valence band edge as a function of the layer composition. In addition, energy levels of the Tb ion are introduced with the ground level coinciding with the valence band edge. The three Tb-RESI energy levels assigned by Lozykowski and Jadwisienczak to the AlN matrix are introduced also (see the three dashes in the right ordinate) and the possible changes with reference to the valence band edge as a function of the In concentration are indicated (green dashed lines). The resonance condition in the transfer of the energy of the decaying exciton bound to the Tb luminescence centre is given in this plot when a dashed line (bound exciton energy) crosses a Tb level line. Two circles coloured violet mark crossings relevant to the present luminescence intensity measurements as a function of the band gap energy (composition). The crossing of the $RESI_{III}$ defect energy line with the $^5D_3$ Tb level indicates the resonance that is correlated with the observed luminescence maximum at the band gap energy $E_g = 4.1$ eV ($x = 0.78$). The second circle at $E_g = 5.6$ eV ($x = 0.96$) denotes resonant interaction with the outer electron states $5d_i$ of the Tb ions. This interaction can tentatively be correlated with the intensity increase observed to be characteristic of the low In end of the $Al_xIn_{1-x}N$ series. The rather broad shape of this intensity increase can then attributed to the comparably broad energy width of the $5d_i$ states that are as outer states very much dependent on the fluctuations in the atomic surroundings in the matrix. A third, empty circle at $E_g = 2.9$ eV ($x = 0.52$) indicates a further resonant interaction. The respective luminescence maximum is, however, not observed in our experiments, very possibly because of thermal quenching (distance of the trap to the conduction band less than, say, 0.4 eV). Luminescence measurements at low temperatures still need to be performed to clarify this aspect.

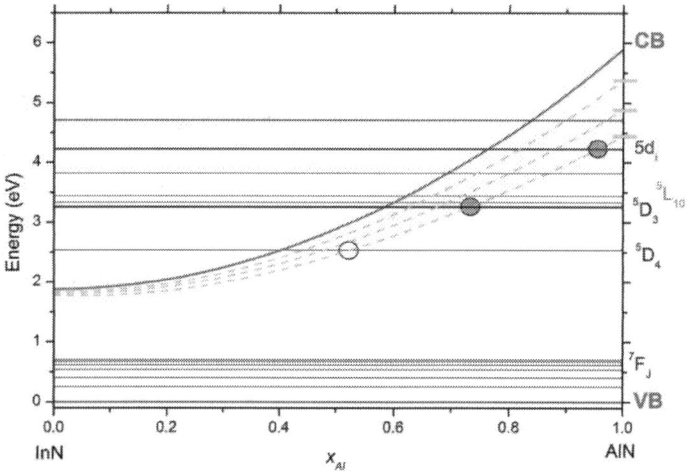

Figure 4. Model of the exciton-Tb resonant energy transfer. Shown are the Tb conduction band CB as a function of the aluminium content x_{Al} with reference to the valence band VB (red lines), the RESI levels of the three exciton states in AlN as treated in ref. 9 (green), and Tb atomic levels, some of them identified by term symbols. The green dashed curves denote the potential positions of the RESI centers with decreasing band gap. Resonant energy transfer can occur where the energetic position of the exciton trap coincides with an atomic Tb level, indicated by circles. For further discussion see text.

3.2. Terbium luminescence as a function of the Terbium concentration

Independent of the detailed mechanism described thus far of the excitation of the Tb ion luminescent centres, insight is helpful into the interaction between bound exciton and Tb luminescent centre. Essentially two different interaction mechanisms were discussed in the literature. Firstly the Förster mechanism which is based on the electrostatic interaction between two electrical dipoles [13], secondly the quantum mechanical exchange interaction [14], see also the summary [15]. Which of these is adequate to describe the present system can in principle be assessed by evaluating the luminescence intensity as a function of the Tb centre concentration. The probability, P_{dd}, for energy transfer according to the dipole-dipole interaction depends on the distance, R_{dd}, between the reacting centres as $P_{dd} \approx R_{dd}^{-6}$. The exchange interaction is characterized by a dependence $P \approx \exp(-R/L)$, which is a very short range interaction (L is the effective Bohr radius [15]). The average spatial distance between centres is just $R = 1/c^{1/3}$ (c is Tb centres per volume), thus $P_{dd} \approx c^{-2}$.

One possibility to evaluate the type of energy transfer consists of using the phenomenon of concentration quenching due to cross-relaxation [15]. It is known for Tb [16] that the transition $^5D_3 \rightarrow {}^5D_4$ is similar in energy to the transition $^7F_6 \rightarrow {}^7F_0$. The respective decay of one Tb ion in the $^5D_3$ excited state can consequently excite a second Tb ion from the ground state to a higher ^7F state, provided this ion is sufficiently near in vicinity, see figure 5. This cross-relaxation diminishes the population of the $^5D_4$ level and thus the green luminescence. Figure 6 displays the presently available experimental results. Specifically the measured green intensities I_{5D4} are plotted as inverse values as a function of the Tb concentration. In order to compensate possible fluctuations in the excitation of the $^5D_3$ level and thus its population, this inverse value is multiplied by the intensity of the transitions $^5D_3 \rightarrow {}^7F_J$. A clear decay of the I_{5D3}/I_{5D4}-values with increasing Tb concentration is evident. Further, the experimental values are approximated with two curves representing the two energy transition models described above. The fit of both the Förster and the exchange model curves to the experimental data is rather good. However, this does not permit us to give a preference for one of the two energy transfer mechanisms. The scatter of the presently available data points is too large, and the investigated Tb concentration range is too small.

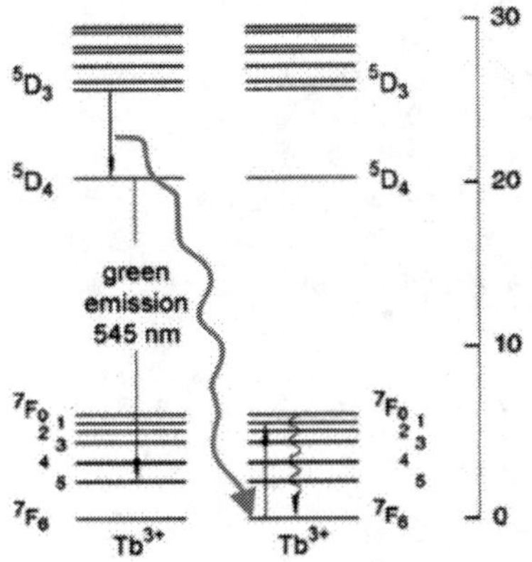

Figure 5. The process of cross-relaxation. Atomic level scheme for two adjacent Tb ions. The energy liberated by the decay of the excited state $^5D_3$ to $^5D_4$ is resonantly transferred to the adjacent Tb ion and excites a higher level ^7F_J state. Sketch after [17].

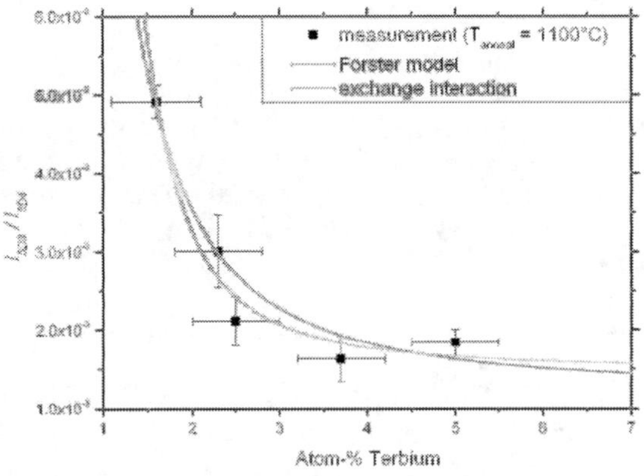

Figure 6. Luminescence intensity as a function of Tb concentration. As quantity for the intensity is used the integrated blue intensity of the $^5D_3$ family divided by the integrated green intensity of the $^5D_4$ family. The red and green curves indicate fit curves according to the dipole-dipole and the exchange mechanisms, respectively.

4. Conclusions

Our first results on luminescence of terbium ions embedded in films of AlInN in various compositions between AlN and InN indicate the advantage of a material system that covers a whole range of band gap energies. The advantage of the group-IIIA-nitrides, AlN, InN and GaN, for the investigation of luminescence properties of rare earth ions is evident. These three compounds are principally miscible, only care has to be taken during deposition to avoid kinetically driven segregation during film deposition, eg. [18]. All the materials can grow with the same, wurtzite structure and are then all direct semiconductors avoiding cumbersome and straggling considerations that are necessary when optical properties of rare earths in matrices with different chemistry and lattice types are to be discussed. The results presented here have indicated that a resonance effect in the energy transfer mechanism can enhance the luminescence intensity very much by appropriately adjusting the band gap of the matrix. Unfortunately the results obtained thus far on the variation of the luminescence intensity with Tb concentration are not sufficient to discriminate whether the Förster or the exchange interaction are dominant in the energy transfer from the matrix into the Terbium luminescence centre.

Acknowledgment

Much support came from the service groups of the Max-Planck-Institute for Metals Research, Stuttgart, the collaborators of which were of indispensable help in different types of measurements, notably S Haug, M Kelsch, T Meissner, R Preuss, U Salzberger, B Siegle, R Völker and M Wieland. HPS acknowledges helpful discussions with M Yang. The transmission electron microscope work has been performed in the Stuttgart Center for Electron Microscopy headed by P van Aken.

References

[1] Fitzgerald E A, Ast D G, Ashizawa Y, Akbar S and Eastman L F 1988 *J. Appl. Phys* **64** 2473

[2] Monemar B and Pozina G 2000 *Progr. Quantum Electronics* **24** 239; Lester S D, Ponce F A, Craford M G and Steigerwald D A 1995 *Appl. Phys. Lett.* **66** 1249

[3] Aldabergenova S B, Albrecht M, Strunk H P, Viner J, Taylor P C and Andreev A A 2001 *Mat. Sci. Eng.* B **81** 29

[4] Mendel H, Aldabergenova S B, Weingärtner R, Frank G, Strunk H P and Andreev A A 2006 *Optical Materials* **28** 79

[5] Gehring P, Weng Y, Wu Z and Strunk H P 2010 *IOP Conf. Ser. Mater. Sci. Eng.* **15** 012007

[6] Gehring P 2009 Diploma Thesis University Stuttgart, Institute of Materials Science

[7] Biggemann D, Mustafa D and Tessler L R 2006 *Optical Materials* **28** 842

[8] Weingärtner R, Erlenbach O, Winnacker A, Welte A, Brauer I, Mendel H, Strunk H P, Ribeiro C T M and Zanatta A R 2006 *Optical Materials* **28** 790

[9] Lozykowski H.J and Jadwisienczak W M 2007 *phys. stat. sol. (b)* **244** 2109

[10] Nepal N, Zavada J, Lee D, Steckl A, Sedhain A, Lin J and Jiang H 2009 *Appl. Phys. Lett.* **94** 111103

[11] Yu P and Cardona M 2005 *Fundamentals of Semiconductors*, 3rd ed. (NewYork Springer-Verlag Inc)

[12] Haynes J 1960 *Phys. Rev. Lett.* **4** 361

[13] Förster T 1946 *Naturwiss.* **6** 166; 1948 *Ann. Phys.* **2** 55; 1959 10th Spiers Memorial Lecture – Transfer mechanisms of electronic excitation *Disc. Faraday Soc.* **27** 7

[14] Dexter D L 1953 *J. Chem. Phys.* **21** 836

[15] Nakazawa E 1999 *Phosphor Handbook*, ed S Shionoya and W M Yen (Boca Raton: CRC Press) chapter 2.7 pp 85-112

[16] Colak S and Zwicker W K 1983 *J. Appl. Phys.* **54** 2156; May P S and Sommer K D 1997 *J. Phys. Chem* A **101** 9571; Lawrence T, Murra K A and May P S 2003 *J. Phys. Chem.* B **107** 4002

[17] Ronda C R 2008 Emission and excitation mechanisms of phosphors *Luminescence* ed C R Ronda (Weinheim: Wiley-VCH) chapter 1 p 20

[18] Albrecht M, Lymparakis L, Neugebauer J, Northrup J E, Kirste L, Leroux M, Grzegory I, Porowski S and Strunk H P 2005 *Phys. Rev.* B **71** 035314

Journal of Physics: Conference Series **281** (2011) 012015 doi:10.1088/1742-6596/281/1/012015

A qualitative semi–classical treatment of an isolated semi–polar quantum dot

Toby D. Young

Instytut Podstawowych Problemów Techniki Polskiej Akademii Nauk
ul. Adolfa Pawińskiego 5b, 02-106 Warszawa, Polska

E-mail: `tyoung@ippt.gov.pl`

Abstract. To qualitatively determine the behaviour of micro–macro properties of a quantum dot grown in a *non*-polar direction, we propose a simple semi–classical model based on well established ideas. We take into account the following empirical phenomena: (i) The displacement and induced strain at heterojunctions; (ii) The electrostatic potential arising from piezoelectric and spontaneous polarisation; and (iii) The localisation of excitons (particle–hole pairs) arising from quantum confinement. After some algebraic manipulation used to cast the formalism into an arbitrarily rotated frame, a numerical model is developed for the case of a semi–polar wurtzite GaN quantum dot buried in a wurtzite AlN matrix. This scheme is found to provide a satisfying qualitative description of an isolated semi–polar quantum dot in a way that is accessible to further physical interpretation and quantification.

1. Introduction

Quantum dot structures are sometimes referred to as 'designer atoms' or 'artificial atoms' and have become an integral part of modern human life in the form of light detectors, chemical sensors, light emitters, and integrated circuits; with applications in medicine, engineering technology, and science. One outstanding property of the wurtzite crystal structure in particular is that it possesses a spontaneous piezoelectric field. The piezoelectric field has been shown to dominate the electronic properties of quantum dot structures; even given the presence of other mechanical fields (*eg.* elastic–electric displacements) [1]. Until now, nanostructures fabricated for scientific investigation and practical devices have been grown along the polar direction. These systems have typically been studied by theoretical modelling with a good degree of success (see for example references [2, 3]).

Recently, three–dimensional quantum dot nanostructures grown in the semi–polar (*i.e. non*-polar) directions and their related physical response have become an area of interest that is enigmatic to both the pure and applied sciences [4, 5]. For this reason in this contribution we investigate the elastic–electric and optical properties of an isolated semi–polar wurtzite GaN quantum dot embedded in a wurtzite AlN matrix that is chosen to be representative of experimental observation. Most such studies tend to make use of the $\mathrm{k}\cdot\mathrm{p}$ Hamiltonian within the effective mass and plane wave approximation [1, 2, 3]; so much so that this tends to be referred to as the 'standard model' of electronic structure theory [6]. While that Hamiltonian has its uses, in this contribution we provide the first few steps toward an alternative form for a semi–empirical model. In the next section the basics of the semi–empirical model and method are described. Following that, in section 3, the model is applied to a quantum dot grown for semi–polar angles $0 \leq \theta \leq \pi/2$. Finally, in section 4, we summarise this contribution and give a tentative outlook for future development.

2. Model and method

The discussion that follows concerns heterostructures examined on the nanoscale that have a built–in spontaneous piezoelectric field, notably nanostructures with a wurtzite crystal phase (hexagonal space symmetry group $P6_3mc$) described by the geometrical parameters a and c. Furthermore, given that we are interested in qualitatively describing the behaviour of micro–macro properties of three–dimensional quantum dot heterostructures, a model based on semi–classical mechanics is proposed that has a classical component and a quantum component.

To describe the elastic–electric properties and particle–hole states quantum confined to a semi–polar quantum dot encapsulated in a polyhedral volume $\Omega \in \mathbb{R}^3$, the following three models are taken to be adequate and sufficient:

(i) The traditional fully-coupled elastic–electric equation set [7, 8] for the elastic displacement u and electric potential V is written as a set of simultaneous equations for the stress state

$$\sigma_{ij} = C_{ijkl}(\partial_l u_k + \Delta_{kl}) - e_{mij}\partial_m V \quad , \tag{1}$$

and the electric displacement state

$$D_i = \epsilon_{ij}\partial_j V_j + e_{ijk}\partial_k u_j + P_i \quad , \tag{2}$$

where C_{ijkl}, e_{ijk}, ϵ_{ij}, and P_i are moduli of elastic, piezoelectric, permittivity, and spontaneous polarisation tensors respectively; and the symmetric gradient of the elastic displacements is defined as $\partial_l u_k := \frac{1}{2}(\partial_l u_k + \partial_k u_l)$. The strain arising from the lattice mismatch in equation (1) is defined by

$$\Delta_{kl} = \begin{cases} \frac{a_{\mathrm{matrix}} - a_{\mathrm{dot}}}{a_{\mathrm{matrix}}}\delta_{kl} & \text{if } k = l \in \{1,2\} \\ \frac{c_{\mathrm{matrix}} - c_{\mathrm{dot}}}{c_{\mathrm{matrix}}}\delta_{kl} & \text{if } k = l = 3 \\ 0 & \text{if } k \neq l \end{cases} \quad , \tag{3}$$

where the indices $\{i, j\}$ run over all degrees of freedom and δ_{kl} is the Kronecker delta function $\delta_{ij} = [i = j]$ for the Iverson bracket $[\cdot]$. Zero boundary constraints are imposed by

$$u(\mathbf{x})|_{\partial\Omega} = V(\mathbf{x})|_{\partial\Omega} = 0 \quad , \tag{4}$$

and equations (1) and (2) are solved for $\partial_j\sigma_{ij} = \partial_i D_i = 0$.

(ii) The Hamiltonian operator for the wurtzite crystal structures is, in the first instance, written as two separable one–band Hamiltonian operators

$$H_{\mathrm{eff}} = \partial_i \frac{1}{2m_i^\star}\partial_i + q\,(V - E_{\mathrm{gap}}) \quad , \tag{5}$$

where the indices $\{i, j\}$ run over all degrees of freedom, $m_i^\star$ is the effective mass of a particle or hole, $q = \mp e$ for particles and holes respectively (where e is the charge of a single electron), and V is the electrostatic potential found from the elastic–electric problem. The operator is substituted into the Schrödinger equation:

$$H_{\mathrm{eff}}\Psi = E\Psi \quad , \quad \Psi = \textstyle\sum_n c_n\psi_n \quad , \tag{6}$$

for normalisation constants c_n and where $E \in \{E_{\mathrm{p}}, E_{\mathrm{h}}\}$ and $\Psi \in \{\psi_{\mathrm{p}}, \psi_{\mathrm{h}}\}$ are the particle and hole eigenenergies and eigenstates respectively. Zero boundary constraints are imposed by

$$\Psi(\mathbf{x})|_{\partial\Omega} = 0 \quad , \tag{7}$$

and equation (5) is then solved $\forall\psi \neq 0$.

(iii) The tensors of moduli of the above two models are written for a fixed–body Cartesian axis $\{x, y, z\} \in \mathbb{R}^3$ from which there exists a rotated–body Cartesian axis $\{\bar{x}, \bar{y}, \bar{z}\} \in \mathbb{R}^3$, where the two spaces are connected by the well–known rotation tensor $\mathcal{R}_{ia}$. The rotation $\mathcal{R}_{ia}$ acts on all tensors of moduli $\mathcal{T}$ such that

$$\bar{\mathcal{T}}_a = \mathcal{R}_{ia}\mathcal{T}_i \quad , \quad \forall \mathcal{T}_i \in \{P_i, m_i^\star\} \quad , \tag{8}$$

$$\bar{\mathcal{T}}_{ab} = \mathcal{R}_{ia}\mathcal{R}_{jb}\mathcal{T}_{ij} \quad , \quad \forall \mathcal{T}_{ij} \in \{\Delta_{ij}, \epsilon_{ij}\} \quad , \tag{9}$$

$$\bar{\mathcal{T}}_{abc} = \mathcal{R}_{ia}\mathcal{R}_{jb}\mathcal{R}_{kc}\mathcal{T}_{ijk} \quad , \quad \forall \mathcal{T}_{ijk} \in \{e_{ijk}\} \quad , \tag{10}$$

$$\bar{\mathcal{T}}_{abcd} = \mathcal{R}_{ia}\mathcal{R}_{jb}\mathcal{R}_{kc}\mathcal{R}_{ld}\mathcal{T}_{ijkl} \quad , \quad \forall \mathcal{T}_{ijk} \in \{C_{ijkl}\} \quad , \tag{11}$$

casting the moduli of the system in a semi–polar direction of growth $\bar{\mathcal{T}}$ defined as the orientation vector g_i which lies parallel to the growth direction. At zero rotation, *ie.* polar direction, the growth axis is $g_i \parallel z \parallel c$. By tracing the rotational motion of the orientation vector $\bar{g}_a = \mathcal{R}_{ia} g_a$ on $\tilde{\Omega}$ the physical properties of "metastable" semi–polar states through a collective path can be found for any arbitrarily chosen orientation vector.

The solution scalars/vectors $\{u_i, V, \Psi\}$ *are determined in the fixed–body axis and are therefore directly comparable among themselves independently of the growth direction.*

2.1. Computational details

The computational procedure involves solving first the elastic–electric problem *cf.* equations (1) and (2); and second the Schrödinger equation *cf.* equation (5). This is repeated for any choice of rotation vector. The system is duly assembled in variational form according to the above prescription and subsequently solved with an application program using a set of standard computational tool kits: In particular, (i) Differential Equations Analysis Library (`deal.II`) [9, 10], which conveniently provides the abstract base for finite element methods; and (ii) Portable, Extensible Toolkit for Scientific Computation (PETSc) [11, 12] and Scalable Library for Eigenvalue Problem Computations (SLEPc) [13] for linear algebra computations and eigensystem computations respectively.

3. A representative half hyper–spheroidal quantum dot

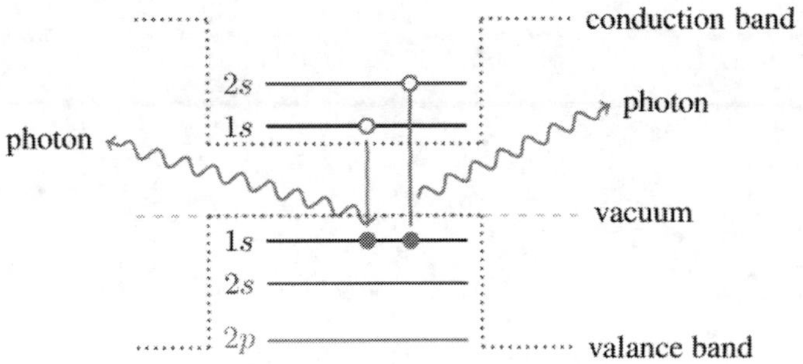

Figure 1. An illustration of photon emission in the particle–hole picture for the two–times–one band Hamiltonian in which a particle collapses to a hole state and emits a photon. Open circles denote particle (unoccupied) states and filled circles denote hole (occupied) states.

It is instructive in the discussion below to consider the two–times–one band Hamiltonian of equation (5) as a particle–hole pair moving in an effective potential generated by the respective energy gap E_{gap} and electrostatic potential V, which is in turn generated by the combined elastic–electric state

Table 1. Table of empirical constants used in this contribution. All data was taken directly from [3].

Parameter	GaN	AlN	Parameter	GaN	AlN
a nm	3.189	3.112	E_g eV	3.475	6.230
c nm	5.185	4.982	$m^\star_z$	0.20	0.28
C_{11} GPa	390	396	$m^\star_{i\in\{x,y\}}$	0.20	0.32
C_{12} GPa	145	137	VBO eV	0	-0.8
C_{13} GPa	106	108			
C_{33} GPa	398	373			
C_{44} GPa	105	116			
e_{13} C/m^2	-0.49	-0.60			
e_{13} C/m^2	-0.49	-0.60			
e_{13} C/m^2	0.73	1.46			
P_{sp} C/m^2	-0.029	- 0.081			
ϵ_z	10.01	8.57			
$\epsilon_{i\in\{x,t\}}$	9.28	8.67			

of the materials. This picture, illustrated in Figure 1 is reminiscent of the quantum oscillator model and the central potential model in atomic physics [14] as well as the Bardeen–Cooper–Schrieffer theory of superconductivity [15]. This picture asserts that we discard many–body effects that normally arise from the underlying atomic structure of the system at hand. Furthermore, in this contribution, many–body effects arising from the interactions of particles–particles, particles–holes, and holes–holes are discarded for simplicity.

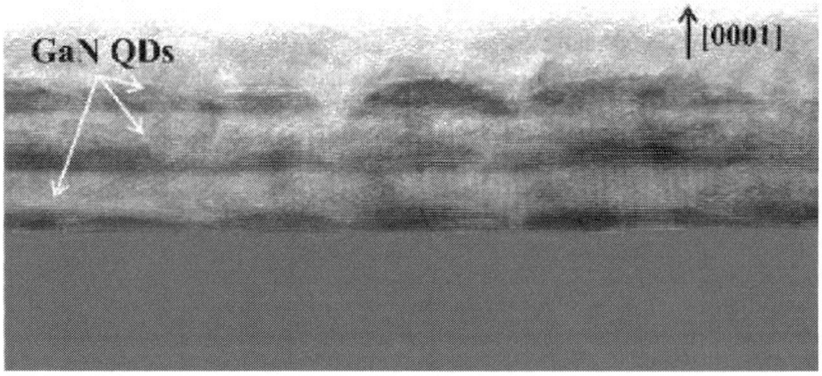

Figure 2. High resolution transmission electron microscopy image along the $[1\bar{1}00]$ zone axis showing self-assembled GaN quantum dots grown by molecular beam epitaxy in an AlN matrix. The heterostructure was deposited on (0001) AlGaN template. The average quantum dot height is 2 nm and their projected radius is 5–8 nm. The projected quantum dot morphology is half–ellipsoidal. Credits: Thanasis Kehagias and George P. Dimitrakopulos, Department of Physics, Aristotle University of Thessaloniki, Greece. (Used with permission).

In this contribution we consider a half oblate quantum dot of radius $A = 7.5$ nm and height $C = 4$ nm *cf* equation (12); which approximates the observations in high resolution transmission spectroscopy for

Table 2. Table of eigenenergies of particle (E_p) and hole (E_h) states and photon emission energies ($E_\gamma = E_p - E_h$) for three selected rotation angles $\theta \in \{0, \pi/6\pi/3\}$ radians. All units are in eV.

State	θ	ψ_{00}	ψ_{10}	ψ_{01}	ψ_{11}	ψ_{20}	ψ_{02}
	0	4.00	4.43	4.43	4.94	5.00	5.00
E_p	$\pi/6$	3.94	4.40	4.42	4.92	4.95	4.99
	$\pi/3$	3.66	4.18	4.33	4.78	4.82	4.90
	0	-0.23	-0.51	-0.51	-0.68	-0.77	-0.77
E_h	$\pi/6$	-0.17	-0.46	-0.50	-0.67	-0.75	-0.76
	$\pi/3$	-0.11	-0.23	-0.40	-0.56	-0.61	-0.65
	0	4.23	4.94	4.94	5.62	5.77	5.77
E_γ	$\pi/6$	4.11	4.86	4.92	5.59	5.70	5.75
	$\pi/3$	3.77	4.41	4.73	5.34	5.43	5.55

quantum dots grown in the polar (z) direction (Figure 2). The GaN material occupied by the quantum dot is then a (partial) spheroid described by the simple equation

$$\frac{x^2}{A^2} + \frac{y^2}{B^2} + \frac{z^2}{C^2} \leq 1 \quad , \quad \text{where } A = B > C \quad , \quad z \geq 0 \tag{12}$$

with the origin at the centre. To emulate semi–polar growth at an angle θ measured from the z axis, we make use of a rotation tensor that operates around the y–axis. The grid on which all computations were performed was a hyper–ball of radius 75 nm containing $\mathcal{O} \sim 10^5$ degrees of freedom. Values for the moduli were taken directly from Reference [3] and are given in Table 1 for completeness.

3.1. Results

The eigenenergies of the quantum dot system for three angles of rotation are given in Table 2. The results we obtain for the low –lying eigenenergies of the polar quantum dot $\theta = 0$ are comparable to alternative schemes described in the literature (Reference [3, 2]); to the author's knowledge there is no currently available data for a semi–polar GaN/AlN quantum dot system. However, we note that our model overestimates the energy gap between particle–hole states. This is most probably due to the lack of coupling between particles and holes. In other words, we have computed the particle–hole states as a vector–valued problem with the Hamiltonian operator

$$H_{ij} = \begin{pmatrix} H_p & H_{ph} \\ H_{hp} & H_h \end{pmatrix}, \qquad \forall H_{ph} = H_{hp} = 0, \tag{13}$$

which ignores interactions between particle–hole states (*ie.* no mixing terms). This can be corrected in a straightforward manner by choosing a correct form of the terms H_{ph} and H_{hp}.

One point of interest to note from the energies in Table (2) is that as the rotation angle increases (that is, as the growth direction diverges from the polar direction of growth) a splitting of energy levels is observed, from the standard degeneracies known from the quantum oscillator problem, to independent energetic states. This symmetry breaking is reflected in the region of increasing angle of rotation in which the particle and hole energies become closer to the vacuum state. The consequence of this is that the energy of photon emission, taken to be the energetic difference between a particle and a hole state ($E_\gamma = E_p - E_h$) tends to decrease as the rotation angle increases.

This feature is emphasised in Figure 3 for hole states which exhibit a more dramatic change in energy than their particle counterparts. The reason for the splitting of eigenenergies can be explained by

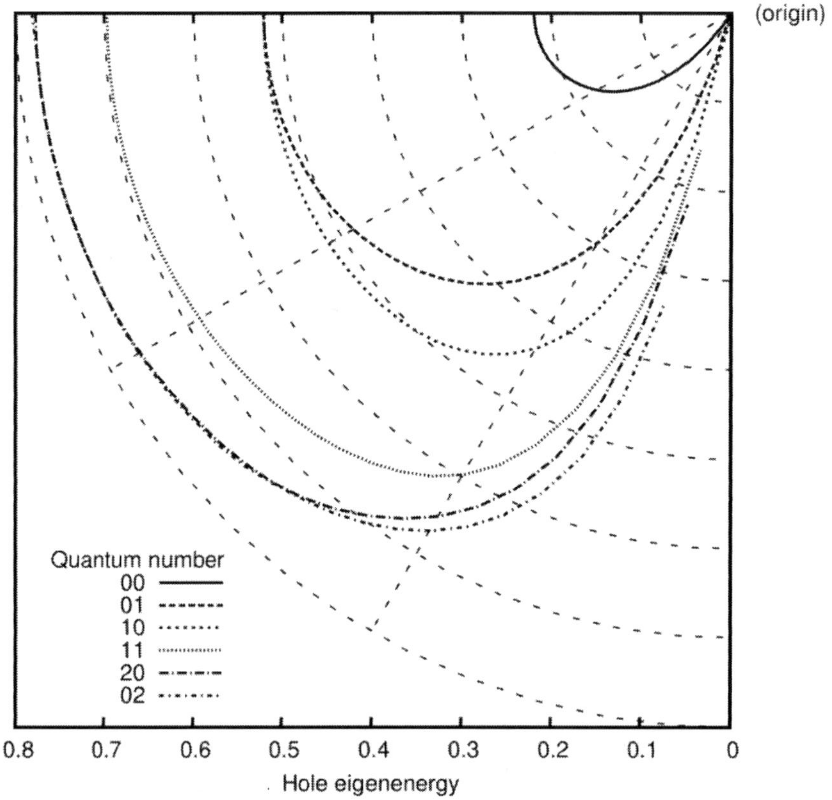

Figure 3. Polar plot of the six lowest hole states. Grid lines mark polar coordinates and the x–axis denotes the modulus of the energy of the holes states. Rotations are performed in an anti–clockwise direction and the eigenenergy is read from the x–axis. The results at specific rotation angles $(0, \pi/6, \pi/3, \pi/2)$ given in Table 2 correspond to radial axis lines.

observing the symmetry of the system. In general terms, the eigenfunctions 'see' material symmetries with respect to the rotation vector g_i. As the system is rotated, eigenfunctions become localised within the corners of the quantum dot and the effect of curvature of the quantum dot surfaces on the localisation of the eigenfunctions becomes more pronounced. In Figure 3, in the limit of an angle of rotation $\pi/2$ radians the eigenenergies are seen to collapse toward the vacuum as particle–hole states are 'squeezed' into a decreased volume at the bottom of the quantum dot. Nevertheless, this effect is possibly not wholly representative of real behaviour in semi–polar quantum dots but rather is a manifestation of the supposition that there are no inter–particle interactions given in our simplified model. If we were to take into account these interactions, we may expect that the energy splitting and divergence would be softened by the Coulomb potential.

Given the discussion above we now focus on the form of the eigenfunctions for two rotation angles $\theta \in \{0, \pi/6\}$ that are given in Figure 4. It is clear for the polar growth direction that the eigenfunction symmetries approximately follow those of the quantum oscillator; namely, we find states with degenerate symmetries, and in the order ψ_{00}, $\psi_{01} \sim \psi_{01}$ (and $\psi_{11} \sim \psi_{20} \sim \psi_{02}$). We find for polar quantum dots (upper panel of Figure4) that particles are located at the top of the quantum dot and holes are located at the bottom of the quantum dot. These results are similar to those of Reference [3]. The picture changes,

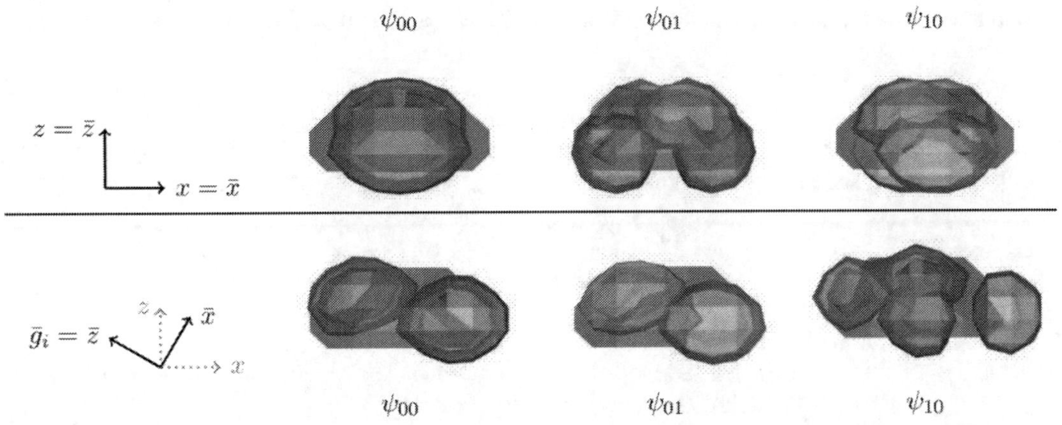

Figure 4. Contour plots of the lowest three eigenfunctions of the model (colour online). Upper panel $\theta = 0$ and lower panel $\theta = \pi/3$. Gold shaded regions denotes particle eigenfunctions and blue shaded areas show hole eigenfunctions. The gray shaded region gives a shadow image of the quantum dot region. All eigenfunctions are l_2–normalised.

as is to be expected, when the rotation vector g_i is aligned at $\pi/6$ radians (lower panel of Figure 4. In those cases the particle–hole separation align themselves along the vector g_i which immediately tends to the conclusion that the spontaneous polarisation vector $P_i \parallel g_i$ is the dominating physical quantity in semi–polar quantum dots as it is in polar quantum dots. This compliments the observations found in Reference [1]

4. Summary

In summary we first note that in all of the above we have surreptitiously invoked the supposition that *the morphology of an isolated quantum dot is unaffected by the given orientation (growth) vector g_i, and vice–versa.* In other words, in this paper we have assumed that a half–spheroidal geometry correctly describes the morphology of a quantum dot in any given growth direction. It is known from high resolution electron microscopy [5] that this supposition is not necessarily a good one! Nevertheless, to obtain a general scope of understanding about the effects of growth orientation, such as strain, the induced electrostatic potential, quantum confinement, and band–gap profile, it is useful to somewhat simplify the geometrical approach as has been done in this investigation.

The semi–classical model introduced in this paper was constructed by solving the elastic–electric equations followed by the two–times–one band Schrödinger equation for stationary bound states in a potential. In essence this is a revival and extension of the atomic theoretical approach to nanoscale structures. It was found that eigenenergies and eigenfunctions of the system followed those of the quantum oscillator for polar grown quantum dots. For semi–polar quantum dots, band splitting was observed due to the fact that the eigenfunctions see a distorted confinement symmetry. This scheme has been found to provide a satisfying qualitative description of an isolated two–level semi–polar quantum dot in a way that is accessible to further physical interpretation and quantification. For example, the inclusion of strain effects in the two–times–one band Hamiltonian operator [16] should improve the quantitative description of semi–polar quantum dots.

Moreover this qualitative scheme is extensible for further elaboration. For instance, one could consider taking into account inter–particle interactions using a Hartree–Fock like method, or to consider the electrostatic potential and particle–hole states within a Schrödinger–Poisson picture. These (and other considerations) are to be the concern of future investigations which, given this promising start, indicate that this is feasible and should be rewarding. That is a good thought on which to end this paper.

Acknowledgements
This research was carried out under the auspices of the development project, "*Finite element analysis of the quantum theory of finite electronic systems*", award number N N519 402837 founded by the Polish Ministry of Science and Higher Education.

References

[1] Young T D and Marquardt O 2009 *Phys. Stat. Solidi (c)* **6** 5557
[2] Andreev A D and O'Reilly E P 2000 *Phys. Rev. B* **62** 15851
[3] Fonoberov V A and Balandin A A 2003 *J. Appl. Phys.* **94** 7178
[4] Paskova T 2008 *Nitrides with nonpolar surfaces: growth, properties, and devices* (Wiley-VCH Verlag GmBH & Co. KGaA, Germany)
[5] Dimitrakopulos G P, Kalesaki E, Kioseoglou J, Kehagias T, Lotsari A, Lahourcade L, Monroy E, Häusler I, Kirmse H, Neumann W, Jurczak G, Young T D, Dłużewski P, Komninou P and Karakostas T 2010 *J. Appl. Phys.* **108** 104304
[6] Zunger A 1998 Electronic–structure theory of semiconductor quantum dots Tech. rep. MRS Bulletin
[7] Pan E and Yang B 2003 *J. Appl. Phys.* **93** 2435
[8] Nye J F 1962 *Własności fizyczne kryształów: w ujęciu tensorowym i macierzowym* (Państwowe Wydawnictwo Naukowe, Warszawa) translated from the original, Physical properties of crystals: their representation by tensors and matrices, Oxford (1957)
[9] Bangerth W, Hartmann R and Kanschat G 2007 *ACM Trans. Math. Softw.* **33** 24/1–24/27
[10] Bangerth W, Hartmann R and Kanschat G `deal.II` *Differential Equations Analysis Library, Technical Reference* `http://www.dealii.org` URL `http://www.dealii.org`
[11] Balay S, Buschelman K, Gropp W D, Kaushik D, Knepley M G, McInnes L C, Smith B F and Zhang H 2009 PETSc Web page http://www.mcs.anl.gov/petsc
[12] Balay S, Buschelman K, Eijkhout V, Gropp W D, Kaushik D, Knepley M G, McInnes L C, Smith B F and Zhang H 2008 PETSc users manual Tech. Rep. ANL-95/11 - Revision 3.0.0 Argonne National Laboratory
[13] Hernandez V, Roman J E and Vidal V 2005 *ACM Transactions on Mathematical Software* **31** 351–362
[14] Lindgren I and Morrison J 1986 *Atomic many–body problem* (Springer–Verlag, Berlin Heidelberg)
[15] Ring P and Schuck P 1980 *The nuclear many–body problem* Theoretical and Mathematical Physics (Springer–Verlag)
[16] Suzuki T 2001 *Phys. Rev. B* **64** 233310

Journal of Physics: Conference Series **281** (2011) 012016 doi:10.1088/1742-6596/281/1/012016

Erbium environments in erbium-silicon/silica light emitting nanostructures

R J Kashtiban[1], U Bangert[1]*, I F Crowe[2], M P Halsall[2]
[1]School of Materials, The University of Manchester, Manchester, M1 7HS, UK
[2]School of Electrical and Electronic Engineering, The University of Manchester, Manchester M60 1QD, UK

E-mail: ursel.bangert@manchester.ac.uk

Abstract. Co-doping of SiO_2 with Si and Er to achieve silica fibre amplifiers has resulted in encouraging levels of light emission, much above those of Er-only doped SiO_2. However, different fabrication methods, i.e., co-implantion and sequential implantation of Er and Si, has led to several factors difference in light levels. This paper looks into the reasons for these differences by establishing structure and local stoichiometry of the created entities via analytical transmission electron microscopy. In both cases Si-nanocrystals (NCs) have formed in the SiO_2 matrix. In the former case Er-ions are co-located with /integrated within the NCs, in the latter case NCs and Er are separate. By assessing the NCs' internal and interfacial structure with the surrounding material, we attempt to identify chemical/structural Er-phases/defects and their effect on the sensitising efficiency in the Er:Si-NCs system; high resolution phase contrast- and high angle dark field imaging as well as nano-scale spatially resolved electron energy core loss- and plasmon-spectroscopy carried out in an aberration corrected dedicated STEM lend valuable support to these studies.

1. Introduction

Si based light sources for Si-integrated technologies (Si-photonics) are a long and much aspired goal, but previous attempts concerned with Er doping of Si (Si:Er laser) and also of SiO_2 (optical fibre amplifiers [1,2]), have poor or insufficient outcomes due to the low absorption cross section of erbium ions in this structure. To improve the optical gain of intraband transitions originating from the 4f shell, erbium-doped silica containing silicon nanocrystals (Si-NCs) has been suggested [3,4]. The internalisation of Si-NCs increases the absorption cross section by ~10^4 times [5,6] and the light emission by ~3 orders of magnitude; the broad continuous absorption band of the Si-NCs allows furthermore use of cheap LEDs as pumping source, making this structure advantageous over the erbium-doped SiO_2 alone. There are various ways of forming the Er-doped Si-NC/Si-oxide structure. In this contribution we compare two fabrication methods, (i) co-implantion and (ii) sequential implantation of Er and Si, the latter method improving the light emission by several factors over the former.

To find the reasons for these differences one needs to address the chemical/structural environment of the Er-ions on the atomic scale in correlation with photoluminescence. In a recent study [7] we investigated highly erbium-doped silica and Si-NCs obtained via method (i) in a dedicated aberration-

* corresponding author

corrected scanning transmission electron microscope (STEM). We demonstrated that co-implantation of erbium and silicon leads to a band of Si-NCs in a silica matrix, where Er is co-located with the NCs, whereas implantation of erbium without silicon leads to large amorphous erbium-rich clusters, presumably Er_2O_3, We suggested that silicon ions might act as agglomeration centres for erbium ions. Here we go further in investigating the detailed Er-environment to better understand the bonding and chemical compound of the Er that gives rise to the improvement.

2. Experimental

Erbium-containing Si-NCs in method (i) were grown by targeting a beam of 85 keV silicon ions providing a dose of 8×10^{16} at/cm^2 together with an Er-ion beam of 320 keV and a dose of $3\text{x}10^{15}$ at/cm^2 onto a 700 nm thick SiO_2 layer, grown by thermal oxidation of a Si (1 0 0) substrate. Post implantation, the samples were annealed at 1100 °C in a N_2 atmosphere for 1 hour, in order to form Si-NCs. The erbium and silicon peak concentrations in the samples were 3×10^{20} cm^{-3} and $8\text{x}10^{21}$ cm^{-3}, respectively. In method (ii) Er was first implanted into a Si-substrate, which was oxidized and subsequently Si-implanted and annealed. Samples (i) and (ii) were compared with only-Er implanted oxides; respective implantation parameters were the same in all samples.

An aberration corrected STEM, at the *SuperSTEM* Labratory Daresbury, fitted with a *Nion* Mark II quadrupole-octupole corrector [8] operating at 100keV was employed to acquire high resolution phase contrast (HRTEM), bright field (BF) and high angle annular dark field (HAADF) images as well as EEL spectrum images (SIs) [9]. The instrument allows Z-contrast lattice images with 1Å resolution to be obtained. An Ultra High Vacuum (UHV) *Gatan Enfina* system was used to obtain spectrum images with energy dispersions of 0.05-0.7 eV /channel, depending on the required energy regime, and achieving energy resolutions down to 0.3 eV.

3. Results and Discussion

3.1. Er-compound determination from HAADF and core loss EELS.

In figure 1a) we show HAADF images of the sample acquired by method (i). Si-NCs (of ~3.5 nm average diameter) are positioned in a narrow band (indicated by orange dashed lines) around the implantation depth. The bright contrast of the NCs suggests co-location of Er and Si-NCs. We have previously proven this [7] by combined HAADF and EELS analysis. In figure 2 a), b) and c) we show that bright patches in the HAADF image and Si and Er signals extracted from EEL spectrum images (procedures are explained elsewhere [7,10]) have indeed identical distributions. We have encircled corresponding features across all panels in fig. 2 by same-colour dashed lines. Panels d) and e) in fig. 2 show results of studies of the O-edge fine structure: in contrast to the single peak of the O K-edge in SiO_2 (at ~538 eV), the rare earth oxide O K-edge has a double peak structure (peaks at ~533 and 538 eV; [11]), hence fitting a single Gaussian function to the O K-edge of an SI taken around Si-NCs would result in a broadening of the fitting function if Er was present as Er-oxide. Simultaneously the fit-peak maximum would shift towards lower energies. This is exactly what we observe in figs. 2 d) and e), which show the spatial distribution of the FWHM, and the peak position of the Gaussian fit, respectively. The O K-edge in locations of Si-NCs has an increased FWHM, as indicated by the yellow/orange colours and a lower peak energy value, as indicated by the black colours. Spatial coincidence of all the features in figure 2 has led us to conclude that an Er-oxide shell of some sort has formed around the Si-NCs.

Coming back to figure 1, we show HAADF images of the sample made by method (ii) in 1b) and of an Er-only implanted sample in 1c). Er is no longer co-located with Si-NCs in sample (ii); a number of large Er-clusters (~average size 15 nm) has formed throughout the oxide layer from the implantation depth right to the top of the sample, similar to the Er-rich clusters which are presumably Er_2O_3, in the Er-only implanted oxide in fig. 2c). Between large Er-rich clusters faint small-scale speckle can be seen in 1b), arising from a band of Si-NCs that has formed similarly to that in sample (i), fig. 1a). Figure 1d) shows an HRTEM image from an enlarged area in 1b), where atomic lattice

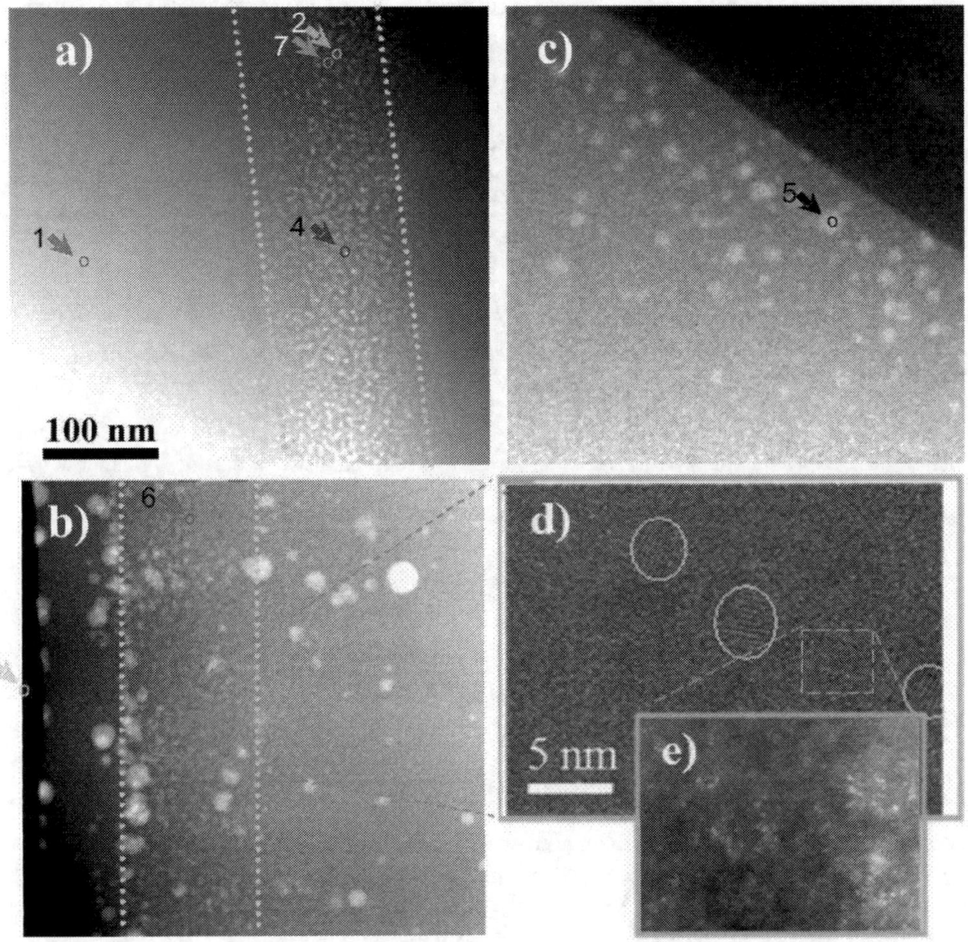

Figure 1. a) – c) HAADF images of co-implanted sample (i), sequentially implanted sample (ii) and of an Er-only implanted SiO_2 layer. d) is a close-up of Si-NCs formed in (b), presenting an HRTEM phase contrast image, which reveals the atomic lattice in NCs and e) is a further close-up of regions in between NCs in (b), showing Er in atomically dispersed or molecular form by its HAADF contrast. Scales in images (a), (b) and (c) are the same. Circles (colour online) with numbered arrows refer to regions, where the low loss EEL spectra in figure 3 have been acquired.

structure is visible in the NCs. Further enlargement of areas in between the NCs in fig. 1e) shows atomic size bright dots in HAADF, indicating atomically dispersed Er, either as single atoms or few-atom clusters, most likely representing Er-oxide molecules.

So importantly in sample (ii) we have two types of appearance of Er, aggregated in large separate clusters, and atomically dispersed around Si-NCs. We think that this is the clue to the high emission efficiency in this sample: in the finely dispersed form Er-ions are close enough to NCs to take part in the energy transfer and sensitisation process. Because Er is highly dispersed there is also less chance of back-transfer or self-excitation, as in the big clusters in the Er-only sample in fig. 1c).

However, an Er-oxide shell surrounding Si-NCs as in sample (i) should also warrant good energy transfer; furthermore ALL Er is consumed in this structure and no 2nd 'waste' Er-phase has formed, contrary to sample (ii). The question arises, why then the efficiency is lower in sample (i). This can be further clarified by finding out, where exactly the Er sits, and whether, due to the high fraction, some Er is incorporated within the NCs in sample (i), e.g., as an Er-silicide phase. This could not be decided from atomic resolution HAADF or BF images, partly due to the fact that the features of interest are extremely small and furthermore embedded in an amorphous matrix.

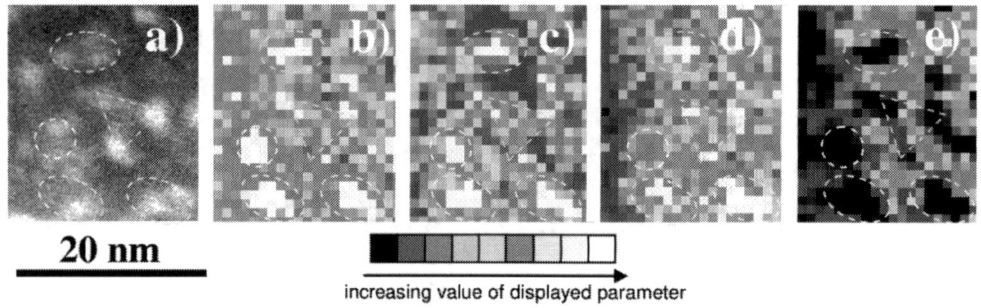

Figure 2. Identical areas within the band of Si-NCs in sample (i) representing a) the HAADF intensity, b) the Si distribution map extracted from an EEL SI using the Si $L_{2,3}$ edge intensity in the energy window 99-101 eV, c) the Er distribution map using the Er $N_{4,5}$ edge intensity at 167–177eV, d) the distribution of the FWHM and e) of the peak energy of a Gaussian function fitted to the O K-edge. The dashed frames in each panel are marking identical features.

3.2. Er-phase determination from low-loss EELS.

EEL spectra in the low loss regime can be treated and interpreted within the dielectric theory. If the loss function can be reliably extracted from EEL spectra, then the dielectric function can be extracted and, by using Kramers-Kronig (KK) relationships, ε_1 and ε_2 can be calculated. Much work has gone into separating the zero-loss energy peak from the loss spectrum [12]; furthermore deconvolution routines [13,14] have successfully been applied to improve the energy resolution in the spectra, so that access to losses down to 2 eV can be obtained.

The dielectric function of $ErSi_2$ is distinct, so if we could extract it from low loss spectra this would provide, maybe the only evidence possible as to whether such a phase exists. Employing the above procedures we have attempted KK analysis, but the outcome so far was inconclusive. Nonetheless, low-loss spectra of the plasmon region should give direct indication as to the materials' phase. However, the contribution of the matrix surrounding the features, of which we want to determine the structural and chemical phase, is dominant in low loss spectra, and needs to be reduced. With the sample thickness lying in the single scattering regime it should be possible to subtract the SiO_2 matrix contribution, which can be estimated from knowledge of the size of the features of interest in relation to the sample thickness. Figure 3 shows *difference* spectra of the plasmon energy region obtained by subtracting an adequate proportion of a matrix spectrum. The arrowed colour circles in figure 1 indicate the regions from which low loss spectra were obtained.

Going from top to bottom through the spectra in figure 3, the purple, uppermost spectrum is from a region between NCs in sample (i) (purple-circled location in figure 1a), after subtraction of an Er-oxide spectrum (obtained from a large Er-cluster in the Er-only implanted sample). The subtraction should have eliminated the Er-oxide contribution, and indeed the purple spectrum shows the pure SiO_2 character by the interband transition at ~ 14 eV and the plasmon at ~23.5 eV. The second spectrum down (red) is from the Si-NC-containing oxide in sample (ii) (red-circled location in fig. 1b), after subtracting a portion of a pure Si-oxide spectrum. The remaining spectrum exhibits mostly Si-oxide characteristics; a small contribution of the Si-plasmon at ~17 eV is present resulting from the large scattering cross sections in the low loss region and proximity of Si-NCs. Significantly there is also an absorption feature from Er5d6s-> O3s and possibly Er4f->5d transitions at ~9 eV. The Er-oxide plasmon at ~15.7 eV, on the other hand, is absent. This confirms that there is atomically dispersed Er in the Si-oxide, as suggested by the HAADF investigations (figure 1b). The 3rd curve down (black) is of a large Er_2O_3 cluster (black-circled region in figure 1c), after subtraction of a Si-oxide spectrum, presenting a typical Er-oxide low loss spectrum with the ~9 eV interband transitions and the Er_2O_3 plasmon [15].

The 4th spectrum down (blue) is of a Si-NC in sample (i) (blue-circled region in figure 1a), after subtraction of an appropriately scaled Si-oxide spectrum, so should reflect the behaviour of an Er-oxide-surrounded Si-NC, and thus give the best evidence concerning a possible Er-silicide phase. The atomic Er-transitions at ~9 eV and the Si-plasmon at ~17 eV are present without evidence of a blue-shift typical for a silicide phase [16]. Importantly, there is also no Er-oxide plasmon, the presence of which would be expected, if the oxide surrounding the Si-NC consisted of a solid shell. So either the surrounding oxide shell shares a highly defective 'interface' with the Si-NC core, or constitutes an unconnected rather 'dendritic' Er-oxide network, or both. In any case the lower light efficiency of sample (i) combined with the lack of a plasmon point towards the existence of high numbers of defects at the Si-NC surface/interface with the Er-oxide, which cause significant non-radiative recombination.

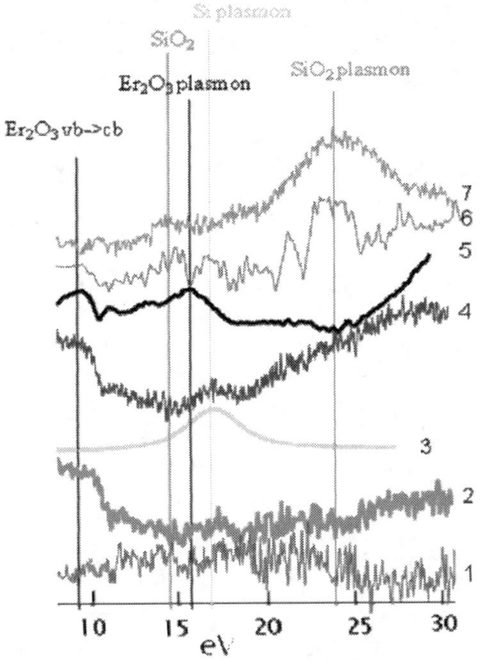

Figure 3. Low EELS difference spectra, taken in locations indicated by the respective numbered arrows and circles (colour online) in figure 1. All curves are a result of a subtraction of a scaled matrix spectrum (either Er-oxide or SiO_2) from the original, raw data. This matrix contribution was estimated from the portion that the matrix would occupy along the e-beam path in relation to the nanofeature of interest, and was derived from sample thickness and NC dimension. Curves in descending numbers (top to bottom): spectrum of the oxide around Er:SiNCs in sample (i) minus Er_2O_3 spectrum from an Er-cluster in the Er-only implanted oxide (i.e., without NCs, purple online); spectrum of NC-containing Si-oxide in sample (ii) minus pure SiO_2 spectrum (red online); spectrum of Er_2O_3 cluster (in the Er-only implanted sample) minus SiO_2 spectrum (black online); spectrum of an Er:SiNC in sample (i) minus SiO_2 spectrum (blue online); Si spectrum (turquoise online); spectrum of Si-oxide between NCs in sample (i) minus pure SiO_2 spectrum (orange online); difference of two pure SiO_2 spectra (green online)

The 5th spectrum (turquoise) is of pure Si, showing the Si-plasmon at ~17 eV, the 6th spectrum (orange) is of the oxide between Si-NCs in sample (i) (orange circle in fig. 1a), after subtraction of a pure Si-oxide spectrum, and the bottom spectrum (green) is the difference between two pure oxide spectra taken in regions near the green circle. Spectrum 6, not surprisingly, has reminiscent features of Er-oxide (at ~9 eV) and of Si (at 17 eV), and the bottom spectrum, again not surprisingly, shows just background noise.

4. Conclusions

Er-environments in SiO_2 matrices (for Er-doped silica fibre amplifiers), sensitised for achievement of enhanced light emission by incorporation of Si-NCs, were investigated by HAADF, HREM, core loss and valence band EELS. Samples, co-implanted with Er and Si before annealing to form Si-NC, gave several times less light emission then samples, which were sequentially implanted with Er (followed by thermal oxidation of Si) and then with Si (followed by annealing to form Si-NC). This was understood in terms of the core shell structures formed in the first case, where Si-NCs act as nucleation centres for Er, which precipitates in form of an Er-oxide shell around them, in contrast to atomically dispersed Er in the SiO_2 matrix surrounding the Si-NCs in the second case. More specifically plasmon EELS results point to a highly defective Er-oxide shell and/or interface with the Si-NC giving rise to non-radiative recombination in the first structure, whereas in the second structure Er-ions are in close proximity but not directly connected to the Si-NCs. There is no indication of Er-silicide phases, which had been considered a possibility with the co-implantation.

Acknowledgments
The authors want to acknowledge M H Gass, previously at Daresbury SuperSTEM Laboratories, for STEM measurements, and R Gwilliam from Surrey University and A J Kenyon from McMasters University (Canada) for ion implantations.

References

[1] Becker P, Olsson A and Simpson J 1999 *Erbium-Doped Fiber Amplifiers* (Elsevier Academic Press) p 1
[2] Kik P G, Brongersma M L and Polman A 2000 *Appl. Phys. Lett.* **76** 2325
[3] Fujii M, Masato Y, Kansawa Y, Hayashi S and Yamamoto K 1997 *Appl. Phys. Lett.* **71** 1198
[4] Kenyon A J, Trwoga P F, Federighi M and Pitt C W 1994 *J. Phys. Con. Mat.* **6** L319.
[5] Kik P G and Polman A, 2000 *J. Appl. Phys.* **88** 1992
[6] Pacifici D, Franzò G, Iacona F, Boninelli S, Irrera A, Miritello M and Priolo F 2003 *Mat. Sci. Eng. B* **105** 197
[7] Crowe I F, Kashtiban R J, Sherliker B, Bangert U, Halsall M P, Knights A P and Gwilliam R M 2010 *J. Appl. Phys.* **107** 044316
[8] Krivanek O L, Dellby N and Lupini A. R. 1999 *Ultramicroscopy* **78** 1
[9] Hunt J A and Williams D B 1991, *Ultramicroscopy* **38**(1) 47
[10] Kashtiban R J, Bangert U, Crowe I F, Halsall M P, Sherliker B, Harvey A J, Eccles J L, Knights A P, Gwilliam R M and Gass M H J. 2010 *Phys. Conf. Ser.* **209** 012043
[11] Schamm S, Scarel G and Fanciulli M 2007 *Rare Earth Oxide Thin Films, Topics Appl. Physics* vol **106**, ed M Fanulli and G Scarel (Berlin Heidelberg: Springer) *p* 153
[12] Bangert U, Harvey A J and Keyse R 1997 *Ultramicroscopy* **68**, 173
[13] Richardson W H 1972 *J. Opt. Soc. Am.* **62** 55
[14] Lucy L B 1974 *Astro. J.* **79** 745
[15] Bertel E, Netzer F P and Matthew J A D 1981 *Surf. Sci.* **103** 1
[16] Cheynet M C and Pantel R 2006 *Micron* **37** 377

Structure and properties of dislocations in interfaces of bonded silicon wafers

M Reiche[1], M Kittler[2], R Scholz[1], A Hähnel[1] and T Arguirov[2]
[1] Max Planck Institute of Microstructure Physics, D-06120 Halle, Germany
[2] IHP Microelectronics, D-15236 Frankfurt (Oder), Germany

Email: reiche@mpi-halle.de

Abstract. The realization of defined dislocation networks by hydrophobic wafer bonding allows the characterization of electrical properties of individual dislocations. The present paper describes the fabrication and characterization of SOI MOFSETs with various dislocations densities in the Si channel. The aim was to investigate the electrical properties in samples containing only 6 dislocations. A drain current of $I_D > 1x10^{-2}$ A induced by a single dislocation was determined by data extrapolation from current measurements in combination with previously analyzed samples containing a varying dislocation density.

1. Introduction

Dislocations in silicon have been extensively studied for more than 50 years. For instance, electron microscopy was applied to investigate the structure of individual dislocations, interactions, and the dynamic behavior of dislocations. The weak-beam method of electron microscopy evolved in the early 1970s [1, 2] demonstrated the dissociation of perfect dislocation into partial dislocations [3]. Modern high-resolution electron microscopy (HREM) and high-angle annular dark field (HAADF) imaging allow a new insight into the atomic structure of individual dislocations and grain boundaries [4-6]. The strain field around individual defects was recently analyzed by using special techniques such as geometric phase analysis (GPA) or peak-pairs analysis (PPA) of high-resolution structure images [7-9], or nano-beam electron diffraction (NBD) [10].

The electronic and optical properties of dislocations were studied by numerous analytical techniques such as the Hall effect, electron paramagnetic resonance (EPR), deep level transient spectroscopy (DLTS), photoluminescence, and electron beam induced current (EBIC) [4, 11-14]. Most of these studies used plastically deformed silicon in order to achieve defined dislocation arrangements and a high density of dislocations to attain the detection limit. Plastic deformation, however, result also in a large number of point defects and defect reactions making it sometimes difficult to interpret experimental data [11]. In order to avoid interactions between dislocations or between dislocations and other defects, methods are required allowing the realization and analyses of only a few dislocations or, in the ideal case, of an individual dislocation. A method to realize defined dislocation arrangements in a reproducible way is semiconductor wafer direct bonding (SWDB) originally developed to produce silicon on insulator (SOI) substrates and three-dimensional micro-electromechanical systems (MEMS) [15]. If two wafers are joined together without any interface layers, a two-dimensional dislocation network is obtained analogous to early experiments on bicrystals [16]. For bicrystals a Czochralski growth process is required allowing only the formation of specific grain boundaries such as $\Sigma = 9$ (rotation 38°56′17′′, boundary plane (122), common [011] axis), $\Sigma = 13$ (rotation 26°37′12′′, boundary plane (510), common [001] axis), and $\Sigma = 25$ large angle grain boundaries (rotation 16°15′36″, boundary plane (710), common [001] axis) [17]. Semiconductor wafer direct bonding, on the other hand, uses commercially available wafers making it possible to realize any grain boundary. Especially small angle grain boundaries having rotational angles $\alpha << 1°$ are of interest allowing dislocation distances of a few hundred nanometers. Applying state of the art techniques for preparation and analyses individual or a small number of dislocations can be characterized.

2. Semiconductor wafer direct bonding

Semiconductor wafer direct bonding (SWDB), or fusion bonding, describes a method to join mirror-polished semiconductor wafers at room temperature without the addition of any glue or external forces. The bonding of silicon wafers was first reported 1985/86. Lasky [18] described the bonding of two oxidized silicon wafers to form silicon on insulator (SOI) substrates. In the same time, Shimbo et al. [19] analyzed the bonding behavior of non-oxidized silicon wafers forming an epitaxial substrate for power devices.

SWDB requires wafers with a high degree of flatness, parallelism and smoothness. Also clean surfaces are necessary which are free of particulate, organic and metallic contaminations. This is important because the surface cleanliness has a direct effect on both the structural and electrical properties of the bonding interface as well as on the resulting electrical properties of the bonded material. After cleaning an activation of the surfaces is required prior to bonding. Then the two mating wafers are brought together face to face in air at room temperature. The top wafer is floating on the other due to the presence of a thin cushion of air between both wafers. When an external pressure is applied onto a small part of the pair to push out the intermediate air, a bond is allowed to be formed by surface attraction forces between the wafers at this location. Depending on the surface conditions, two bonding mechanisms are distinguished:

(i.) Wafer bonding using hydrophilic surfaces is the most common technique [15]. Silicon surfaces are covered with an oxide layer under room temperature conditions. If oxidized surfaces are bonded, an oxide layer results in the interface. A model for hydrophilic wafer bonding was first described by Stengl et al. [20] using the analogy of surface chemistry of silica and oxidized silicon. Based on results of infrared spectroscopy, a three-dimensional hydrogen bonded network of water molecules was assumed. The water is primarily bonded via Si–OH groups on the silica surface. During heating above 180°C the adsorbed water molecules desorb under atmospheric pressure leaving a hydroxylated silica surface, on which most of the SiO groups are linked via hydrogen atoms. OH groups are bonded more stably with increasing temperature. Annealing above 800°C results in the formation of strong Si-O bonds via the interface.

(ii.) When the oxide layer from a crystalline silicon substrate is removed with HF, a hydrophobic surface with unique properties is obtained, i.e. having a good resistance to chemical attack and a low surface recombination velocity, which means a surface with a very low density of surface states. Bonded hydrophobic wafers are characterized by completely different interfaces. The removing of the oxide results in two silicon lattices that are in contact, and Si-Si bonds are formed via the interface. Crystal defects (dislocations) are generated forming a two-dimensional network in order to match both crystal lattices. The structure of the dislocation network depends on the surface orientation of both wafers. Figure 1 shows transmission electron microscope (TEM) images of dislocation networks formed by bonding of (100)/100)-, (110)/(110)-, and (111)/(111)- wafer pairs. Bonding of Si(100) wafers, for instance, cause a $\Sigma = 1$ (100) small angle grain boundary characterized by a square-like mesh of screw dislocations expected from theory [21]. These dislocations are formed by the rotational misfit (twist) between both crystal lattices. There is, however, an additional tilt component caused by the deviation on the [001] axis of real wafers (cut-off). The tilt component is compensated by a periodic array of 60° dislocations. The spacing between dislocations, S, in both networks are indirectly proportional to the misalignment angle and are given by

$$S_{twist} = \frac{a}{2\sqrt{2} \cdot \sin \frac{\alpha_{twist}}{2}} \tag{1}$$

for the screw dislocation network. On the other hand, the relation between dislocation distance and tilt angle of the network formed by 60°-dislocations follows as

$$S_{tilt} = \frac{a}{2 \cdot \tan \alpha_{tilt}} \tag{2}$$

In both equations a represents the lattice constant (a = 0.543 nm for Si) and α_{twist} and α_{tilt} are the angles of misorientation of the twist and tilt components, respectively.

Both dislocation fractions were investigated for hydrophobic wafers bonded under environmental conditions [22-25] and under ultra-high vacuum (UHV) conditions [26, 27]. There are different observations for wafer pairs bonded under UHV conditions. Some reports present dislocation networks already after bonding at room temperature [28], while dislocation networks are observed by other authors only after a subsequent annealing above 800°C [27]. The reason probably is that the boundary is unrelaxed after bonding at room temperature, while relaxation occurs for energetic reasons only at higher temperatures [29].

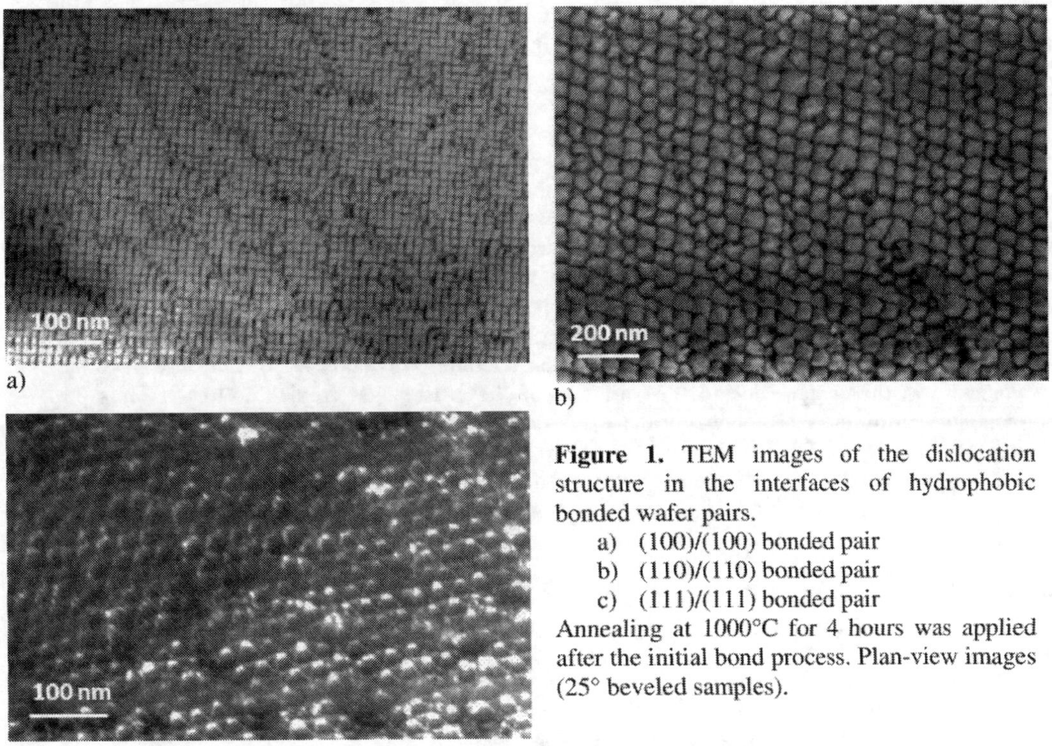

Figure 1. TEM images of the dislocation structure in the interfaces of hydrophobic bonded wafer pairs.
a) (100)/(100) bonded pair
b) (110)/(110) bonded pair
c) (111)/(111) bonded pair
Annealing at 1000°C for 4 hours was applied after the initial bond process. Plan-view images (25° beveled samples).

Numerous investigations on (100)/(100) bonded wafer pairs showed that there are no differences in the structure of the interfacial dislocation networks produced by bonding under UHV and environmental conditions. Differences are only obtained in the dislocation spacing (caused by the different twist angles) and in the types of interaction between screw dislocations and 60°-dislocations (caused by different twist and tilt angles as well as different annealing processes). An almost undisturbed mesh of the screw dislocations exists on {100}-terraces while at steps on the surface (characterized by the 60°-dislocations) interactions between both dislocations portions are found. Different interactions were discussed for hydrophobic bonded wafers [30] and grain boundaries in sintered silicon [31].

Dislocation reactions modify the existing dislocation network especially at very low twist and tilt angles ($\alpha_{tilt} \cong \alpha_{twist} < 0.1°$) where the mesh size of the screw dislocation network reaches the same values as the distance of the 60°-dislocations. At $\alpha_{tilt} \cong \alpha_{twist} \cong 0.07°$ dislocation networks with hexagonal meshes appear [25]. Twist and tilt angles below 0.1° can be realized experimentally only by aligned wafer bonding allowing the control of the twist angle down to $\alpha_{twist} = \pm 0.005°$ [24, 32]. At $\alpha_{twist} \leq 0.001°$ dislocation networks were not observed. Instead, segments of individual dislocations of different geometries were found. Straight segments of screw dislocations are observed first, while – probably at lower angles – loops of various geometries are dominant [24].

3. Properties of dislocations in bonded interfaces

Dislocations in the interface of bonded wafers possess numerous remarkable properties which may be used for different applications [33, 34]. The electrical properties of bonded hydrophobic silicon wafers were studied for the first time by Bengtsson et al. [35]. The measurement of the capacitance-voltage (CV) characteristics on bonded unipolar wafers were interpreted on the assumption of two distributions of interface states, one of acceptors and one of donors, causing a potential barrier at the bonded interface. The origin of the interface states was assumed to be impurities and crystal defects. More recent analyses by EBIC proved barrier heights generally smaller than 100 meV for different types of bonded hydrophobic wafers [36]. The concentration of deep levels along the interface was determined to be a few 10^5 per cm. Low concentrations of deep levels at the interface are also the reason for low dark currents, an improved CV-characteristics, and fast rise times of pin-diodes prepared on bonded hydrophobic wafers [37].

The luminescence properties of dislocation networks were also studied. Figure 2 shows the luminescence spectra of different bonded samples. The spectra are obtained from samples having different misorientation. Detailed photoluminescence and cathodoluminescence measurements provide direct evidence that the wavelength of light emitted from the dislocation network could be tailored to some extent by misorientation of the wafers during the bonding procedure. D1 or D3 lines have the largest intensity in the spectra due to the variation of the twist angle from 8.2° to 9°. Thus the luminescence spectrum can be tailored by the misorientation angles in a controlled manner and the dominance of either D1 or D3 radiation can be attained. Furthermore, in some special cases the D1 emission could completely dominate the spectrum, even at room temperature (RT) [33].

An external bias voltage applied across the bonded interface can significantly enhance the luminescence intensity from the bonded interface. The reason for this effect can be that, on the one hand, the external bias could reduce the potential barrier at one side of the bonded interface and therefore increase the occupancy of dislocation related states responsible for the D-lines; on the other hand, the bias could change the distribution of minority carriers close to the bonded interface. This results in the variation of carrier recombination velocity at the bonded interface and in relevant variation of luminescence. The application of the proper external bias voltage enhances the total luminescence at any excitation level, but the enhancement is more pronounced for lower excitation conditions. A maximum enhancement factor of 130 was achieved.

4. Analyses of individual dislocations

The study of individual dislocations or only of a small number of dislocations is an important issue because all of the interactions between defects can be eliminated. This results in more precise data about the electronic structure of dislocations and, combining these investigations with microscopic observations, conclusions about the correlation between the structure and properties of dislocations can be derived. A combination of bonded wafer pairs with preparation methods to separate individual dislocations or a small number of dislocations allow such experiments. Twist angles between two bonded Si wafers below 0.1° results in dislocation distances of more than 100 nm (equation 1). Using photolithography and etching techniques, individual dislocations can be separated and measured.

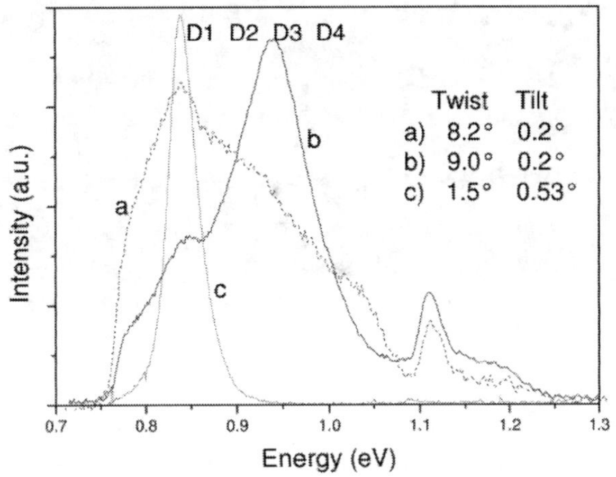

Figure 2. The impact of the misorientation/dislocation structure on the luminescence spectra of dislocation networks. The misorientation (tilt and twist components) are indicated in the figure. Cathodoluminescence spectra recorded at 77K.

Diodes and metal-oxide-semiconductor field-effect transistors (MOSFETs) were prepared on such bonded wafers. In order to avoid the effect of the bulk material, dislocation networks were realized in SOI wafers having only a thin device layer. The substrates were prepared by hydrophobic wafer bonding of commercially available SOI wafers having the following specification of the silicon device layer: CZ-grown silicon, 150 mm diameter, p-type, resistivity ρ = 13.5 – 22.5 Ωcm, <100>-orientation, thickness of the buried oxide (BOX) layer was 60nm or 1000 nm. The final device layer thicknesses of 15 nm and 40 nm were obtained by thinning down the initially 260 nm or 600 nm thick layers, respectively, by thermal oxidation. After dipping in diluted HF the wafers were immediately bonded under hydrophobic conditions in an atmospheric environment. Various twist angles between $0.01° < \alpha_{twist} < 0.65°$ were realized. The bonded wafer pairs were annealed at 1050°C for 4 hours in nitrogen. Finally one of the handle wafers was removed by a combination of mechanical grinding and chemical etching (spin etching) followed by chemical etching of the oxide layer. This process results in SOI wafers having 2-dimensional dislocation networks in their thin device layers (figure 3).

SOI MOSFETs were prepared on such substrates using lithographic techniques and dry reactive ion etching (RIE). The channel region was defined first. Because dislocations are parallel to <110>-directions in Si, they are parallel to the channel. In order to study the effect of the dislocation density, channel width and length, respectively, are varied between 1 μm and 10 μm. Source and drain contacts were formed by As^+ implantation (5 keV, $1 \cdot 10^{15} cm^{-2}$) combined with a RTA step (950°C, 60 sec.). A thin gate oxide of about 6 nm was formed by thermal oxidation. The device gates were prepared by low-pressure chemical vapour deposition (LP-CVD) of polycrystalline silicon (100 nm thick) followed by As^+ implantation (30 keV, $1 \cdot 10^{15} cm^{-2}$) and a RTA step (950°C, 60 sec). Finally, contacts were formed by Al deposition and annealing at 420°C for 30 minutes in hydrogen.

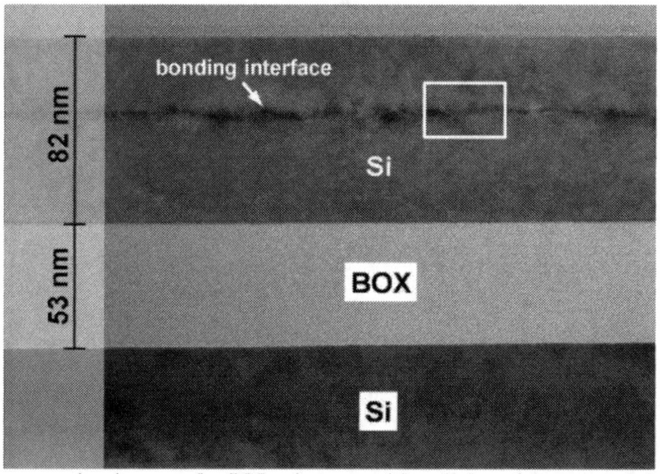

Figure 3. TEM cross-section image of a SOI substrate with a dislocation network in the device layer. The dislocation distance is about 15 nm corresponding to a twist angle of about 0.7°.

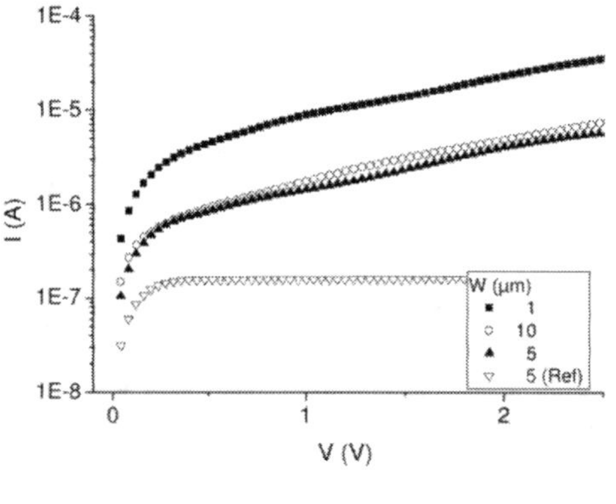

Figure 4. I-V characteristics of SOI-MOSFETs with and without dislocations (V_G = 0V). The distance between contacts is 1 µm, while widths of 1, 5, and 10µm are used. Because the dislocation spacing is about 34 nm, 30, 150, and 300 dislocations, respectively, are present in the channel. The reference is an SOI wafer having the same device layer thickness but without dislocations.

Figure 4 shows the conductance of dislocation layers by measuring MOSFETs at gate voltages V_G = 0V. A characteristic increase of the current I is obtained for devices containing dislocations compared to a reference sample prepared on a SOI wafer with the same device layer thickness (40 nm). The current increase depends on the dislocation density and probably on the type of the dislocation. The reason is that dislocations form channels of higher conductance in the silicon layer resulting in a higher carrier concentration in the channel [38] but increase also the effective carrier mobility which is consistent with quasi-ballistic transport along dislocations [39]. From figure 4, we can see that the carriers transported through dislocations are electrons [40]. Furthermore, it can be also seen from the figure that the current increases as the number of dislocations decreases. The highest current is obtained if only 30 dislocations are present.

The I-V characteristics of nMOSFETs with and without a dislocation network in the channel are shown in figure 5. The thickness of the device layer was about 80 nm. The channel length was 1 μm. Typical output and transfer characteristics are obtained for the reference sample without a dislocation network in the device layer (figure 5 a,b). The devices are characterized by a sub-threshold slope $S = 100$ mV/dec. and a threshold voltage $V_T = -150$ mV. The output and transfer characteristics of a device with a dislocation network in the channel are shown in figures 5c and 5d. It can be seen that higher drain currents (I_D) are measured at the same gate (V_G) and drain (V_D) voltages, compared to the devices without a dislocation network. The increase of the drain current even at very low gate voltages

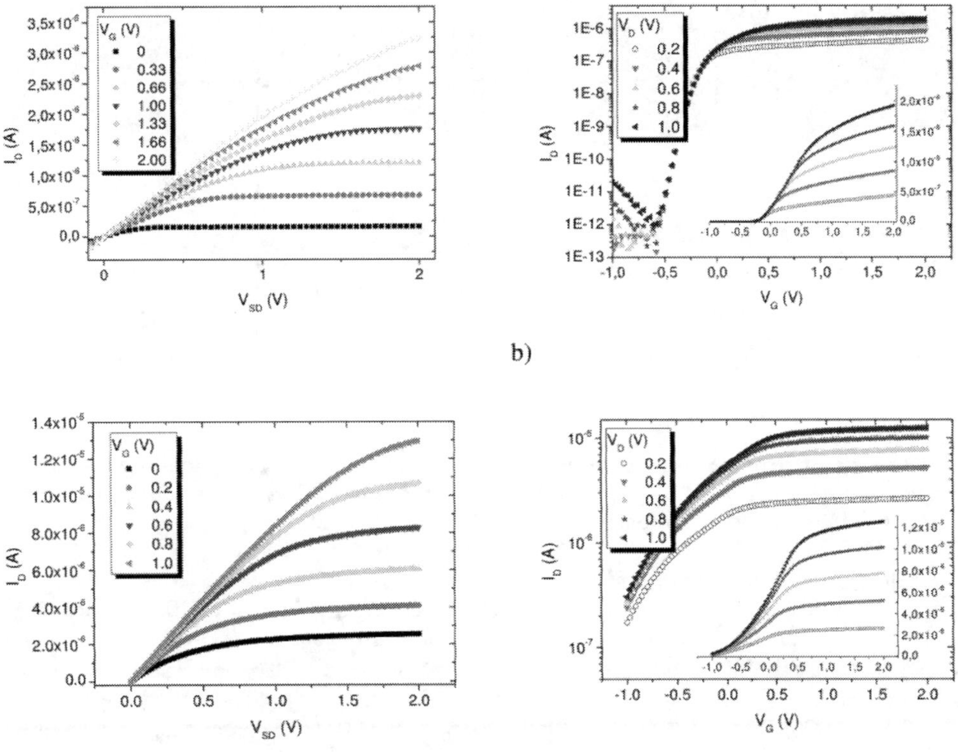

a) b)

c) d)

Figure 5. Output and transfer characteristics of nMOSFETS without dislocations (a,b) and with a dislocation network (c,d). The insets represent the transfer characteristic ($I_D = f(V_G)$) in linear scale for a better current comparison. The device layer thickness was 80 nm. The channel length and width, respectively, for both devices are 1 μm. The sub-threshold slope of the dislocation MOSFET is in this example 550 mV/dec.

is about one order of magnitude. Similar results were also obtained by other authors [38] and are ascribed to the presence of dislocations. The relatively high source-drain current even at $V_G = 0V$, in contrast to the reference sample, indicates the presence of charged carriers on the dislocations.

The analysis of the device data clearly proved that the number of dislocations in the channel characteristically affects the device parameter. Indications are found by measurements on devices prepared on wafers having dislocation networks with different dislocation density. As described above, such networks are realized by varying the twist angle during the wafer bonding process. Besides devices prepared on a dislocation network with $\alpha_{twist} = 0.31°$ (resulting in a dislocation spacing of about 35 nm) analogous samples were prepared on a wafer having a dislocation network characterized by $\alpha_{twist} = 0.035°$. Here, the dislocation spacing is about 150 nm. Using channel widths, W, between 1 µm and 10 µm, devices having about 660 dislocations (at a dislocation spacing of 15 nm and W = 10 µm) up to 6 dislocations per channel at a dislocation spacing of 150 nm and W = 1 µm were prepared. The $I_D - V_D$ curves of these devices show that the drain current depends on the number of dislocations in the channel. At $V_D = 2$ V, a drain current of $3 \cdot 10^{-3}$ A is obtained if there are only 6 dislocations in the channel. On the other hand, a value of $I_D = 2 \cdot 10^{-6}$ A is measured under the same conditions if the channel includes about 660 dislocations.

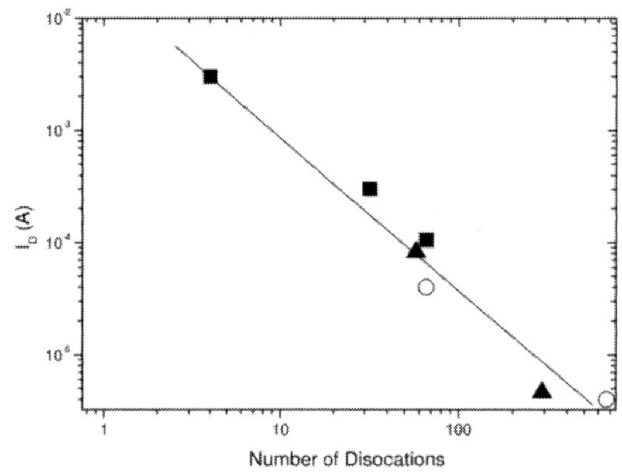

Figure 6. Source-drain current I_D as a function of the number of dislocations in the channel. Measurements at $V_D = 2V$ and $V_G = 0V$. Different symbols characterize measurements on various wafers having different dislocation densities.

Besides I_D also the threshold voltage (V_T) and the sub-threshold swing of the MOSFETs dependent on the dislocation density. Increasing the number of dislocations by a factor of 10, by increasing the channel width from 1µm to 10 µm, results in a decrease of V_T from about -400 mV down to -150 mV. An explanation could be the dependence of V_T and the sub-threshold voltage shift (ΔV_T) on the effective channel length for MOSFETs [41]. Decreasing the effective channel length results in an increase of ΔV_T. If dislocations are present in the channel, the effective channel length is defined by the number of dislocations as the effective transport channels. Therefore, reducing the number of dislocations in the channel results in an increase of V_T and ΔV_T.

On the other hand, the sub-threshold slope increases as the dislocation density increases. An analogous increase of the sub-threshold swing is generally interpreted as a thickness effect of the device layer for short channel SOI-MOSFETs and is caused by an inhomogeneous electron concentration in the layer [42].

Measurements on MOSFETs prepared on wafers with different dislocation networks proved a dependence of the drain current on the dislocation density (figure 6). The figure clearly shows that the current depends only on the number of dislocations where the current increases as the number of dislocations decreases. Because the data were measured on different wafers prepared in different experimental runs, impurities on the dislocations would modify the results. Fitting the data allows to extrapolate the current given by one dislocation of more than 10^{-2} A. This corresponds to a current density of more than 100 A/cm. Further measurements are required to confirm this result.

5. Conclusions

The analysis of the structure and properties of dislocations is a fundamental issue in materials science. Most of the hitherto existing investigations used dislocation arrangements having a high density of dislocations. In order to avoid interactions between dislocations or between dislocations and other defects, methods are required allowing the realization and analyses of only a few dislocations or, in the ideal case, of an individual dislocation.

A method to realize defined arrangements in a reproducible way is hydrophobic wafer bonding. Two-dimensional dislocation networks are produced if two wafers are bonded. The length and distance of the dislocations in a network is defined by the mesh size and can be up to several micrometers. Combining wafer bonding with pattern formation techniques (photolithography and dry etching), devices were realized allowing the measurement of the electrical characteristic of only a few dislocations.

The presented data clearly show an indirect behaviour of the drain current on the number of dislocations in the channel. The fact that the highest current is obtained if only a few dislocations are present allows the conclusion that electrically active centres in the dislocation core of the straight dislocation segments are responsible for the electron transport while dislocation nodes and dislocation segments oriented orthogonal to the channel direction act as "scattering centres" and reduce the carrier transport.

Acknowledgment
This work was financially supported by the German Federal Ministry of Education and Research in the framework of the SiGe-TE project (contract no. 03X3541B).

References

[1] Cockayne DJH, Ray ILF and Whelan MJ 1969 *Phil. Mag.* **20** 1265
[2] Cockayne DJH, Ray ILF and Whelan MJ 1970 *Proc. 7th Intern. Conf. Electron Microscopy (Grenoble)* vol 2 (Paris: Société CIB) p 277
[3] Ray ILF and Cockayme DJH 1971 *Proc. R. Soc. Lond* A **325** 543
[4] Holt DB and Yacobi BG 2007 *Extended Defects in Semiconductors* (Cambridge: Cambridge University Press)
[5] Narayan J and Nandedkar AS 1991 *Phil Mag.* B **63** 1181
[6] Spence JCH 2007 Experimental studies of dislocation core defects *Dislocations in Solids* (vol 13) ed FRN Nabarro and JP Hirth (Amsterdam: Elsevier) pp 419-452
[7] Hytch MJ, Putaux JL and Penisson J-M 2003 *Nature* **423** 270
[8] Galindo PL, Kret S, Sanchez AM, Caval J-Y, Yanez A, Pizarro J, Sanchez AM, Ben T and Molina SI 2007 *Ultramicroscopy* **107** 1186
[9] Sanchez AM, Galindo PL, Kret S, Falke M, Beanland R and Goodhew PJ 2006 *Microsc. Microanal.* **12** 285
[10] Béché A, Rouviére JL, Clément L and Hartmann JM 2009 *Appl. Phys. Lett.* **95** 123114-1
[11] Alexander H and Teichler H 1991 Dislocations *Materials Science and Technology* (vol 4) ed RW Cahn, P Haasen and EJ Kramer (Weinheim: VCH) pp 249-319
[12] Schröter W and Cerva H 2002 *Solid State Phenom.* **85-86** 67
[13] Kittler M, Ulhaq-Bouillet and Higgs V 1995 *J. Appl. Phys.* **78** 4573
[14] Kveder V, Kittler M and Schröter W 2001 *Phys. Rev.* B **63** 115208
[15] Tong QY and Gösele U 1999 *Semiconductor Wafer Bonding* (New York: Wiley)
[16] Thibault-Desseaux J, Putaux JL, Bourret A and Kirchner HOK 1989 *J. Phys. France* **50** 2525
[17] Aubert JJ and Bacmann JJ 1987 *Rev. Phys. Appl.* **22** 515
[18] Lasky JB 1896 *Appl. Phys. Lett.* **48** 78
[19] Shimbo M, Furukawa K, Fukuda K and Tanzawa K 1986 *J. Appl. Phys.* **60** 2987
[20] Stengl R, Tan T and Gösele U 1989 *Jpn. J. Appl. Phys.* **28** 1735
[21] Bollmann W 1970 *Crystal Defects and Crystalline Interfaces* (New York: Springer Verlag)
[22] Gafiteanu R, Chevacharoenkul S, Gösele U and Tan TY 1993 *Inst. Phys. Conf. Ser.* **134** 87

[23] Benamara M, Rocher A, Sopéna P, Claverie A, Laporte A, Sarrabayrouse G, Lescouzéres L and Peyre-Lavigne A 1996 *Mat Sci. Eng.* B **42** 164

[24] Wilhelm T, Mchedlidze T, Yu X, Arguirov T, Kittler M and Reiche M 2008 *Solid State Phenom.* **131-133** 571

[25] Reiche M 2008 *Mat. Sci. Forum* **590** 57

[26] Akatsu T, Scholz R and Gösele U 2004 *J. Mat. Sci.* **39** 3031

[27] Reznicek A, Scholz R, Senz S and Gösele U 2003 *Mat. Chem. Phys.* **81** 277

[28] Plößl A, Scholz R and Akatsu T 1999 Room temperature covalent bonding: effect on interfacial properties *Semiconductor Wafer Bonding: Science, Technology, and Applications (ECS Proc. Vol. 99-35)* ed CE Hunt, H Baumgart, U Gösele and T Abe (Pennington: The Electrochem. Soc.) pp 232-243

[29] Schober T and Baluffi RW 1970 *Phil. Mag.* **21** 109

[30] Benamara M, Rocher A, Laporte A, Sarrabayrouse G, Lescouzéres L, Peyre-Lavigne, Fnaiech M and Claverie A 1995 *Mat. Res. Soc. Symp. Proc.* **378** 863

[31] Föll H and Ast D 1979 *Phil. Mag.* A **40** 589

[32] Fournel F, Moriceau H, Aspar B, Rousseau K, Eymery J, Rouviére JL and Magnea N 2002 *Appl. Phys. Lett.* **80** 793

[33] Kittler M, Yu X, Mchedlidze T, Arguirov T, Vyvenko OF, Seifert W, Reiche M, Wilhelm T, Seibt M, Voß O, Wolff A and Fritzsche W 2007 *Small* **3** 964

[34] Kittler M and Reiche M 2009 *Adv. Eng. Mat.* **11** 249

[35] Bengtsson S, Andersson GL, Andersson MO and Engström O 1992 *J. Appl. Phys.* **72** 124

[36] Yu X, Arguirov T, Kittler M, Seifert W, Ratzke M and Reiche M 2006 *Mat Sci. Semicond. Proc.* **9** 96

[37] Reiche M, Hiller E and Stolze D 2002 New substrates for MEMS *Proc. IEEE Sensors 2002, vol 1* (Piscataway NJ: IEEE) pp 607-612

[38] Ishikawa A, Yamauchi K, Yamamoto C and Tabe M 2005 *Mater. Res. Soc. Symp. Proc.* **864** E6.5.1

[39] Mchedlidze T, Wilhelm T, Arguirov T, Trushin M, Reiche M and Kittler M 2009 *Phys. Stat. Sol.* (c) **6** 1817.

[40] Reiche M, Kittler M, Buca D, Hähnel A, Zhao Q-T, Mantl S and Gösele U (2010) *Jpn. J. Appl. Phys.* **49** 04DJ02-1

[41] Liu Z-H, Hu C, Huang J-H, Chan T-Y, Jeng M-C, Ko PK and Cheng YC 1993 *IEEE Trans. Electron Dev.* **40** 86

[42] Rauly E, Potavin O, Balestra F and Raynaud C 1999 *Solid-State Electron.* **43** 2033

Deformation slip and twinning in bulk and nanocrystalline semiconductors

Maryam Zahiri-Azar, Pirouz Pirouz[*] and K Peter D Lagerlöf
Department of Materials Science and Engineering Case Western Reserve University
Cleveland, OH 44106-7204 U.S.A.

[*]E-mail: pxp7@case.edu

Abstract. Yielding and twinning in bulk and nanocrystalline semiconductors are considered based on the activation of perfect or partial dislocation loops. Assuming that the yield stress of the crystal is proportional to the critical stress necessary for the activation of a perfect Frank-Read source, it is shown that for a range of crystal orientations with respect to the applied shear stress, the Hall-Petch relationship for a silicon polycrystal fails when the average grain diameter is of the order of 3 nm or less. It is further shown that, assuming the double-cross-slip mechanism of twin formation in silicon, twinning is easier than slip for a range of crystal orientations in nanocrystals.

1. Introduction

The double-cross-slip mechanism of twinning, proposed in 1987 [1], arose out of two experimental observations about semiconductors. The first was that glide dislocations in all semiconductors that had been investigated up to then had been shown to be dissociated [2-4] and, secondly, the experiments of Wessel and Alexander [5] had indicated that the mobility of the leading and trailing partials in Si and in Ge [6] were different. This was quite unexpected because, unlike the case of compound semiconductors, where the core of partial (or perfect) dislocations can be different thus leading to so-called α and β dislocations with different dislocation velocities, the core of partial dislocations in elemental semiconductors is the same atom and one would not expect a mobility difference. It was thus interesting to see what difference the dissociation of dislocations and the difference in mobility of partials would make to the classical mechanism of Frank and Read [7] for dislocation multiplication that was based on the expansion of a pinned segment of perfect dislocation under the action of a shear stress $\tau_{z\varphi}$ (figure 1). In particular, it was interesting to consider the case where the mobility of the leading partial was sufficiently larger than that of the trailing partial such that the former expands to the unstable semicircular position while the latter lags behind. This led to the interesting situation that the leading partial forms a *faulted* loop that surrounds the trailing partial. Clearly, as the leading partial – that is now behind the trailing partial – approaches the latter and the separation distance, d, between them decreases, the repulsive force between the two partials increases rapidly thus preventing a recombination of the two partials into a perfect dislocation. However, even if such a recombination could occur, say under an immensely large stress, a second dissociation could not be envisaged because of the resulting high energy AA-type stacking configuration. In such a case, the applied stress $\tau_{z\varphi}$ could not be relieved by the operation of the Frank-Read source and the latter would come to a stop after the formation of only one faulted loop. On the other hand, at the two pinning points X and X' (figure 1), the two partials are naturally recombined into a perfect dislocation.

One could thus envisage that if the perfect dislocation had a screw character, it would be able to relieve the applied stress $\tau_{z\varphi}$ by double cross-gliding on the cross-slip plane (csp) and repeating the formation of a second faulted loop on a glide plane parallel to the original one. In fact this process could be repeated many times and n faulted loops could form on n parallel planes until the cross-slip stress is not enough for further double cross-glide of the screw dislocation. Assuming that these planes are all adjacent to one another, the result would be a twin band with a thickness nd_{111} [1], where d_{111} is the inter-planar spacing for the {111} set of slip planes and n is an integer. It was later realized [8] that the cross-slip stage of the twinning, could be described much more naturally and conveniently by the Escaig-Friedel [9-11] model. Also, since it was very likely that the mobility difference of partial dislocations would increase with decreasing temperature, it could be concluded that while nucleation and glide of perfect dislocations is the predominant mode of deformation at moderate to high temperatures in semiconductors, twinning would become more important at low temperatures or higher strain rates [1]. This model was actually more applicable to compound semiconductors because the core of two 30° partials of a screw dislocation are known to have different identities and their velocities can be appreciably different [12]. Further extension of this model could also explain polytypic transformation in SiC [13], and twinning in BCC crystals [14], basal [15-17] and rhombohedral [18] twinning in sapphire and in the Laves phases [19, 20].

2. Dislocation Plasticity of Semiconductors

Recently Estrin *et al.* [21] considered the role of Frank-Read sources in the plasticity of nanomaterials. Following the suggestion of other investigator [22, 23], these authors attributed the breakdown of the classical Hall–Petch relationship in nanocrystalline materials to a non-monotonic grain-size dependence of the required stress for the operation of Frank-Read sources. The length λ of the source dislocation segment acting as the F-R source in figure 1 is assumed to scale with the grain size d, and the origin of this non-monotonic dependence is the logarithmic term in the self-energy or elastic energy, E_{el}, of a dislocation line of length λ, see e.g. Ref. [24, 25]:

$$E_{el} = \frac{Gb^2\lambda}{4\pi(1-\nu)}\left[\left(1+\frac{\nu}{2}\cos^2\theta-\frac{3\nu}{2}\sin^2\theta\right)\ln\left(\frac{\lambda}{b}\right)+\left(2.0\nu\sin^2\theta-0.4\nu\cos^2\theta\right)-2.05\right] \quad (1)$$

where b is the magnitude of the dislocation Burgers vector and G and ν are, respectively, the elastic modulus and Poisson's ratio of the material. The line tension T of the dislocation line is simply the self-energy per unit length of dislocation, i.e.:

$$T = \frac{dE_{el}}{dl} = \frac{Gb^2}{4\pi(1-\nu)}f(\lambda) \quad (2a)$$

where:

$$f(\lambda) = \left[1-\frac{\nu}{2}\left(3-4\cos^2\theta\right)\right]\ln\left(\frac{\lambda}{b}\right)-1+\frac{\nu}{2} \quad (2b)$$

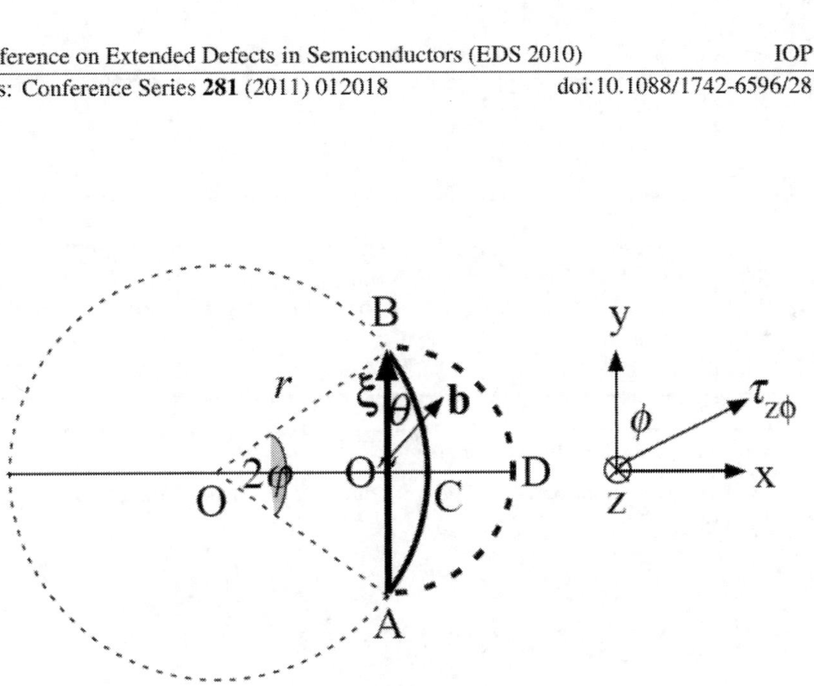

Figure 1. Straight dislocation segment AB with line direction ξ and Burgers vector **b** bows out under the action of the resolved shear stress $\tau_{z\varphi}$ that makes an angle φ with respect to the dislocation line direction lying in the plane of the paper, normal to the z-axis.

According to the Peach-Kohler equation [26], the glide force $\mathbf{F}_g$ per unit length of a dislocation with line direction ξ and Burgers vector **b** caused by the stress tensor $\underline{\sigma}$ is [25]:

$$\mathbf{F}_g = \frac{[(\mathbf{b}\cdot\sigma)\times\xi]\cdot[\xi(\mathbf{b}\times\sigma)]}{|\mathbf{b}\times\xi|} \tag{3a}$$

If we denote the shear stress component of the stress tensor acting at an angle φ to the dislocation line direction by $\tau_{z\varphi}$ (figure 1), this equation reduces to:

$$F_g = \tau_{z\phi} b(\sin\theta\sin\phi + \cos\theta\cos\phi) \tag{3b}$$

where θ is the angle between the Burgers vector **b** of the source dislocation segment and ξ (figure 1). Note from Eq. (3b) that the glide force on a screw dislocation segment is zero when $\varphi = 90^o$.

Under the applied shear stress $\tau_{z\varphi}$, the pinned dislocation segment bows to a radius of curvature r subtending an angle ϑ at O (figure 1), a process that increases the dislocation length – from AB to ACB - and thus its line tension T. The bow out of the dislocation line is also opposed by the lattice resistance, or Peierls stress τ^P, given by [25]:

$$\tau^P = \frac{2G}{K}\exp\left(-\frac{2\pi d_{111}}{Kb}\right) \tag{4}$$

where d_{111} is the interplanar spacing of the $\{111\}$set of slip planes in semiconductors; in this equation, $K \approx 1$ for a dislocation with a screw character and $K = 1 - \nu$ for one with an edge character [25]. Thus, balancing the bow-out force F_g with the opposing forces due to the line tension T and the Peierls stress τ^P, we get (figure 1):

$$F_g = \frac{T\sin\vartheta}{r} + b\tau^P \tag{5}$$

Substituting (2a), (3b) and (4) in Eq. 5 gives the radius of curvature of the bowed out segment under the resolved shear stress $\tau_{z\varphi}$:

$$r = \frac{Gb\sin\vartheta}{4\pi(1-\nu)\left[\tau_{z\phi}(\sin\theta\sin\phi + \cos\theta\cos\phi) - \tau^P\right]} f(\lambda) \tag{6}$$

Clearly as the stress $\tau_{z\varphi}$ increases, the radius of curvature r of the bowed-out segment decreases until the pinned dislocation segment attains a semi-circular shape (ADB in figure 1) at which point it becomes unstable leading to the formation of a loop and re-generation of the original straight dislocation segment [25]. Thus the critical stress $\tau^*_{z\varphi}$ required to form an unstable configuration is obtained by the condition $2r = \lambda$ at $\varphi=\pi/2$, i.e.

$$\tau^*_{z\phi} = \frac{1}{(\sin\theta\sin\phi + \cos\theta\cos\phi)}\left[\frac{Gb}{2\pi(1-\nu)\lambda} f(\lambda) + \tau^P\right] \tag{7}$$

$\tau^*_{z\varphi}$ is plotted for silicon as a function of λ and φ in figure 2; in these calculations, it is assumed that $G = 52$ GPa and $\nu = 0.28$.

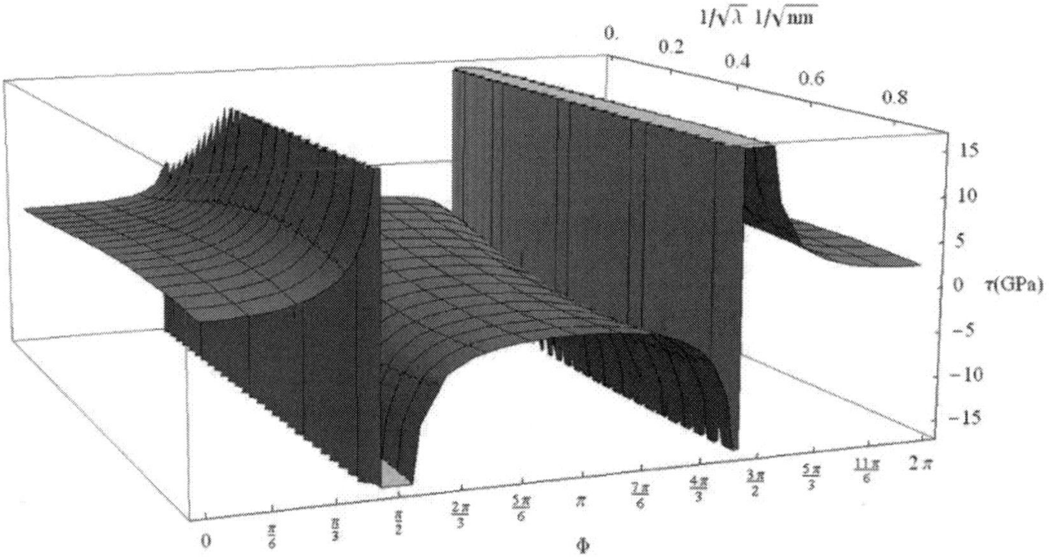

Figure 2. Variation of the critical resolved shear stress $\tau^*_{z\varphi}$ required to activate a perfect screw dislocation loop from a Frank Read source as a function of $1/\sqrt{\lambda}$ (where λ = source length) and φ (the angle between the direction of the applied stress with the dislocation line direction). The negative values of $\tau^*_{z\varphi}$ between $\varphi = \pi/2$ and $\varphi = 3\pi/2$ indicate a bow-out in the opposite direction.

As expected, $\tau^*_{z\varphi}$ is a minimum along the line where the applied stress is parallel to the Burgers vector **b**, i.e. when $\varphi = 0°$. Notice that $\tau_{z\varphi}$ increases with decreasing source length (Hall-Petch effect) but reaches a maximum in the small nanometer length scale after which it decreases. This is more clearly shown in figure 3, which is a section of the three-dimensional plot at $\varphi = 0°$.

Assuming that λ scales with the average grain size diameter d and that the yield stress of polysilicon varies as $\tau^*_{z\varphi}$, figures 2 and 3 show that - similar to polycrystalline metals [21] - its yield stress goes through a maximum. For silicon, this occurs at $\lambda = 2.22$ nm, indicating that for a silicon polycrystal with a grain size smaller than 2.22 nm, the inverse Hall-Petch relationship holds and the crystal softens as the grain size become smaller. According to this figure, the stress required to activate a Frank-Read source in bulk silicon is about 17 MPa. This stress increases to about 2.6 GPa at $\lambda = 2.22$ nm before it starts to decrease. The large stress value is not surprising considering the calculations do not take temperature into account, i.e. they are 0 K calculations.

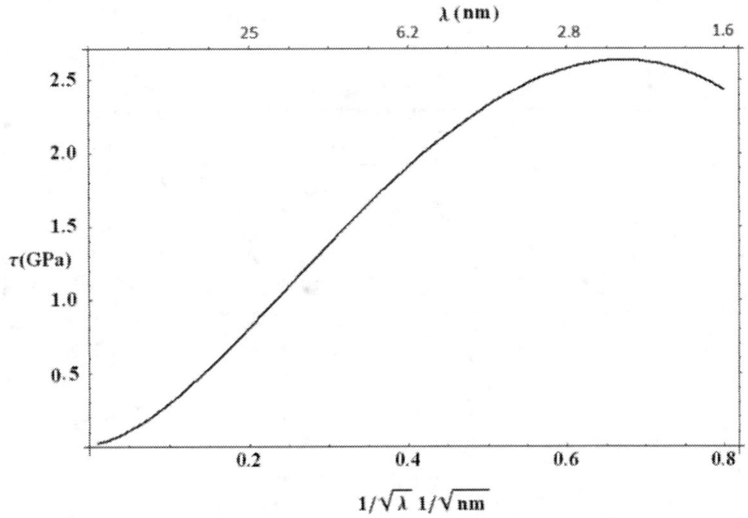

Figure 3. Variation of the critical resolved shear stress $\tau^*_{z\varphi}$ required to activate a perfect screw dislocation loop from a Frank-Read source as a function of $1/\sqrt{\lambda}$ when the stress acts in a direction parallel to the dislocation line direction (i.e., $\theta = \varphi = 0°$ in figure 1).

3. Deformation twinning in semiconductors

Let's use the same methodology as in section 2 to study the effect of the source length in the double cross-slip model of twinning [1]. The source dislocation segment now has a screw character and is dissociated into two 30° partials with Burgers vectors $\mathbf{b}_l$ and $\mathbf{b}_t$ making angles of θ_l and θ_t with respect to their line direction ξ (the sub/superscripts l and t denote leading and trailing, respectively). We shall again assume that the applied shear stress makes an angle φ with respect to ξ (figure 4). As before, both partial dislocations bow out under the action of the applied shear stress, but since the Burgers vectors of the two partials, $\mathbf{b}_l$ and $\mathbf{b}_t$ are different (even though $b_l = b_t$), they bend to different radii r_l and r_t.

We can now write a similar equation to (5) except that we need to incorporate the forces exerted by the stacking fault energy γ, generated by the leading partial dislocation, and the interaction force $\mathbf{F}_{lt}$ between the two partials. The latter is given by [27]:

$$\begin{aligned} \mathbf{F}_{lt} &= \frac{G}{2\pi d}\left[(\mathbf{b}_l \cdot \xi)(\mathbf{b}_t \cdot \xi) + \frac{1}{1-\nu}(\mathbf{b}_l \times \xi)(\mathbf{b}_t \times \xi)\right] \\ &= \frac{G b_l b_t}{2\pi d}\left(\cos\theta_l \cos\theta_t + \frac{\sin\theta_l \sin\theta_t}{1-\nu}\right) \end{aligned} \tag{8}$$

The force balance for the two partials is as follows [27, 28]:
– for the leading partial:

$$F_g + F_{lt} = \frac{\sin \vartheta}{r_l} T + \gamma + \tau_l^P b_l \tag{9a}$$

i.e.

$$\tau_{z\phi}^l b_l \left(\sin\theta_l \sin\phi + \cos\theta_l \cos\phi\right) + \frac{Gb_P^2}{2\pi d}\left(\cos\theta_l \cos\theta_t + \frac{\sin\theta_l \sin\theta_t}{1-\nu}\right) = \frac{Gb_l^2 \sin\vartheta}{4\pi(1-\nu) r_l} f_l(\lambda) + \gamma + \tau_l^P b_l \tag{9b}$$

– for the trailing partial:

$$F_g + \gamma = \frac{\sin \vartheta}{r_t} T + F_{lt} + \tau_t^P b_t \tag{10a}$$

i.e.:

$$\tau_{z\phi} b_t \left(\sin\theta_t \sin\phi + \cos\theta_t \cos\phi\right) + \gamma = \frac{Gb_t^2 \sin\vartheta}{4\pi(1-\nu) r_t} f_t(\lambda) + \frac{Gb_p^2}{2\pi d}\left(\cos\theta_l \cos\theta_t + \frac{\sin\theta_l \sin\theta_t}{1-\nu}\right) + \tau_t^P b_t \tag{10b}$$

where r_l and r_t are the radii of curvature of the leading and trailing partials, $b_l = b_t = b_p$, and $f_l(\lambda)$ and $f_t(\lambda)$ are the same as Eq. (2b) with the Burgers vector of the dislocation replaced by b_l or b_t.

Solving for $2r_l = \lambda$ and $2r_t = \lambda$ when $\vartheta = \pi/2$, we get the critical resolved shear stresses $\tau_{z\varphi}^{l*}$ and $\tau_{z\varphi}^{t*}$ for activating the leading and trailing partials, respectively:

$$\tau_{z\phi}^{l*} = \frac{1}{b_l\left(\sin\theta_l \sin\phi + \cos\theta_l \cos\phi\right)} \times \left\{\frac{Gb_l^2}{2\pi\lambda(1-\nu)} f_l(\lambda) + \frac{Gb_p^2}{2\pi d}\left(\cos\theta_l \cos\theta_t + \frac{\sin\theta_l \sin\theta_t}{1-\nu}\right) - \gamma + \tau_l^P b_l\right\} \tag{11a}$$

and:

$$\tau_{z\phi}^{t*} = \frac{1}{b_t\left(\sin\theta_t \sin\phi + \cos\theta_t \cos\phi\right)} \times \left\{\frac{Gb_t^2}{2\pi\lambda(1-\nu)} f_t(\lambda) - \frac{Gb_p^2}{2\pi d}\left(\cos\theta_l \cos\theta_t + \frac{\sin\theta_l \sin\theta_t}{1-\nu}\right) + \gamma + \tau_t^P b_t\right\} \tag{11b}$$

A 3D plot of the critical resolved shear stresses $\tau_{z\varphi}^{l*}$ and $\tau_{z\varphi}^{t*}$ to activate the leading and trailing partials, respectively, is presented in figure 5.

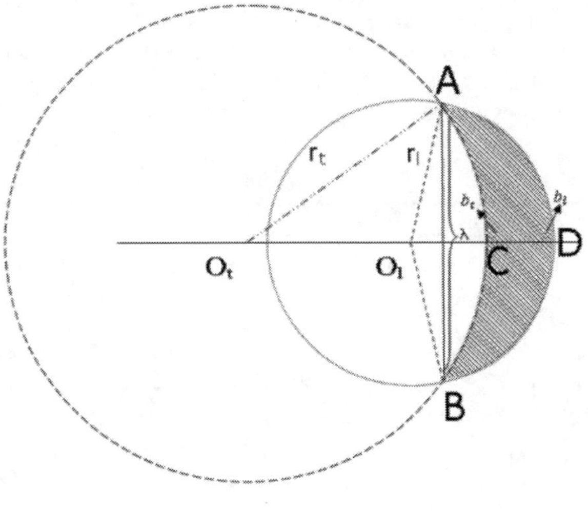

Figure 4. Bow out of pinned leading and trailing partial dislocation sources under an applied shear stress $\tau_{z\varphi}$. Under this stress, the leading and trailing partials with Burgers vectors $\mathbf{b}_l$ and $\mathbf{b}_t$ expand to radii r_l and r_t, respectively.

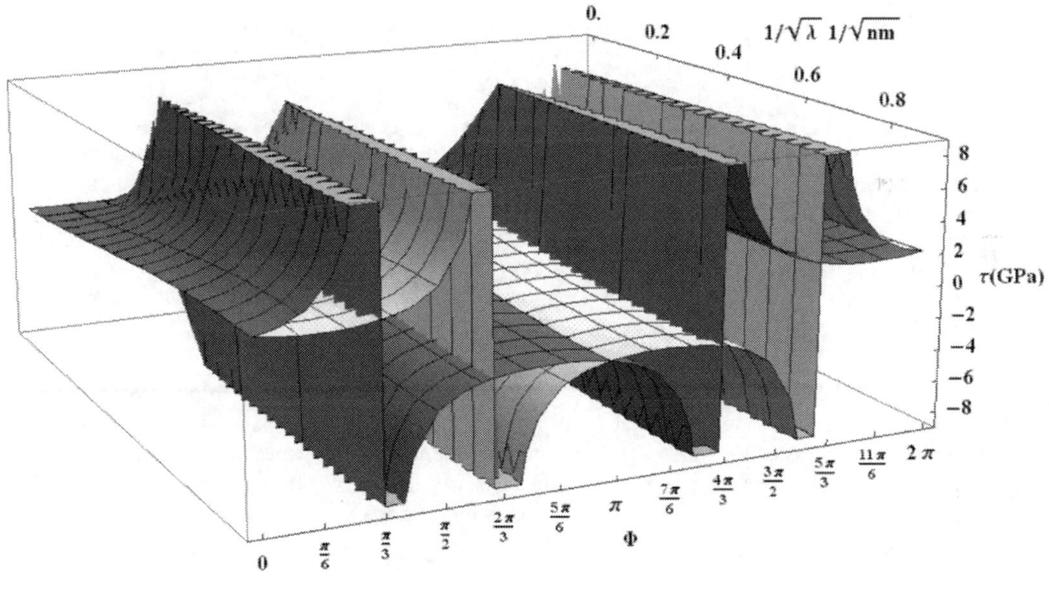

Figure 5. Variation of the resolved shear stresses, $\tau_{z\varphi}^{l*}$ and $\tau_{z\varphi}^{t*}$, required to activate the leading (blue – light - surface) and trailing partial (red – dark - surface) dislocation loops, respectively, from a dissociated dislocation source as a function of $1/\sqrt{\lambda}$ (where λ = source length) and φ (angle between the direction of the applied stress with the dislocation line direction). For clarity, the surface for the perfect dislocation is not shown in this figure.

Note that in different regions of the plot $\tau_{z\varphi}^{l*}$ may be larger than $\tau_{z\varphi}^{t*}$ or vice versa, i.e. whether $\tau_{z\varphi}^{l*} > \tau_{z\varphi}^{t*}$ or, conversely, $\tau_{z\varphi}^{t*} > \tau_{z\varphi}^{l*}$ depends on the source length λ and the direction of the applied shear stress with respect to the source dislocation ξ - this is because as φ is varied, the angles between the direction of the applied stress $\tau_{z\varphi}$ and the partial Burgers vectors $\mathbf{b}_l$ and $\mathbf{b}_t$ vary. In particular, when the applied shear stress is parallel to the Burgers vector of the leading partial, $\tau^{l*}_{z\varphi}$ is much smaller than both $\tau^{t*}_{z\varphi}$ and the critical stress required for activation of a *perfect* F-R loop. Under such conditions, twinning is favoured over slip. This is shown in the 2D plot of figure 6 where $\varphi = \theta_l = 30°$.

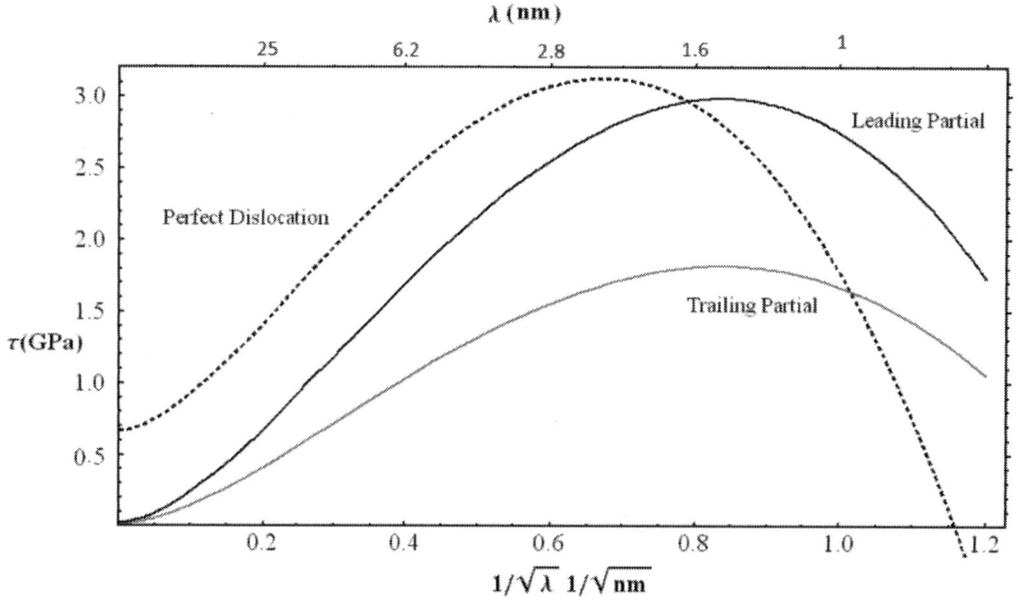

Figure 6. Variation of the critical resolved shear stresses, $\tau^{l*}_{z\varphi}$ and $\tau^{t*}_{z\varphi}$, and $\tau^{p}_{z\varphi}$, required to activate the leading partial (continuous dark line - blue), trailing partial (continuous light line - purple), and perfect (dashed line - yellow) dislocation loops, respectively, from a Frank Read source as a function of $1/\sqrt{\lambda}$ (where λ = source length) when the shear stress is parallel to $\mathbf{b}_l$ ($\varphi = \theta_l = 30°$). Note that under this condition, twinning is favored at all length scales (elasticity equations, upon which the above figures are based, are not valid when λ has atomic scale dimensions, say below 1 nm).

A number of approximations have been used in our calculations that can be improved in future work. As an example we have taken the separation distance d between the two partials to stay constant at the equilibrium separation d_{eq}. This effectively eliminates the two terms γ and F_{lt} from equations (9a) and (10a). In reality, as d increases, the repelling force between the two partials, F_{lt}, decreases while the attractive force due to the stacking fault energy γ does not change.

4. Conclusions

The critical stress for the activation of a pinned segment of perfect and dissociated dislocation in silicon has been evaluated as a function of the source length λ and the angle φ between the applied stress and the dislocation line. For different values of φ and the source length, initiation of either perfect dislocation or the leading partial dislocation is favored. In the first case, this leads to plastic deformation by slip while in the second case, the first stage of twinning by the mechanism of double cross-glide is satisfied. The variations of the critical stress with the source length indicates that while the crystal strengthens with decreasing grain size for bulk polycrystals of silicon (the Hall-Petch relationship), softening of the crystal can occur when the dimensions of the grain is in the small nanoscale regime. Moreover, for a wide range of stress orientations, the critical resolved shear stress for activation of the leading partial dislocation, $\tau^{l*}_{z\varphi}$, is much less than that required for the activation of the trailing partial, $\tau_{z\varphi}^{J^*}$, or for the perfect dislocation $\tau_{z\varphi}^{P^*}$, indicating that in such orientations, twinning is favored. Moreover, the stress difference $\tau_{z\varphi}^{J^*} - \tau_{z\varphi}^{J^*}$ increases with decreasing grain size indicating that twinning is generally more likely in nanocrystals than in bulk silicon.

References

[1] Pirouz P 1987 *Scripta Met.* **21** 1463
[2] Gottschalk H, Patzer G and Alexander H 1978 *phys. stat. sol. (a)* **45** 207
[3] Takeuchi S and Suzuki K 1999 *phys. stat. sol. (a)* **171** 99
[4] Pirouz P, Cockayne D J H, Sumida N, Hirsch P B, and Lang A R 1983 *Proc. Roy. Soc. Lond. A* **386** 241
[5] Wessel K and Alexander H 1977 *Phil. Mag.* **35** 1523
[6] Gottschalk H 1979 *J. de Physique Colloque* **40** C6-127
[7] Frank F C and Read W T 1950 *Phys. Rev.* **79** 722
[8] Pirouz P and Hazzledine P M 1991 *Scripta metall. mater.* **25** 1167
[9] Friedel J 1957 *Proceedings of the Dislocations and Mechanical Properties of Crystals (New York)* p 330
[10] Escaig B 1968. In *Dislocation Dynamics*, ed A R Rosenfield, G T Hahn, A L Bement Jr and R I Jaffee (MacGraw-Hill Book Company, London) pp. 655-77
[11] Escaig B 1968 *J. de Physique* **29** 225
[12] Pirouz P 1989 *Scripta Met.* **23** 401
[13] Pirouz P and Yang J W 1993 *Ultramicroscopy* **51** 189
[14] Lagerlöf K P D 1993 *Acta metall. mater.* **41** 2143
[15] Bilde-Sørensen J B, Lawlor B F, Geipel T, Pirouz P, Heuer A H and Lagerlöf K P D 1996 *Acta mater.* **44** 2145
[16] Pirouz P, Lawlor B F, Geipel T, Bilde-Sørensen J B, Heuer A H and Lagerlöf K P D 1996 *Acta mater.* **44** 2153
[17] Geipel T, Bilde-Sørensen J B, Lawlor B F, Pirouz P, Lagerlöf K P D and Heuer A H 1996 *Acta mater.* **44** 2165
[18] Geipel T, Lagerlöf K P D, Pirouz P and Heuer A H 1994 *Acta metall. mater.* **42** 1367
[19] Hazzledine P M and Pirouz P 1993 *Scripta Met.* **28** 1277
[20] Hazzledine P M 1994 in *Twinning in Advanced Materials*, ed M H Yoo and M Wuttig (TMS Publications, Pittsburgh) pp. 403-14
[21] Estrin Y, Kim H S and Nabarro F R N 2007 *Acta mater.* **55** 6401
[22] Scattergood R O and Koch C C 1992 *Scripta Met.* **27** 1195
[23] Lian J and Baudelet B 1993 *Nanostructured Materials* **2** 415
[24] Hull D and Bacon D J 2001 *Introduction to Dislocations*, Fourth edition, (Oxford: Butterworth Heinemann)
[25] Hirth J P and Lothe J 1982 *Theory of Dislocations* (New York: J. Wiley & Sons, Inc)
[26] Peach M and Koehler J S 1950 *Phys. Rev.* **80** 436
[27] Byun T S 2003 *Acta Mater.* **51** 3063
[28] Lagerlöf K P D, Castaing J, Pirouz P and Heuer A H 2002 *Phil. Mag. A* **82** 2841

Journal of Physics: Conference Series **281** (2011) 012019 doi:10.1088/1742-6596/281/1/012019

Luminescence of SiC films grown by a vapor phase reaction

E A Steinman[1], E B Yakimov[2], K N Filonov[1] and A N Tereshchenko[1]
[1] Institute of solid state physics RAS, Chernogolovka, Moscow distr., 142432 Russia
[2] Institute of Microelectronics Technology and High Purity Materials RAS, Chernogolovka, Moscow distr., 142432 Russia

E-mail: steinman@issp.ac.ru

Abstract. SiC films grown by vapor phase reaction consist of multicrystalline grains. The size of grains depends on growth parameters, such as temperature and the relative concentration of the Si and C components. A small fraction of large grains demonstrate very efficient luminescence. Several techniques have been used to study the origin of the luminescence. Cathodoluminescence and photoluminescence revealed flat regions as sources of emission. Diffraction of back scattered electrons and X-ray analysis suggests that efficient emission is observed for grains with a small fraction of the rhombohedral phase. On the basis of the experimental data we assumed that the layers of rhombohedral phase or the interface between these layers and the host material are the origin of the luminescence.

1. Introduction

SiC, as a wide band gap semiconductor with unique properties, has attracted considerable attention as a material for high temperature electronics and short wavelength optical applications. In electronics great success has been achieved in the field of high power transistors and hostile-environment devices [1]. However due to its indirect energy gap, SiC has a significant drawback in optical device applications: light emission is a phonon mediated process with a low probability. To overcome this difficulty, two approaches can be used: reducing the non-radiative channels or making the radiative process much faster. One of the solutions for the second approach is to exploit the quantum confinement effects [2] to increase the radiative probability. As an alternative to the epitaxial growth of quantum well structures, defect engineering looks rather promising for this purpose. This work is devoted to an analysis of efficient radiation arising from individual grains of SiC films.

2. Experimental method

Samples were grown by a vapor phase reaction. This technique has the advantage of being cheap and producing a potential to use different types of substrates such as Si, Al_2O_3 and graphite. A vertical graphite crucible was employed as a reactor. Si was placed at the bottom of the reactor while natural gas was discharged from the top. The pressure inside the reactor was held in the range of 2–10 mbar. The reactor was placed into a resistance heating furnace and heated to 1600°C. Decomposition of hydrocarbons started in the upper hot part of the reactor. Due to the opposite gradients of Si and hydrocarbons, their relative concentration along the height of the reactor was height dependent.

The morphology of the surface was analyzed by optical and scanning electron microscopy (SEM). X-ray diffraction (XRD) and diffraction of back scattered electrons (EBSD) were used for structure analysis. Photoluminescence (PL) was measured using 325 nm excitation of a He-Cd laser with a conventional phase sensitive registration. The photomultiplier F79 was used as a detector. Cathodoluminescence (CL) was measured by SEM.

3. Results and discussion

Figure 1 shows an example of SEM images for films grown on graphite. The size of grains and film morphology strongly depend on the growth parameters. Figure 1 illustrates the influence of relative Si concentration on morphology. Large grains were created in the zone of high Si concentration (left image); quite a different structure was formed at a lower Si concentration (right image). For the sake of clarity we will call these films type 1 and type 2. Typically large grains are well shaped, have an arbitrary orientation and their dimensions can reach several hundred microns. In the case of a low Si concentration we have small rounded grains. Their dimensions are about 1 micron and the particles consist of smaller granules.

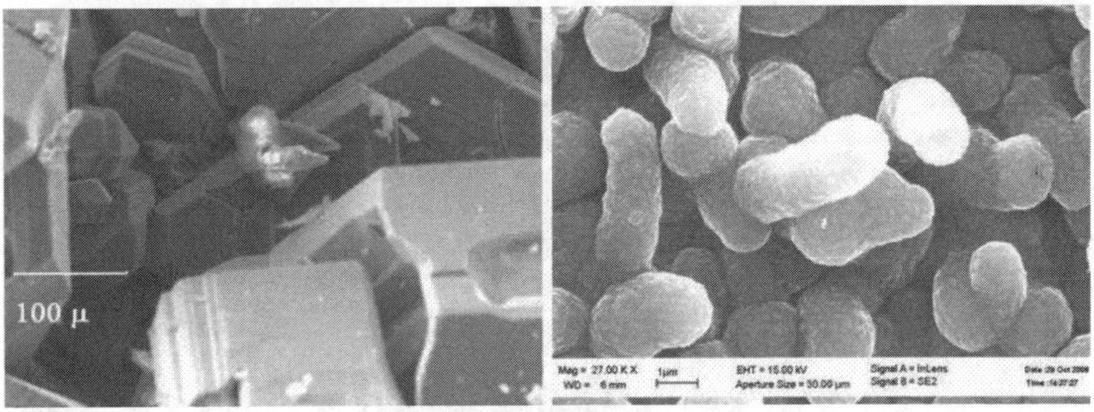

Figure 1. SEM images of SiC films illustrating the influence of the relative relation of Si concentration and C vapors. The left and right films were grown with the excess of Si and C, respectively.

It turned out that the PL properties of these types of films are quite different. While type 2 films demonstrate a weak but uniform broad luminescence band with a maximum around 400 nm, the luminescence of type 1 films is strongly non-uniform. Most of the grains show the same weak luminescence as type 2 films with an additional band in the red part of the spectrum, but there is a limited number of grains with a very efficient luminescence. We will concentrate on type-1 grains with efficient luminescence.

Analysis of PL by optical microscopy revealed that even within a single grain the emission is non-uniform (figure 2). There are some small luminescent parts and the light emitted from them penetrates into the grain, is scattered by the imperfections on the surface and in the bulk giving an illusion that the whole grain glows.

Figure 3 shows a spectral distribution of PL for bright grains of type 1. The spectra are normalized to the spectral sensitivity. Generally emission consists of two bands, blue and yellow. The intensity of the blue band grows slightly with decrease of temperature, but remains unresolved. The yellow part of the luminescence is most active at low temperature and disappears completely at 150K. Therefore grains with different visual appearance have the same spectral features but with different relative intensity.

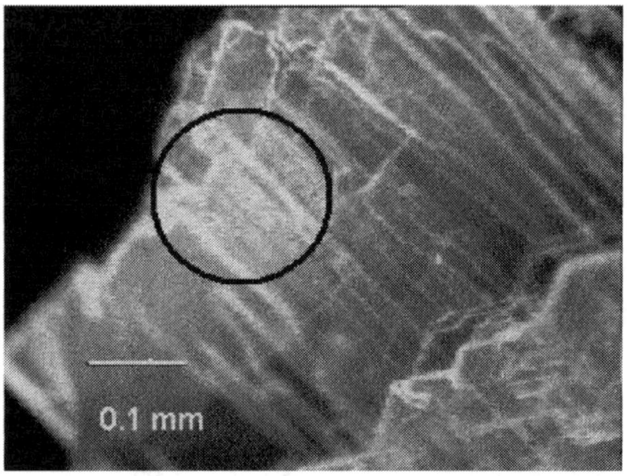

Figure 2. Photoluminescence excited by 325 nm radiation of HeCd laser. The excitation spot is shown by the circle.

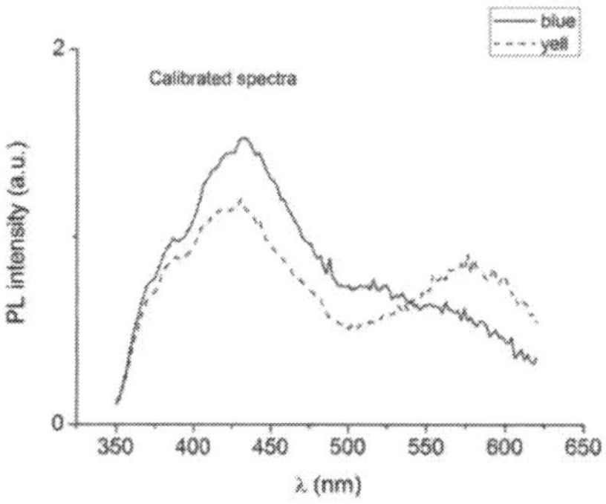

Figure 3. PL spectra of grains with different visual appearance of the glow. Solid line for blue grains, dashed for yellow grains.

Cathodoluminescence, measured in the monochrome mode, clearly shows (figure 4) that there is a sharp boundary between dark and white parts of the grain. The white lines on the SEM image show an approximate location of the luminescent part. Its shape replicates some features of the grain relief. It appears that the last layer grown gives the origin of luminescence.

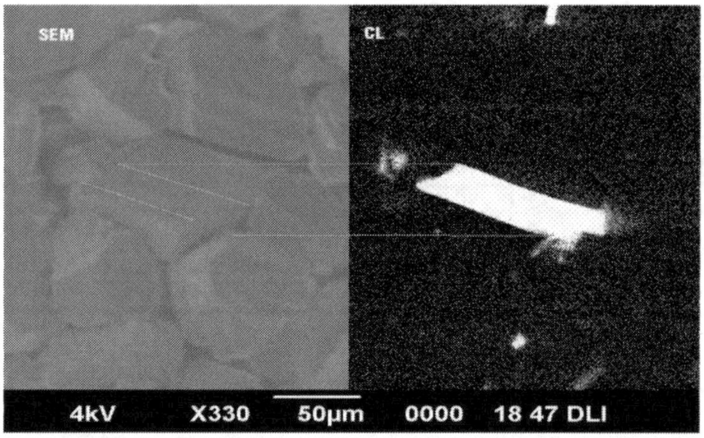

Figure 4. SEM image and CL of the same part of film.

In the case of a more complicated lamellar structure shown in figure 5, several active layers are seen simultaneously. Taking into account that the grain surface morphology is defined by the growth direction we can assume that the luminescent parts have a flat shape oriented perpendicular to the growth direction. So far as a distance of the electron penetration into the sample is about 1 micron, we cannot say definitely whether the active planes are located at the surface or inside the grain. The dependence of integrated CL intensity on acceleration voltage (figure 6) implies that the active layer is located under the surface. Here the growth of CL intensity can be explained by the contribution of the surface recombination, which affects the initial increase of CL intensity with increasing penetration depth. Further decrease of CL at high voltage implies that electrons passed through the active layer due to a deeper penetration of electrons.

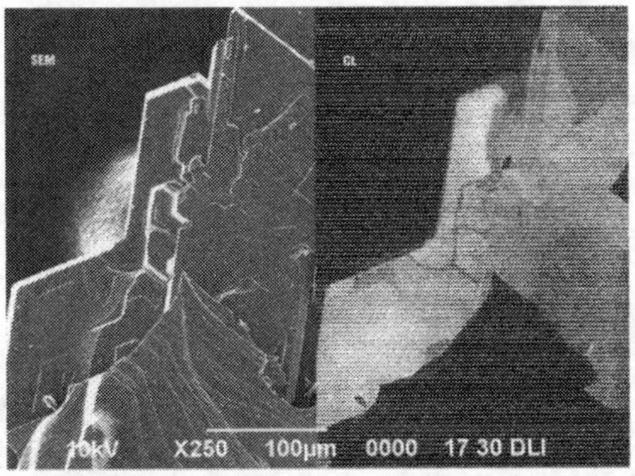

Figure 5. SEM image and CL of grain with lamellar structure.

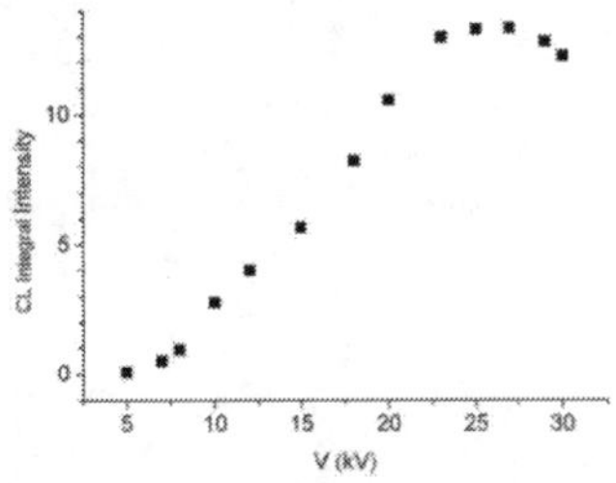

Figure 6. Normalized dependence of CL integrated intensity upon accelerating voltage.

The two possible reasons can be suggested to explain the difference of particular grains from the remaining material: it can be either local segregation of some impurity or some structural features. We checked the composition of the material in dark and light places. No significant difference was found. On the other hand, the comparison of diffraction of backscattered electrons for dark and luminescent grain segments clearly shows the structural difference of these parts (figure 7).

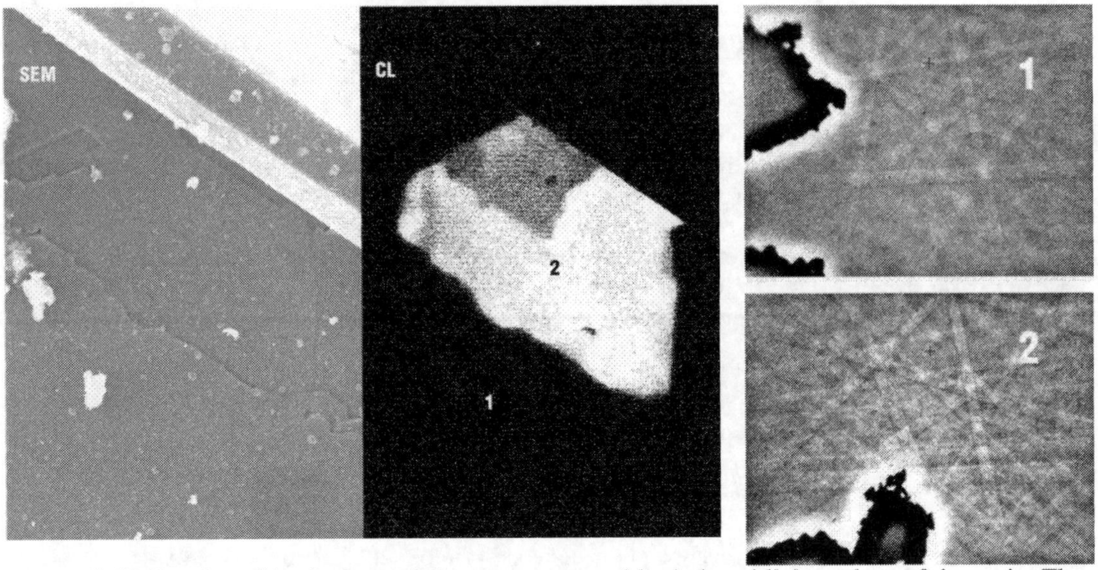

Figure 7. Diffraction of backscattered electrons measured in dark and light regions of the grain. The left part is SEM and CL images of the grain. The particular spots marked as 1 and 2 in the CL image for EBSD analysis are shown in the figure on the right.

Therefore we can make a conclusion that the PL and CL active parts are layers of different polytypes, or the interface between the host material and the different polytype. From the data of figure 7 we cannot say unambiguously what structure corresponds to the two patterns, but we can definitely state that the origin of the intense luminescence is caused by the different structure of the small layer regions.

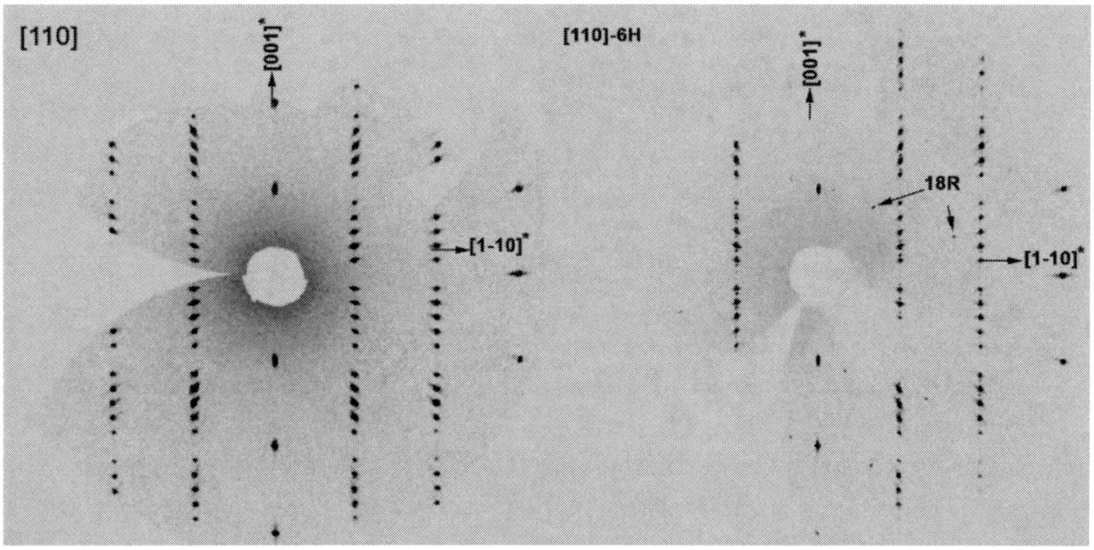

Figure 8. XRD patterns of dark grain (left image) and grain with efficient luminescence (right image). The additional reflections of a rhombohedral phase are shown by arrows.

To elucidate the nature of luminescent layers, we studied dark grains and grains with luminescent fragments separately by the XRD technique. It was found that dark grains have a pure hexagonal structure while PL active grains have an additional small fracture of the rhombohedral phase. Figure 8 illustrates the two XRD images for different types of grains, which clearly shows additional reflections for luminescent grains, corresponding to the rhombohedral phase. Taking into account the above consideration, we can infer that thin interlayers of the rhombohedral phase created during the growth of some grains are the origin of luminescence. It should be noted that the possibility of creating a quantum well like structure due to different polytypes in SiC was suggested by Iwata *et al* [3]. One possible mechanism of creation of a different polytype layers is the motion of partial dislocations in adjacent basal planes [4]. Identification of the growth conditions responsible for creation of these interlayer fragments is yet to be investigated.

Acknowledgments

This work was supported by the Russian academy of Sciences Grant "Fundamental Research of Nanotechnology and Nanomaterials".

References

[1] Baliga B J, 2006 *Silicon Carbide Power Devices* (World Scientific Press)
[2] Pavesi L, Dal Negro L, Mazzoleni C, Franzo G and Priolo F 2000 Nature **408** 440
[3] Iwata Y P, Lindefelt U, Öberg S and Briddon P R 2003 *Phys. Rev. B* **68** 24309
[4] Pirouz P and Yang J W 1993 *Ultramicroscopy* **51** 189; Pirouz P 1997 *Solid State Phys.* **56** 107

Radiation properties of dislocations created by plastic deformation of Si under high pressure and low temperature

A N Tereshchenko[1], E A Steinman[1] and J Rabier[2]

[1] Institute of Solid State Physics RAS, Chernogolovka, Moscow distr., 142432 Russia
[2] Laboratoire de Métallurgie Physique, UMR CNRS 6630, Université de Poitiers, SP2MI, BP 30179, 86962 Futuroscope-Chasseneuil Cedex, France

E-mail: tan@issp.ac.ru

Abstract. Dislocation structures produced at room temperature under confining pressure (5 GPa) have been investigated by transmission electron microscopy and photoluminescence. Following annealing treatments in the temperature range 300°C - 600°C the initial DRL spectrum changes to a well-known spectrum of dislocation luminescence of dissociated glide set dislocations. In spite of the fact that no transformation of perfect shuffle dislocations to dissociated glide ones was observed by transmission electron microscopy, experimental photoluminescence data analysis does not exclude the possibility either of such transformation or nucleation of glide dislocations following annealing treatments at temperature as low as 300°C.

1. Introduction

Dislocation related luminescence (DRL) in silicon consists of four bands, D1 to D4, with their maxima located at low temperature at 0.81 eV, 0.87 eV, 0.93 eV and 0.99 eV, respectively [1]. While it is widely accepted that D4 and its TO phonon replica D3 originate from recombination at straight segments of dissociated 60° dislocations, the D1/D2 lines arise most probably from recombination on some structural defects on dislocation lines [2]. However, at the same time the exact origin of luminescent D1/D2 bands is still unknown.

Recently D1 luminescence has been demonstrated to be a promising candidate for light emitting diodes based on silicon technology [3, 4]. In this connection, further investigation of the microscopic nature and the generation process of these centres is required. One of the steps for solving this problem is investigation of different microstructural configurations related to generation of these centres. It was shown earlier that structural defects in or near the glide dislocation core are a source of low-energy bands D1/D2 of DRL [5]. It was assumed that these defects result from moving dislocations and dislocation reactions.

In the Si crystal the glide planes for moving dislocations are {111}. However, there are two possible {111} sets of glide systems for dislocations: the shuffle set when the dislocation core is between widely spaced {111}-planes and the glide set when the core lies between narrowly spaced {111}-planes. At present it is widely accepted that in silicon most dislocations generated at a relatively low stress and moderate temperature are a glide set [6]. Investigation of dislocation structure created by plastic deformation of Si under high pressure and low temperature shows the presence of a shuffle set of dislocations [7]. This dislocation type does not contain D1/D2 centres [8]. In this context, looking at nucleation of glide dislocations or transformation of the dislocation core from shuffle to glide can be useful to understand the origin of the D1/D2 luminescence bands. It should be noted that the core of perfect shuffle dislocations and dissociated glide dislocations can be transformed from one form to the other through two basic mechanisms: cross slip and climb. The first mechanism can be achieved under stress with the help of thermal activation. Since it does not require diffusion of point defects, the cross slip mechanism can operate at lower temperatures as compared to the climb mechanism. Nevertheless, no transformation of a shuffle set to a glide set and back was observed in ref. [9]. In this work the samples similar to those used previously [9] were investigated by transmission electron microscopy (TEM) and photoluminescence (PL) methods.

2. Experimental details

Plastic deformation of float zone grown (FZ) n-type Si single crystals (ρ=1800 Ωcm) was performed under a hydrostatic confining pressure of 5 GPa at room temperature. For this purpose a multi-anvil apparatus (see details in Ref. 10) was used. TEM observations were performed using a JEOL 200CX electron microscope. Annealing treatments were performed in situ using a hot-stage experimental setup.

Subsequent isochronal anneals (10 minutes) of deformed samples were carried out in the temperature range 300°C-600°C. Each thermal treatment of samples was performed in a quartz furnace with Ar flow and each step of that was followed by PL measurements. The PL spectra were recorded at T=6K using a conventional lock-in technique with a cooled Ge detector. PL was excited by a GaAs laser with P_{exc} of about 1W/cm^2 and λ=920nm. The area of the laser spot was estimated to be about 1mm^2.

3. Results and discussion

TEM investigations of as deformed samples show perfect dislocations that were nucleated at the lateral surfaces of cracks and others that propagated in the bulk material and accumulated in areas with very high dislocation densities. Weak-beam dark field contrast studies revealed that dislocations nucleated at crack edges were of 1/2 <110> (111) type (figure 1a). These half-loops were elongated along the [321] direction. However, in the bulk the same dislocations were found to lie along the screw orientation, as well as along specific <112>/30° and <123>/41° crystallographic directions (figure 1b).

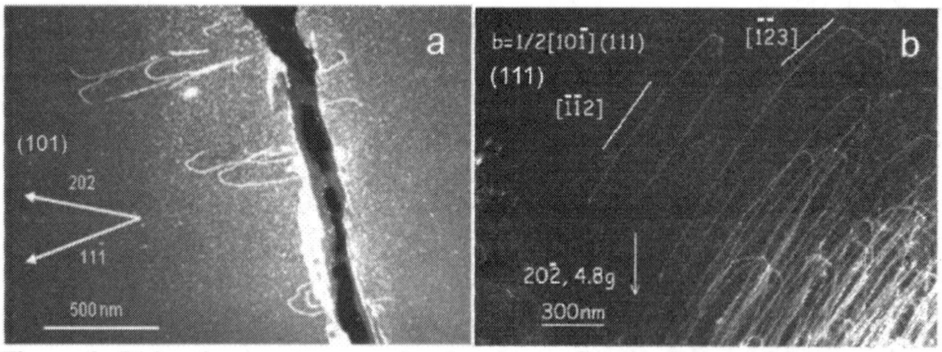

Figure 1. Dislocation structures produced under confining pressure of 5 GPa at room temperature. (a) dislocations nucleated at the lateral surfaces of cracks, and (b) in the bulk.

Figure 2a shows PL spectra of the sample after deformation at room temperature under 5GPa and consequent annealing at 300°C and 400°C (10 minutes). Since all PL measurements were performed in identical conditions, the intensity of luminescence is proportional to the concentration of the correspondence centres. The PL spectrum of the sample after deformation is rather different from that typical of glide dislocations. It consists of a broad emission in the whole DRL range with some features in the high energy part. Note that the observed broad background is a common component of the DRL spectrum at low temperature deformation [11-13].

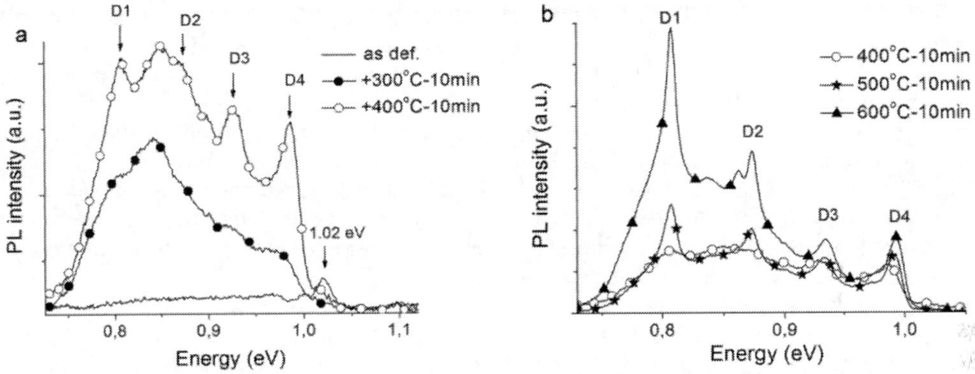

Figure 2. Modification DRL spectra due to annealing treatments.

One can see that after 300°C annealing the PL intensity grows considerably and can be interpreted as unresolved dislocation related luminescence (DRL) due to the presence of the typical features at D1 to D4 line positions. At the same time the band with the maximum at 1.02 eV decreases rapidly.

Further annealing at 400°C leads to an increase of DRL and a relative increase of the D4 and D3 bands which are known to originate from recombination of excess carriers at straight segments of dissociated 60° dislocations. The considerable increase of D1 band intensity starts after subsequent annealing at 500°C and it dominates in the spectrum after annealing at 600°C (figure 2b). This means that the generation process of these centres is very effective at annealing of the investigated dislocation structure.

Figure 3 illustrates the gradual change of the spectral distribution of intensity in the range of D4 band as a result of consequent isochronous annealing. The first curve (without symbols) shows the spectrum after deformation at room temperature and an applied pressure of 5GPa. Actually the D4 region of the spectrum consists of some broad bands. One of them has a clear maximum of about 1.02 eV. This line was observed earlier in an analogous sample by Pizzini *et al.*[8]. It was assumed that the 1.02 eV band was related to the point defects in the deformed Si crystal.

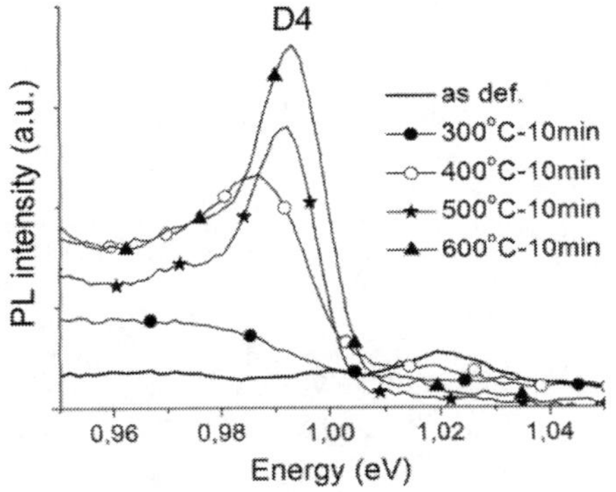

Figure 3. Influence of annealing on the D4 region of the DRL spectrum.

Journal of Physics: Conference Series **281** (2011) 012020 doi:10.1088/1742-6596/281/1/012020

At the same time another interpretation of this line origin can be considered. It has been shown in Refs. [14, 15] that two bands arise as a result of anomalous splitting of 60° dislocations. The low energy band corresponds to narrow splitting, while the high energy band relates to widely dissociated dislocations. Therefore, we can assume the presence of some amount of glide dissociated dislocations in our sample after deformation.

It was observed [14, 15] that non-equilibrium dislocation structure relaxes to its equilibrium state starting upon annealing at 265°C. In our case after annealing at 300°C the 1.02 eV band disappears while the low energy band in the 0.96–1 eV range increases and shifts to the conventional position of the D4 band. It should be noted that the transformation of the low and high energy sides of the spectrum to D4 takes place gradually in intensity as well as in position. This observation supports the glide dislocation hypothesis better than the model of unknown point defects of Ref.[8].

Therefore, the appearance and development of the DRL spectrum can be presumably interpreted as an emergence of glide set dislocations. We can imagine two possibilities for generation of such dislocations. It can be either reconstruction of the existing defects to glide dislocations or emission of new dislocations due to frozen internal stress.

In situ annealing in the TEM was performed in order to study the possible transformation of perfect dislocations into dissociated glide dislocations. The sample having a microstructure built by perfect dislocations with their common <112>/30° orientation as well as cracks was annealed in the microscope heating stage at temperatures up to 685°C. Figure 4a,b shows a TEM image of the same sample region before and after annealing for 90 min. Evidently there is no significant evolution in the shape of the perfect dislocation glide loops. Only the front of the loops moves, the dislocation segments staying locked in the <112>/30° and screw Peierls valleys. This shape evolution provides no evidence of transformation.

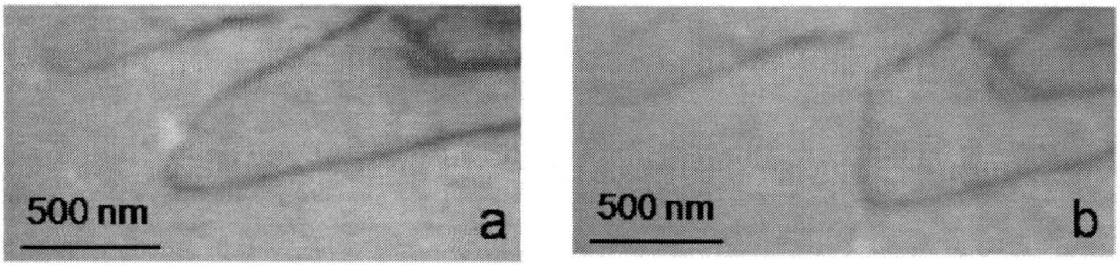

Figure 4. Result of in situ annealing of undissociated dislocations up to 685°C in TEM: (a) initial dislocation structure, and (b) the same region after annealing for 90 min.

However, in the same thin foil, in an area initially free of dislocations, the same annealing treatment results in nucleation of glide loops with a hexagonal shape characteristic of glide set dislocations (figure 5).

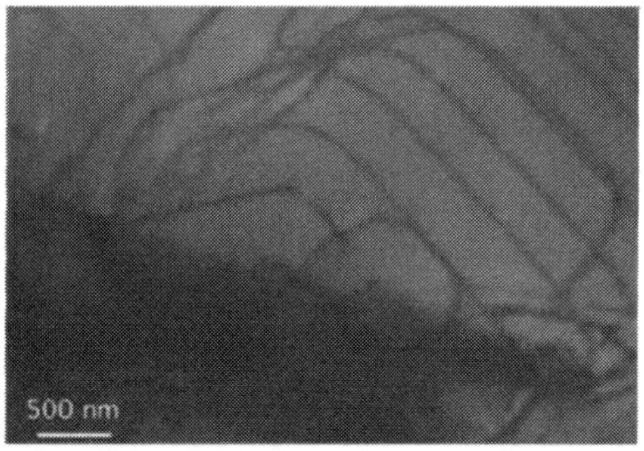

Figure 5. Nucleation of glide dislocations from lateral surfaces of cracks during in situ annealing.

While the increase of the D1/D2 bands after 600°C (figure 2b) annealing can be explained by nucleation of glide dislocations, it is hardly the case at lower temperatures. In figure 2a it is seen that the appearance and increase of D-bands occurs already upon short time 300-400°C annealing. This is typical for DRL of glide dislocations. Taking into account that D1/D2 centres are of the point type, we can assume that there are two contributions to the growth of D1/D2 intensity. At low temperature the relaxation process in the core of perfect shuffle dislocations leads to generation of some D1/D2 centres. This type of transformation can hardly be revealed by the TEM technique. At temperature above 500°C generation of new glide dislocations is possible, which provides a further significant increase in D1/D2 intensity.

On the other hand, the observation of the 1.02 eV line (figure 3) can be interpreted as the presence of regular segments of widely dissociated 60° dislocations just after deformation. The rapid decrease of the 1.02 eV line after primary annealing and simultaneous growing of the conventional D4 line supports this conclusion. However, this conflicts with TEM data. The following explanation can be put forward to explain this discrepancy. TEM can reveal local events with no statistical significance, whereas PL spectra are sensitive to modifications in the bulk material or to premonitory mechanisms associated with nucleation of glide dislocations. Therefore, we cannot exclude the possibility of either nucleation of glide dislocations or transformation of existing defects to glide dislocations. Additional effort is necessary to elucidate the problem.

Acknowledgments
This work was partially supported by the RAS grant "Quantum Physics of Condensed Matter".

References

[1] Drozdov N A, Patrin A A and Tkachev V D 1976 *Sov. Phys. JETP Lett.* **23** 597
[2] Kveder V V, Steinman E A, Shevchenko S A and Grimmeiss H G 1995 *Phys.Rev. B* **51** 10520
[3] Kveder V, Badylevich M, Steinman E, Izotov A, Zeibt M and Schröter W 2004 *Appl. Phys. Lett.* **84** 2106
[4] Kveder V, Badylevich M, Schroter W, Seibt M, Steinman E and Izotov A 2005 *Phys. Stat. Sol. (a)* **202** 901
[5] Steinman E A, Vdovin V I, Yugova T G, Avrutin V S and Izyumskaya N F 1999 *Semicond. Sci. Technol.* **14** 582
[6] Schroter W and Cerva H 2002 *Solid State Phenomena* **85-86** 67
[7] Rabier J and Demenet J L 2001 *Scripta Materialia* **45** 1259
[8] Pizzini S, Binetti S, Donne A, Marzegalli A and Rabier J 2005 *Appl. Phys. Lett.* **88** 211910
[9] Rabier J and Demenet J L 2005 *Phys. stat. sol. (a)* **202** 944
[10] Rabier J, Cordier P, Demenet J L and Garem H 2001 *Mater. Sci. Eng.* A **309–310** 74
[11] Drozdov N A, Patrin A A and Tkachev V D 1977 *Phys. Stat. Sol. (b)* **83** K137
[12] Alexander H, Kisielowski-Kemmerich C and Weber E R 1983 *Physica B* **116** 583
[13] Suezawa M and Sumino K 1983 *Phys. stat. sol. (a)* **78** 639
[14] Izotov A N and Steinman E A 1987 *Phys. Stat. Sol. (a)* **104** 777
[15] Sauer R, Weber J and Stolz J 1985 *Appl. Phys.* **A 36** 1

Journal of Physics: Conference Series **281** (2011) 012021 doi:10.1088/1742-6596/281/1/012021

Room temperature luminescence from Germanium

T Arguirov [1,2], M Kittler [1,2], and N V Abrosimov[3]

[1] IHP, Im Technologiepark 25, 15236 Frankfurt (Oder), Germany
[2] Joint Lab IHP/BTU Cottbus, Erich-Weinert-Str. 1, 03046 Cottbus, Germany
[3] Leibniz Institute for crystal Growth, Max-Born-Str. 2, 12489 Berlin, Germany

E-mail: arguirov@tu-cottbus.de

Abstract. We analyze the band to band radiative transitions in germanium thin films deposited on silicon and compare them with those in bulk material. A significant down shift of the direct transition related peak was observed from the thin films samples, caused by the preparation formed tensile strain in the film. A comparison between the ratios of the direct to indirect transitions peak intensities showed that those are very similar for the thin films and the bulk material, when the self absorption of the emission is accounted for. The observed similarity in the spectral shape indicates that the strong direct transition luminescence detected in strained germanium films is mainly due to the improved light extraction in the energy range of about 0.8 eV, rather than an increase in the probability of band to band direct transition. We find a feature at around 0.72 eV in the spectrum of the germanium luminescence, which correlates with the presence of dislocations in the crystal. We discuss the origin of this feature in view of one dimensional dislocation bands, split off from the Gamma valley of the conduction band due to the dislocations local strain field.

1. Introduction

Efficient light emitters compatible with standard CMOS technology have been a subject of research for more than a decade [1]. On-chip integration of light emitters would on one side significantly boost the performance of the integrated circuits by allowing fast optical data transfer, and on the other would add new functionalities to microelectronic devices for versatile optical applications [2]. A variety of concepts have been proposed [3-5], but a real application of those in process practice is still missing. Two main groups of light emitter concepts can be defined: one group embraces those which are incorporated in the silicon substrate on the so called device layer; the other covers the hybrids which in a simplistic view represent a construct consisting of direct gap materials glued onto a fully processed integrated circuit [3]. It is obvious that the compatibility requirements for the first group of devices are far more restrictive than those for the second, because they have to sustain unchanged the entire process sequence. In regard to those compatibility requirements, the material of choice for light emitters is limited to silicon, germanium and silicon-germanium alloys. Indeed, germanium has been proven as a suitable material for implementation in microelectronics and an instance of such application is the hetero-junction bipolar transistor used for high frequency digital data processing [6].

Recent reports present a new exciting application of germanium in light emitting devices. Thin, strained films grown on silicon exhibit luminescence, which maximum lies at about 0.8 eV [7-10]. It is generated by a direct transition between the conduction and the valence band. Indeed, germanium is regarded a quasi-direct semiconductor with two closely situated energy minima in the conduction band: a direct one at the Gamma valley and an indirect one at the L valley [11]. The energy position of the Gamma valley is just 140 mV above the L one and it can become partially filled at room temperature if the Fermi level for electrons is close to the upper band gap edge (for example in highly doped n-type materials). Partial occupation of the Gamma valley usually causes direct transition luminescence which is much more efficient than the phonon mediated indirect one, involving the L valley electrons [9].

It has been reported that at low temperatures the occupation of the Gamma valley almost vanishes due to the sharper energy distribution of the carriers. The observed intensity of the direct transition decreases with temperature corresponding to the decreasing occupation of the Gamma valley [10].

Recently it was shown that high n-type doping concentrations favor direct band luminescence, ensuring higher occupation of the Gamma valley [7, 12]. Besides doping, tensile strain in the germanium films additionally improves the probability for direct band gap transitions due to the modification of the semiconductor band structure [13]. Strain lowers the energy difference between the two minima in the conduction band and also decreases the band gap by shifting the maximum of the valence band light holes to a higher energy [14]. It has been shown that germanium becomes a direct semiconductor at strains of about 2% [15]. It is obvious that the "more direct" strained germanium films will exhibit stronger direct luminescence than the fully relaxed material.

The influence of strain and doping concentration on the efficiency of the direct luminescence has been already studied [13]. Even light amplification in thin germanium films has been reported [16]. However those studies consider changes in the luminescence in a very restricted spectral range involving only the direct band transitions related peak. Thus for example the role of luminescence self absorption in films of different thickness has been neglected. Moreover, the growth of thin germanium films on silicon substrates is related with formation of crystal defects, which also exhibit luminescence in the energy range 0.45-0.58 eV [17, 18]. Although there is no doubt that the reported luminescence enhancement effects are primarily due to the doping and strain of the germanium layers, the role of their material quality and its influence on the luminescence properties has not yet been clarified sufficiently.

Here we report on the luminescence of undoped and differently doped p- and n- type bulk germanium samples and make a comparison with the luminescence obtained from non-doped thin germanium films prepared by chemical vapor deposition (CVD) on silicon substrates [19]. For a bulk reference sample containing dislocations we used a wafer from a poly-crystalline block cast grown crystal. The reported spectra are recorded over a broad spectral range, so that indirect band to band transitions as well as defect related luminescence can be analyzed. We consider the role of the self absorption of the direct luminescence in bulk germanium and discuss the stricture of the indirect band transition peak. Moreover, spectral features in the low energy range are detected and identified as dislocation related peaks. We confirm the correlation between the appearance of those features and modification in the fine structure of the indirect band to band transition luminescence, which was already reported in the work of Kittler *et al* [19].

2. Experimental Method

2.1. Samples

Eight wafers of Czochralski ‹100› grown (Cz) bulk germanium were investigated. Four of them were n-type doped by Sb and the other four were p-type doped by Ga. Doping concentration in each case was varied from 10^{15} to 10^{16} cm^{-3}. In addition, undoped germanium wafers and undoped thin germanium layer samples were studied. The thin germanium films were prepared by CVD of germanium on a silicon substrate. The dislocation density for the CVD samples was estimated to be 10^{7} cm^{-2}. The germanium layer was 5 µm thick and was not intentionally doped. To understand the influence of the dislocations on the luminescence from the thin films we examined the luminescence from wafers of poly-crystalline block-cast-grown-germanium and compared its spectrum with that of the thin film sample. Details of the poly-crystalline samples can be found in [20].

2.2. Setup for luminescence measurements

For the PL study reported below we used a standard spectroscopic setup. The Germanium samples were excited by means of a frequency doubled Nd:YAG diode-pumped solid-state laser, emitting continuous-wave at wavelength of 532 nm. The spot of the laser beam at the surface of the sample was about 100 μm in diameter. The laser power used for excitation was in the range of 100 mW to 2 W. The luminescence was collected by a parabolic mirror and spectrally analyzed by a monochromator (0.64 m focal length, F/5.2 aperture) equipped with a 300 l/mm grating and an InSb detector. The excitation beam was chopped at 133 Hz and a lock-in technique was used to process the signals.

3. Results and discussion

The luminescence from the bulk samples shows two distinct peaks with maxima at about 0.7 and 0.8 eV. Examples of the as-measured spectra on samples with different doping type but similar doping concentration are shown in figure 1.

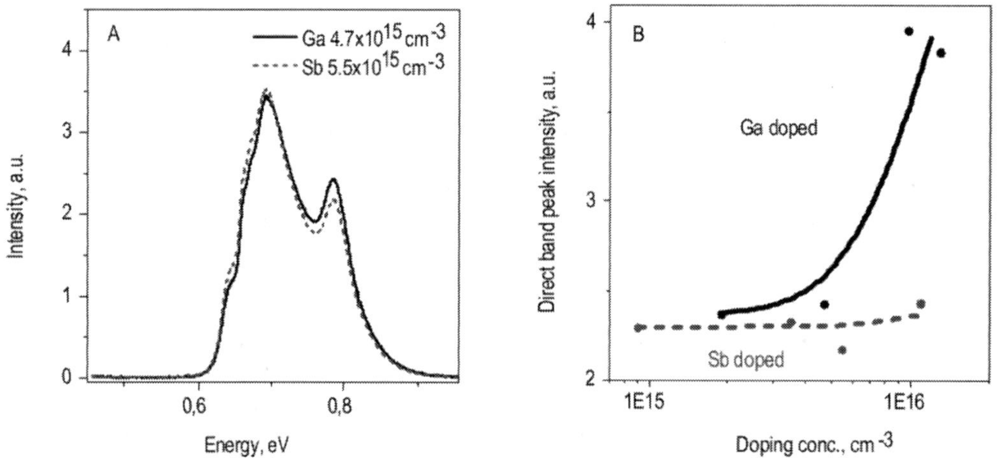

Figure 1. Luminescence from bulk Germanium samples.
(A) Spectra measured on p and n type doped samples with similar doping concentrations ($10^{15}cm^{-3}$).
(B) Dependence of the direct band transition peak intensity on the doping concentration.

The maximum of the intensity of those spectra appears in the energy range of the indirect band to band transition. The shape of the spectra for both doping types is alike and no indication for a stronger luminescence from the Sb doped n-type samples was found. Surprisingly, the luminescence from the p-type Ga doped samples appears even stronger than that for Sb doped ones at concentration of about 10^{16} cm^{-3}. Indeed, the doping concentration used here is small, and one cannot expect that the position of the quasi Fermi levels at high excitation would significantly be influenced by it, but the results imply that the luminescence intensity is also influenced by factors related to the quality of the material, especially when the doping concentration is low. We should mention that detectable direct band recombination is recorded in low n-doped and even p-doped samples so a high n-doping is not necessarily the reason for its appearing.

All measurements presented in figure 1 are carried out at equal excitation conditions and the intensity ratio of the radiation due to direct and indirect related transition, from all samples is same. Upon increasing the excitation the direct band transition related peak increases. Figure 2 shows spectra recorded at different power of the excitation beam from a non-intentionally doped sample. The curves are normalized to the maximum of the indirect transition peak.

Figure 2 shows the dependence of the direct, I_d and indirect, I_i transitions intensities on the excitation power P. Both data sets can be well described by a power low I~P^n, where n=1.4 for the indirect and n=2.1 for the direct transition. Considering that the material is non-doped, the intensity of the luminescence should be proportional to the square of the injected carriers, because each emitted photon is generated from a recombination of a pair of excess electron and hole. The recombination rate is then proportional to the product of the concentrations of electrons and holes and each of those is proportional to the excitation power [21, 22].

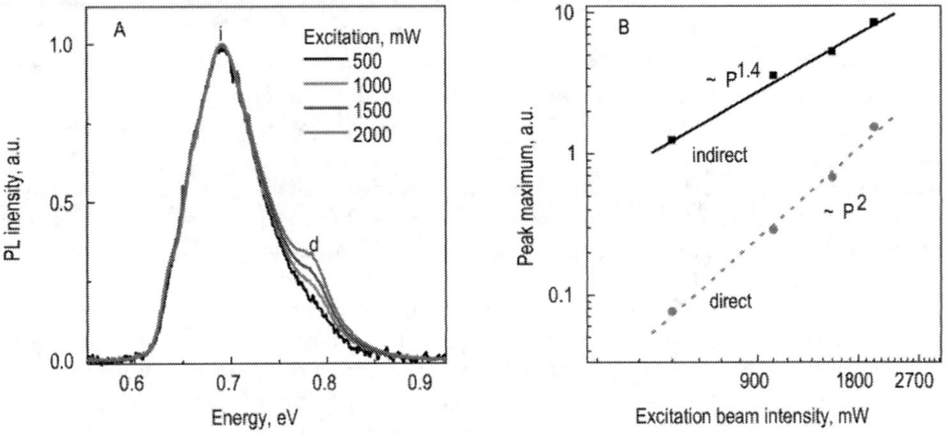

Figure 2. Intensity of direct and indirect recombination peaks dependence on the excitation level. (A) Spectra normalized on maximum, taken at different powers of the excitation laser beam. (B) Log-log plot for estimating excitation dependence of the direct and indirect radiative transition peaks.

The near quadratic dependence observed for the direct transition luminescence matches well the expected one, but there is no such match in the case of indirect band luminescence. A possible reason for the sub-quadratic dependence of the indirect band luminescence is a competitive process of non-radiative recombination, which effectively restricts the increase of electron concentration in the L valley of the conduction band upon increasing excitation power. A candidate for such a competitor is the Auger recombination. Indeed, the concentration of the electrons in the L valley is significantly higher than that in the Gamma valley and the probability of a three carrier mediated Auger process with energy transferred to a second electron should also be relatively high. One can assume that the Auger recombination (Cn^3) is limiting the concentration of electrons, n in the L valley. The generation rate (~ excitation power), G, is equilibrated by the Auger recombination ($G=Cn^3$), so that the concentration of carriers n in the L valley should be proportional to the excitation to the power of 1/3.

The indirect radiative recombination rate is then given by the product of the electron density in the L valley ($\sim P^{1/3}$) and the hole density in the valence band (~P). This leads to an indirect transition rate proportional to the power of 1.3, which is close to the experimentally obtained result.

We use the above quadratic dependence of the direct recombination on the excitation to derive the correction for the luminescence-self-absorption in bulk germanium. The spectra shown above represent the wavelength dependent intensity distribution of the luminescence which escaped from the sample after passing through a significant amount of material. To obtain the truly generated luminescence spectrum one should account for the self-absorption in germanium itself. Indeed, the intensity of the indirect radiation is barely influenced, while a significant part of the direct radiation is absorbed. To correct the spectra for self absorption we assumed that the luminescence intensity is proportional to the square of the generated excess carriers.

In the case of photo-excitation with a green laser (532 nm) the light is absorbed in a few nm beneath the surface and the carriers generated there diffuse into the bulk. Their density exponentially decreases in depth with a scaling factor corresponding to their diffusion length.

Similarly the luminescence $I(h\nu)$ generated at a certain depth passes a layer with absorption coefficient $\alpha(h\nu)$ and is attenuated exponentially.

Assuming one dimensional distribution and integrating all contributions from recombinations at different depths we obtained a relation for correcting the spectra due to the self absorption:

$$I_{meas}(h\nu) = I(h\nu) \cdot \left(\propto (h\nu) + \frac{2}{L_d}\right). \tag{1}$$

In the equation above $I_{meas}(h\nu)$ denotes the luminescence intensity escaped from the sample, $I(h\nu)$ denotes the intensity of the generated luminescence, $\alpha(h\nu)$ is the absorption coefficient, and L_d stands for the diffusion length of the excess carriers. In all parameters the term $h\nu$ marks their photon energy dependence.

To obtain the spectra corrected for self absorption we measured the spectrum of the absorption coefficient of the samples, utilizing the same spectrometer and detector, which we used for luminescence measurements. The data were normalized to fit those available in the literature [23, 24]. The diffusion length was taken to be 100 μm, a typical value for the investigated material.

Figure 3 shows a spectrum of an Sb doped sample with correction for the self absorption. A non corrected spectrum is added for comparison. It is seen that the intensity of the direct transition exceeds the indirect one about 10 times. A similar value has been reported in an earlier study [25]. Moreover, the position of the direct transition related peak in the corrected spectrum is at 0.8 eV, exactly corresponding to the reported value for the direct band gap energy.

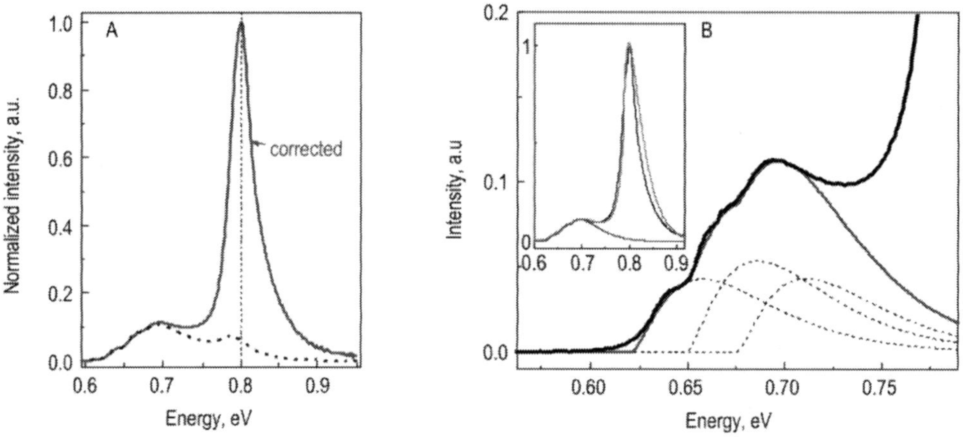

Figure 3. Spectra of band to band transition of an Sb doped sample normalized on maximum. (A) Corrected for self-absorption. (B) Fit of the spectral features in the indirect transition peak. The inset shows the fit of the whole spectrum with the indirect- and direct-transition related peaks.

The diagrams in figure 3b present a fit of the corrected spectra with the known dependences for direct and indirect allowed transitions in a semiconductor.

The peak shapes according Bebb and Williams [26] are given by:

$$I_d\,(h\nu) \sim \sqrt{E_{gd} - h\nu} \cdot exp\left(\frac{E_{gd}-h\nu}{kT}\right) \tag{2}$$

for a direct transition and

$$I_i\,(h\nu) \sim (E_{gi} - h\nu)^{\,2} \cdot exp\left(\frac{E_{gi}-h\nu}{kT}\right) \tag{3}$$

for the indirect one. Here $I_d\,(h\nu)$ and $I_i\,(h\nu)$ denote the intensity of the luminescence due to direct and indirect transitions, E_{gd} and E_{gi} are the direct and the indirect band gaps, kT is the mean thermal energy term at room temperature, and $h\nu$ is the energy of the emitted photons.

For the indirect related feature three peaks of similar intensity were used to fit its fine structure. The difference in the position of those peaks is 28 meV, which corresponds to the energy of the longitudinal acoustic phonons in germanium at the L point [27]. Those peaks we relate to phonon mediated radiative recombinations, namely, one phonon emission, one phonon absorption, and zero phonon transitions.

To estimate the real enhancement of the direct luminescence in germanium thin films we use the intensity of the indirect transition as a reference. Figure 4 compares the spectra of a thin germanium sample with that of the bulk one, but corrected for the self-absorption. A correction for the thin sample spectra is not necessary, because of the negligible attenuation of the luminescence inside the thin film.

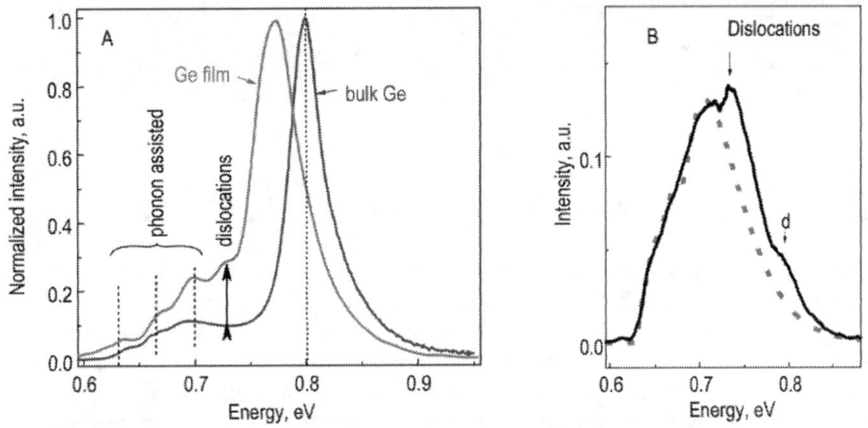

Figure 4. (A) Band to band luminescence spectra from thin strained samples compared with that from bulk material. (B) Band to band luminescence feature in the spectrum of a poly-crystalline germanium sample containing dislocations.

Figure 4A shows that the luminescence spectra from the thin film sample and the corrected bulk samples are similar. The spectrum from the thin film sample shows a well pronounced shift of the direct transition related feature, which we consider caused by the strain induced band gap shrinkage. The strain originates from the thermal expansion coefficient mismatch between the germanium film and the silicon substrate. The germanium is deposited on the substrate at a temperature of about 800°C. By cooling down to room temperature it undergoes a stronger contraction than silicon and the thin film becomes tensile strained.

It can be seen that the ratio between direct and indirect transitions features in both spectra is similar. Considering the enhancement of the direct luminescence in the thin strained films, one can state that the major effect is due to the better light extraction from the film. In fact we do not observe any improvement of the direct transition related luminescence due to the strain, because the ratio between direct and indirect transition peaks is similar to those for the bulk sample (when corrected for the self-absorption).

But let us consider the spectra in more detail. A peak at 0.72 eV is observed in the spectrum of the thin film sample, which is not present in the spectrum of the bulk material. A similar peak is also detected in the poly-crystalline samples containing dislocations (figure 4B). We relate that peak with the presence of dislocations in the crystalline lattice. Indeed, well known dislocation related features are observed in the luminescence spectra of plastically deformed germanium samples in the range 0.45-0.58 eV [28]. Figure 5 shows the spectra of epitaxially grown thin germanium films recorded at room temperature and at 80 K exhibiting the above mentioned dislocation spectral features. Similar spectra were recorded on the poly-crystalline samples as well.

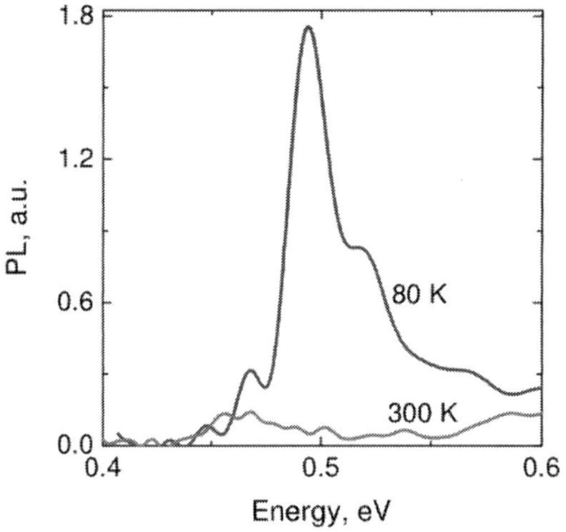

Figure 5. Dislocation related features in the spectrum of a 5 μm thin germanium layer on silicon.

The presence of a peak at around 0.72 eV in the spectra of samples with dislocations and its absence in the high crystal quality Cz samples suggests that it originates from extended defects. The origin of the dislocation radiation is still not fully understood and is a subject of intensive research [29]. However, some of the characteristic features can be related to formation of one dimensional dislocation bands due to the modification of the semiconductor three dimensional energy bands by the local dislocation strain field. A general expression for such modification is given by Lelikov et al [30].

$$E_{dis} = E_g - A\left(\frac{b}{a}\right)^2 \tag{4}$$

where E_{dis} is the energy of the radiation, E_g is the band gap, a is the lattice constant, A is a constant accounting for the deformation potential and the effective mass of the carriers, and b is the edge component of the Burgers vector of the dislocation.

A modification of the direct as well as the indirect band gaps could be expected. We think that the observed feature at around 0.72 eV is a result of the dislocation related direct gap modification of germanium. A shift of the dislocation related bands of about 80 meV below the direct band edge is expected by an evaluation of the material specific parameters [31].

4. Conclusions

In this work we analyzed the band to band luminescence of germanium bulk material and compared with that from thin germanium films deposited on silicon. Spectral features in the region of direct and indirect transitions were discussed. A strong direct transition luminescence peak was found in the spectra from the thin film germanium samples. The peak was slightly downshifted from its expected position at 0.8 eV, indicating tensile strain in the film. A comparison of the spectra from the bulk and from the thin film germanium samples, excluding the role of self absorption, showed that both are very similar in shape. This observation let us to conclude that the main reason for the observed strong luminescence from the thin films is the improved light extraction of the direct transition radiation.

We showed that the samples containing dislocations have an additional peak at about 0.72 eV in their spectrum, which is not present in the spectrum of the defect free germanium crystals. We related this peak to the dislocations and their long range strain field. We suggested that the dislocation induced strain modifies the direct band gap of germanium locally, forming a one dimensional band along the dislocation line, shifted by about 80 meV below the direct band of the bulk material.

Acknowledgements
We would like to thank Dr. Yuji Yamamoto (IHP Frankfurt (Oder)) for providing the thin film germanium samples and Dr. Ch. Frank-Rotsch (IKZ Berlin) for the poly-crystalline ones. This work has been partially supported by the German Ministry of Education and Research in the frame of SiliconLight project.

References
[1] Liang D and Bowers J E 2010 *Nature Photonics* **4** 511
[2] Vogel U, Kreye D, Reckziegel S, Toerker M, Grillberger C, Amelung J 2007 *Proc. SPIE* **6477** 647703
[3] Coffa S, Franzò G, Priolo F, Polman A, and Serna R 1994 *Phys. Rev. B* **49** 16313.
[4] van Campenhout J, Romeo P R, Regreny P, Seassal C, van Thourhout D, Verstuyft S, di Cioccio L, Fedeli J-M, Lagahe C, and Baets R 2007 *Optics Express* **15** 6744
[5] Arguirov T, Kittler M, Seifert W, Yu X, and Reiche M 2006 *Materials Science and Engineering B* **134** 109
[6] Heinemann B, Knoll D, Barth R, Bolze D, Blum K, Drews J, Ehwald K-E, Fischer G, Köpke G, Kurps R, Rücker H, Schley P, Winkler W, and Wulf H-E 2001 *Proc. IEDM* 348
[7] Cheng S-L, Lu J, Shambat G, Yu H-Y, Saraswat K, Vuckovic J, and Nishi Y 2009 *Optics Express* **17** 10019
[8] de Kersauson M, Jakomin R, El Kurdi M, Beaudoin G, Zerounian N, Aniel F, Sauvage S, Sagnes I, and Boucaud P 2010 *Journal of Applied Physics* **108** 023105
[9] Chen T-H, Ko C-Y, Chen C-Y, Peng K-L, Luo G-L, Liu C W, and Tseng H-H 2010 *Applied Physics Letters* **96** 091105
[10] Kasper E, Oehme M, Aguirov T, Werner J, Kittler M, and Schulze J 2010 *Proc. GEP2010* – accepted
[11] Mataré H 1971 *Defect Electronics in Semiconductors* (New York: John Wiley & Sons)
[12] Klingenstein W and Schweizer H 1978 *Solid-State Electron.* **21** 1371
[13] Sun X, Liu J, Kimerling L C, Michel 2009 J *Optics Letters* **34** 1198
[14] Zhang F and Crespi V H 2009 *Physical Review Letters* **102** 156401
[15] Soref R, Kouvetakis J, and Menendez J, *Mat. Res. Soc. Symp. Proc.* 2007 **13** 958
[16] Liu J, Sun X, Camacho-Aguilera R, Kimerling L C, and Michel J 2010 *Optics Letters* **35** 679
[17] Newman R 1957 *Physical Review* **105** 1715
[18] Steinmann E A 1981 *Crystal Research and Technology* **16** 247
[19] Kittler M, Arguirov T, Oehme M, Yamamoto Y, Tillack B, and Abrosimov N V 2010 *Physica Status Solidi (c)* – accepted
[20] Holla M, Arguirov T, Jia G, Kittler M, Frank-Rotsch C, Kiessling F, Rudolph P, 2010 *Solid State Phenomena* **156-158** 483
[21] Varshni Y P 1967 *Physica Status Solidi* **19** 459
[22] Schlangenotto H, Maeder H, Gerlach W 1974 *Physica Status Solidi (a)* **21** 357
[23] Palik E D 1985 *Handbook of optical constants of solids* (New York: Academic Press Handbook Series)
[24] http://www.ioffe.ru/SVA/NSM/Semicond/Ge/
[25] Haynes R 1955 *Physical Review* **98** 1866
[26] Barry Bebb V H and Williams E W 1972 *Semiconductors and Semimetals* **8** 239
[27] Weber W 1977 *Phys. Rev. B* **15** 4789
[28] Shevchenko S and Tereshchenko A 2007 *phys. stat. sol. (c)* **4** 2898
[29] T. Mchedlidze, O. Kononchuk, T. Arguirov, M. Trushin, M. Reiche and M. Kittler 2010 *Solid State Phenomena* **156-158** 567
[30] Lelikov Y, Rebane Y, Ruvimov S, Tarhin D, Sitnikova A, and Shreter Y 1992 *Mater. Sci. Forum* **83-87** 1321
[31] Kamieniecki E 1976 *J. Phys. C: Solid State Phys.* **9** 1211

Propagation and interactions of cracks in Si induced by H supply into He-filled cracks

S Reboh[1,2,3], J F Barbot[2], M F Beaufort[2], and P F P Fitchner[1]

[1] PGCIMAT, Universidade Federal do Rio Grande do Sul, Porto Alegre, Brasil
[2] Institut Pprime, UPR3346, Université de Poitiers - CNRS - ENSMA, SP2MI, BP 30179, 86962 Futuroscope, France
[3] Groupe nMat, CEMES-CNRS, 29 rue J. Marvig, 31055 Toulouse, France

E-mail: Jean.francois.barbot@univ-poitiers.fr

Abstract. The phenomena of interaction and propagation of cracks under the contribution of hydrogen were studied in (001) silicon substrate in which an array of scattered over-pressurized He-plates was previously introduced at a given depth. Their propagation under subcritical regime was activated through diffusional supply of H atoms introduced by implantation/annealing. Interactions between the tips of non coplanar cracks take place in a nanometric scale; they can be of plastic-type leading to the formation of extended defects or of elastic-type resulting in deviations of crack-tip propagation. While the planar interactions facilitate the propagation of cracks, those of non coplanar-type stop them. The observations were carried out by transmission electron microscopy and the results were discussed and modelled by using concepts of elasticity and fracture mechanics.

1. Introduction

Hydrogen implantation induced cracking in semiconducting material has been extensively studied in connection with thin layers transfer technology. This process is a key issue for the fabrication of semiconductor-on-insulator (SOI) substrates – the so-called Smart-Cut® process [1]. This ion cutting method drives the lateral propagation of cracks over the whole wafer area resulting in a SOI structure when bound onto another wafer. A significant amount of research on this hydrogen ion-cut process in silicon has thus been carried out to clarify the underlying mechanisms of layer splitting [2-5]. Platelets resulting from H implantation grow during post-thermal annealing to form large microcracks in the narrow depth region around the projected range of the H ions. Their propagation and interactions lead to exfoliation of a thin layer from the wafer. Later it has been shown that the co-implantation of H and He ions allowed blistering at much lower total fluence than H implantation alone [6]. The blistering facilitation is understood as due to helium capture by the previously introduced H platelets, He acting as an overpressurizer of platelets. Even in silicon, despite significant efforts, many physical aspects of the evolution of platelets to microcracks and then their interaction into large cracks are still unclear.

This paper investigates the propagation and interactions of cracks into implanted silicon. Specific two-steps experiments were designed to allow controlled crack propagation. H atoms were introduced into pressurized He-filled cracks previously formed under particular conditions of He implantation. This original approach produces rather efficient, flat and controlled growth of cracks [7]. In this paper previously the results obtained [7] are more detailed and completed. The resulting microstructure was examined by transmission electron microscopy (TEM) and discussed in terms of elasticity and fracture mechanics.

2. Experimental

The experiments were performed using p-type (001) Si Cz-wafers with a resistivity of 1-25 Ω cm. In the first step sample pieces were implanted at room temperature with He^+ ions accelerated at 45 kV to a fluence $\Phi = 1x10^{16}$ cm^{-2}, and then annealed at 350 °C during 900 s. These experimental conditions are known to lead to the formation of over-pressurized He-platelike cavities (or He-plates) located at the mean projected range of He atoms (≈ 400 nm underneath the surface) [8]. In the second step of the process, the samples were implanted at 200°C with 30 keV H_2^+ ions to a fluence $\Phi = 0.5x10^{16}$ cm^{-2}. The predicted mean depth range of the H^+ implantation is approximately 200 nm according to SRIM calculations [9]. The as-He/H co-implanted samples as well as a reference sample containing only the He-plates were finally subjected to a final thermal treatment at 300°C for 900 s. Microstructure characterization was performed by transmission electron microscopy (TEM) in cross-sectional specimens prepared by mechanical polishing and ion milling. The observations were carried out in a JEOL JEM 2010 microscope operating at 200 kV.

3. Results

Figure 1 shows the microstructure after the first step of the procedure, i.e. the He implantation followed by the 350°C annealing [7]. As expected scattered He-plates form close to the mean projected range of He. The average distance between He-plates exceeds their mean diameter of ≈ 150 nm, resulting in a discrete planar array of plates oriented along the (100) lattice planes parallel to the surface. The surrounding fringes of the two edge-on He-plates seen on the cross sectional TEM image are ascribed to the strain-related contrast induced by the high gas pressure inside the over-pressurized He-plates [8]. The elastic strain field of the He-plate can be calculated in 2D by modelling the He-plate by a dipole of edge dislocations with an effective Burger's vector [10]. The local strain field can be used to favour sublocal organized arrangement of H precipitates [11]. No modification of both the arrangement and the size of He-plates are observed after an annealing at 350°C suggesting that the He-plates are quite stable with respect to 350°C.

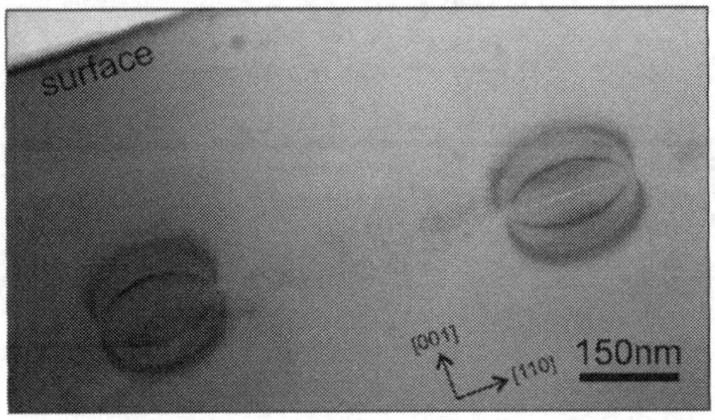

Figure 1. Cross-sectional bright-field TEM image showing two edge-on overpresurized He-plates observed in He-implanted silicon ($1x10^{16}$ cm^{-2}, RT) and then annealed (350°C, 900s) [7].

After the formation of the He-plates, the samples were implanted with H_2^+ ions at 200°C and then annealed at 300°C during 900 s. A general view of the developed microstructure is provided in figure 2. Nano-sized H-platelets are observed around a depth of 200 nm. However, at a depth of ≈ 400 nm, where the He-plates were originally located, a chain of long and quite flat cracks is now observed. This shows that some H atoms diffused towards the He-plates initiating a process of rectilinear crack propagation.

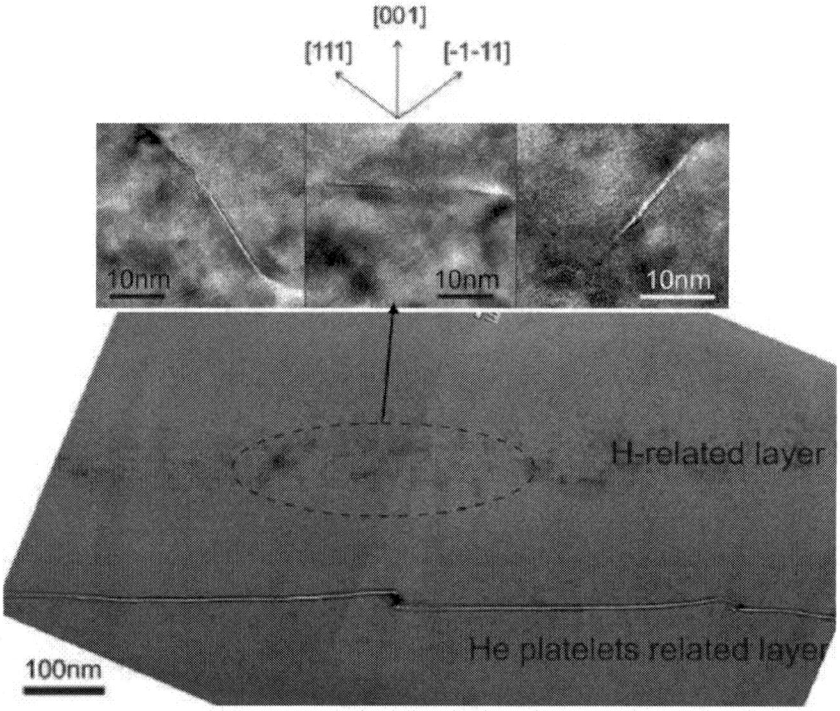

Figure 2. Bright field cross sectional TEM image of sample submitted to the total process, i.e., after He-implantation (45 keV,$1x10^{16}$ cm^{-2}, RT, annealed 350°C for 900s), and then H_2-implantation (30 keV, $0.5x10^{16}$ cm^{-2}, 200°C, annealed 300°C for 900s) [7]. The upper part shows in detail the different types of H-platelets as formed.

Various types of interaction between the long cracks are observed, see figures 3 and 4. These interactions can be of plastic or elastic type. In case of plastic interactions dislocations and nanotwin structures connecting the crack tips are observed, figure 3. For interactions of elastic type, deviations of crack propagation are observed; the two approaching cracks merge by redirecting the direction of propagation against each other, figure 4. Experimentally the elastic-type interaction seems to be effective for distances of approach lower than 50 nm; the spatial extension of the interaction is thus limited around the tip of crack [7].

4. Discussion

Most of the He-plates formed during the first step of the process form along (001) plans parallel to the surface due to the implantation in-plane stress [10]. For simplicity the He-plate was treated as a penny-shaped crack of radius d in an infinite elastic medium subjected to an internal pressure p according to the linear elastic fracture mechanism. The He-crack stability up to 350°C shows that the stress intensity factor at the crack tip, K_I, does not overcome the fracture toughness, K_{IC}, of the medium (=$1.19x10^6$ $Nm^{-3/2}$ in silicon [12]) required to propagate the crack in unstable manner. Therefore, this suggests that:

$$p < p_{\max} = K_{IC}\left(\frac{2}{\pi d}\right)^{1/2} \quad (1)$$

Journal of Physics: Conference Series **281** (2011) 012022 doi:10.1088/1742-6596/281/1/012022

where d is the diameter of the crack. Conversely the pressure is larger than the pressure for a crack in equilibrium given by [13]:

$$p > p_{\min} = \left(\frac{2\pi\mu\gamma}{(1-\nu)d} \right)^{1/2} \quad (2)$$

where μ is the shear modulus, ν the Poisson's ratio and γ the surface energy. So, the catastrophic propagation of a 150 nm He-crack (introduced by the first step of the process) controlled by the fracture toughness would require an additional overpressure of maximal value $\Delta p_{\max} = p_{\max} - p_{\min} \approx 200MPa$ resulting in the second step of the process (H_2 diffusion facilitated by the out-of plane strain of the He-plates). According to the ideal gas law the increase of pressure in the He-plate is directly proportional to the amount of added molecular hydrogen. Thus only few percents of implanted H atoms would require diffusing towards the He-plates to initiate the crack propagation via the increase of the inner gas pressure. However the role of hydrogen can be of chemical nature by breaking the strained Si-Si bonds at the crack-tip. In this case the Si-H formed should influence the internal structure of the crack and therefore should assist the crack propagation. This last scenario is more likely because it was shown that He not only leads to a high internal pressure but also induces the reconversion of molecular H_2 into bound Si–H [14]. Similarly recent dynamical simulations confirmed the growth of H-induced platelets via subcritical stress-corrosion mechanism rather than any significant H_2 pressure buildup [15]. In this last case the propagation of cracks takes place under subcritical conditions (i.e., with $K_I < K_{IC}$). The additional contribution of H is thus an effective process for the growth of large cracks. Moreover the rectilinear cracks suggest flat final surface in the ion cutting process.

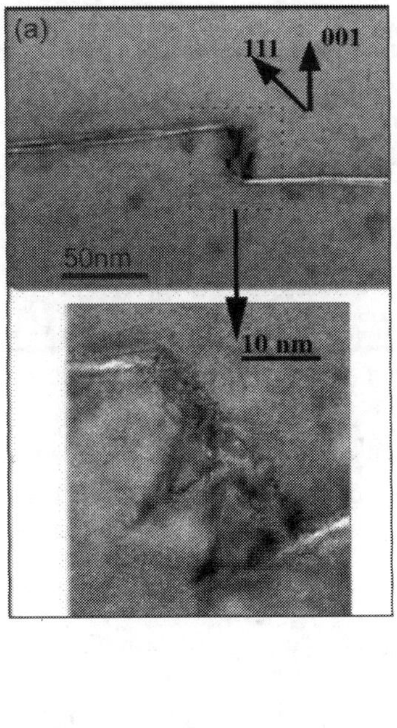

Figure 3. Example of plastic-type interaction between two cracks propagating towards each other. The HRTEM shows the microtwin-like defect between the crack-tips [7].

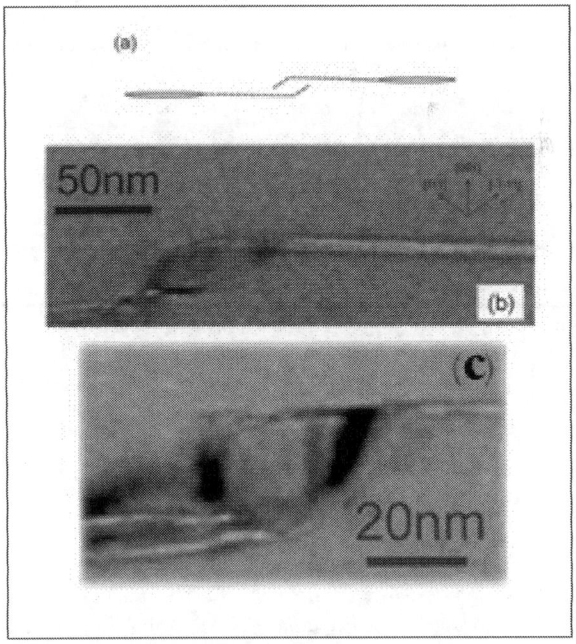

Figure 4. Example of elastic interaction where both crack tips run around each other (a) schematic of the process [14], (b) and [7] (c) TEM examples

During the growth of cracks, when the distance between two parallel crack tips becomes small enough complex phenomena of interaction are observed (figure 2). According to the different TEM observations, when two cracks get closer, they can interact in a plastic way blocking so brutally their growth, figure 3. This mechanism of relaxation of elastic strain by plastic deformation is the origin of the formation of structural defects such as nanotwins. The formation of structural defects requires high stresses at the crack-tips. TEM observations showed that the nucleation of perfect dislocations requires stresses larger than 1 GPa [16]. However, when the stress between a pair of parallel cracks is not quite important for satisfying the plastic deformation conditions, the cracks continue their expansion while remaining subjected to the interaction of elastic type, leading to a deviation of their path, figure 4. This behaviour of attractive-type can be explained by taking into account modifications of the stress intensity factor K of a given crack-tip due to the elastic forces induced by the approaching second collinear crack-tip. According to Gross and Seelig [17] and assuming equilibrium conditions (i.e, equation (1) ~ equation (2)) the K-factors, K_I (mode-I loading) and K_{II} (mode-II loading), of two parallel cracks of midpoint distance L and of interaction angle θ are given by:

$$K_I \approx (\pi\beta)\left[1+\frac{1}{8}\alpha^2(2\cos 2\theta - \cos 4\theta\right] \quad (3)$$

$$K_{II} \approx \frac{\pi}{8}\beta\alpha^2\left[-\sin 2\theta + \sin 4\theta\right] \quad (4)$$

where $\beta = \left(\frac{\gamma\mu}{1-\nu}\right)^{1/2}$ is a constant characteristic of the matrix and $\alpha = \frac{d}{L}$ an adimensional term characterizing the distance of interaction. For large distance L and for coplanar cracks ($\theta = 0$) K_I prevails (mode-I loading) and cracks develop in a rectilinear way to form long straight cracks as observed (figure 2). However, for cracks of vertical separation Z, when interacting, deviations from a straight path occur; the angle at which a propagating crack is deviated is mainly controlled by the K_{II} factor [17]. Figures 5 and 6 show the variations of the reduced factors K_I/β and K_{II}/β respectively as function of both α and θ.

As seen on figures 5 and 6 when two crack-tips approach (for $Z \neq 0$), both α and β increase and a first phase of slight increase is observed for both K_I and K_{II} resulting in minor modifications of propagating parameters. However, with decreasing interaction distance K_{II} becomes negative and increases in absolute value (for $\theta < \pi/6$) which means a change (attraction) in the prospective direction of crack growth [18]. Both crack-tips run thus around each other and each of them merges with the other crack as observed in figure 4. Conversely, along their interaction, K_I decreases suggesting a reduction in the propagation kinetics of cracks. The dotted lines in figures 5 and 6 show two possible paths of interaction. Furthermore K_{II} reaches maximal values when cracks are close and for $\theta \approx \pi/3$. Larger values of Z result in larger values of θ for a given value of α causing thus a reduction in the interaction strength and therefore limiting the cutting effects for the layer separation. More work is needed to determine more precisely the transition stress at the crack-tips between plastic and elastic interaction.

The formation of coplanar cracks being ideal to form straight cracks but not realistic due to the implantation straggling. This limitation can be overcome by introducing well localized sites of nucleation as observed in strained interfaces [19].

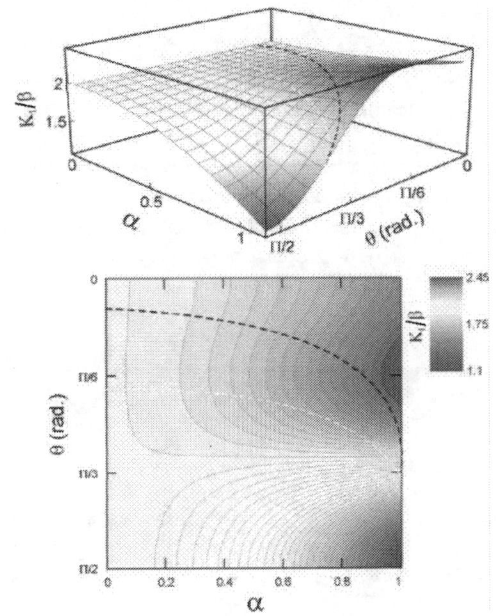

Figure 5. Plot of the reduced factor K_I/β, equation (3), as function of α and θ. The dotted lines show different examples of interaction.

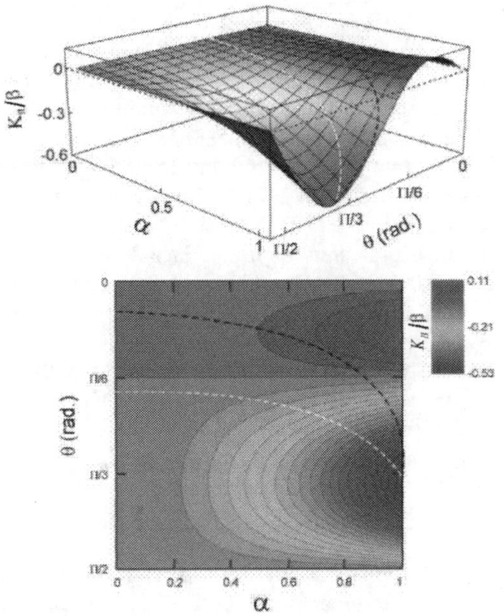

Figure 6. Plot of the reduced factor K_{II}/β, equation (4), as function of α and θ. K_{II} shows a change in the sign for $\theta = \pi/6$ which means a qualitative change in the prospective direction of crack growth.

The diffusion of H can also be performed by using a plasma hydrogenation process. Plasma hydrogenation based techniques have also been studied for their use in the transfer of thin films [20] Figure 7 shows a long and straight crack formed by the diffusion of H during the plasma hydrogenation into the previously introduced He-plates. This last point confirms that the He-plates can be used as precursors of cracks whose propagation under subcritical regime is activated through diffusional supply of H atoms.

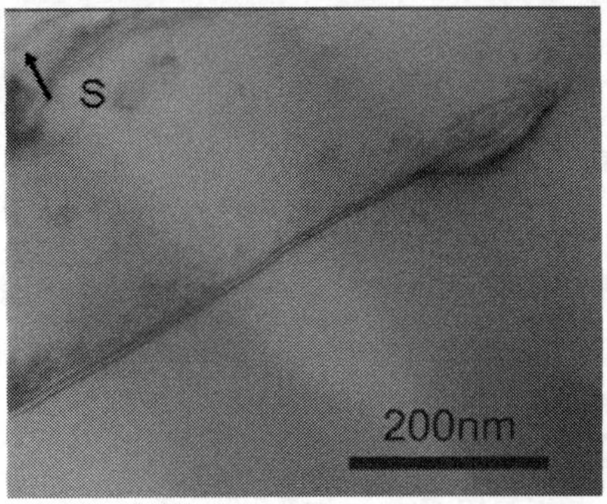

Figure 7. Cross-sectional bright-field TEM image showing a long and sharp crack observed after He-implantation (45 keV,$1\text{x}10^{16}$ cm^{-2}, RT, annealed 350°C for 900s) and plasma hydrogenation at 300°C during 1h. (S denotes the surface)

5. Conclusion

The propagation and interactions of cracks in Si were observed within a nanoscopic size scale by using a specific co-implantation process. Starting from overpressurized He-plates, subcritical propagation of cracks was activated through additional H atoms and micrometer sized rectilinear cracks were obtained. This type of He-based engineering could be used to reduced the amount of H required for the propagation of cracks and thus for the layer transfer. Plastic or elastic interactions of these cracks were observed and studied. Plastic interactions result in the formation of dislocations and twin-like extended defects due to a high shear stress of the two interacting cracks. Elastic interactions cause deviations in their path resulting in an interlacement of their tips as predicted by the calculations. The roughness of the as-split layer should be controlled by the interactions of propagating cracks.

Acknowledgments

The authors acknowledge the support from the CAPES-COFECUB program

References

[1] Bruel M 1995 *Electron. Lett.* **31** 1201
[2] Weldon M K, Marsico V E, Chabal Y J, Agarwal A, Eaglesham D J, Sapjeta J, Brown W L, Jacobson D C, Caudano Y, Christman S B and Chaban E E 1997 *J. Vac. Sci. Technol.* B **15**, 1065
[3] Reboh S, de Mattos A A, Barbot J F, Declemy A, Beaufort M F, Papaléo R M, Bergmann C P and Fichtner P F P 2009 *J. Appl. Phys.* **105** 093528
[4] Hochbauer T, Misra A, Nastasi M, Mayer J W 2002 *J. Appl. Phys.* **92** 2335
[5] Reboh S, Schaurich F, Declemy A, Barbot J F, Beaufort M F, Cherkashin N and Fichtner P F P 2010 *J. Appl. Phys.* **108** 023502
[6] Agarwal A, Haynes T E,. Venezia V C, Holland O W and Eaglesham D J 1998 *Appl. Phys. Lett.* **72** 1086
[7] Reboh S, Barbot J F, Beaufort M F and Fichtner P F P 2010 *Appl. Phys. Lett.* **96** 031907
[8] Fichtner P F P, Kaschny J R, Yankov R A, Mücklich A, Krei*B*ig U and Skorupa W 1997 *Appl. Phys. Lett.* **70** 732
[9] Ziegler J F, Biersack J P 2008 *SRIM computer code at the URL* http://www.srim.org
[10] Reboh S, Beaufort M F, Barbot J F, Grilhe J and Fichtner P F P 2009 *Proceedings of the First Workshop on the Use of in situ TEM/ion Accelerator techniques in the Study of Radiation Damage in Solids (Salford, UK, 18-20 June 2008)* edited by J A Hinks and S E Donnelly pp 13-21
[11] Reboh S, Beaufort M F, Barbot J F, Grilhe J and Fichtner P F P 2008 *Appl. Phys. Lett.* **93** 022106
[12] Tsai Y L and Mecholsky Jr J J 1991 *J. Mater. Res.* **6** 1248
[13] Hartmann M and Trinkaus H 2002 *Phys. Rev. Lett.* **88** 55505
[14] Weldon M K, Collot M, Chabal Y J, Venezia V C, Agarwal A, Haynes T E, Eaglesham D J, Christman S B, Chaban E E 1988 *Appl. Phys. Lett.* **73** 3721
[15] Moras G, Ciacchi L C, Elsässer C, Grumbsch P and De Vita A 2010 *Phys. Rev. Lett.* **105** 075502
[16] Rabier J, Pizzagalli L and Demenet J L 2010 *Dislocations at high stress in Dislocations in solids* vol 16 edited by J P Hirsch and L Kubin pp 49-98
[17] Gross D and Seilig Th 2006 *Fracture Mechanisms* (Springer Netherland)
[18] Seilig Th and Gross D 1999 *J. of Mechanical and Physic of solids* **47** 935
[19] Shao L, Di Z, Lin Y, Jia Q X, Wang Y Q, Nastasi M, Thompson P E, Theodore N D and Chu P K 2008 *Appl. Phys. Lett.* **93** 041909
[20] Chen P, Chu P K, Höchbauer T, Lee J K, Nastasi M, Buca D, Mantl S, Loo R, Caymax M, Alford T L, Mayer J W, Theodore N D, Cai M, Schmidt B and Lau S 2005 *Appl. Phys. Lett.* **86** 31904

Journal of Physics: Conference Series **281** (2011) 012023 doi:10.1088/1742-6596/281/1/012023

Point and extended defects in chemical vapour deposited diamond

H Pinto[1], R Jones[1], J P Goss[2] and P R Briddon[2]

[1]School of Engineering, Mathematics and Physical Sciences, University of Exeter, Stocker Road, Exeter, EX4 4QL, United Kingdom

[2] School of Electrical, Electronic and Computer Engineering, Newcastle University, NE1 7RU, United Kingdom

E-mail: pinto@excc.ex.ac.uk

Abstract. Homoepitaxially grown single crystal CVD diamond has a dislocation arrangement not seen in natural IIa diamond. Whereas in the latter, dislocations lie on (111) planes along $[1\bar{1}0]$ directions, dislocations in CVD diamond lie nearly along the [001] growth axis and are arranged in bundles with almost four fold rotational symmetry. Their large strain is easily seen in birefringence and we model the arrangements in the bundle to account for the singular behaviour of the birefringence when the polarisers are rotated with respect to the crystal. The optical absorption spectrum of brown CVD diamond displays a continuum similar to that found in brown IIa natural diamonds but in addition broad point defect induced bands. We investigate the suggestion that the 2.38 eV (560 nm) peak is related to *V*NH. We suggest that the breadth of these bands is related to the large structural change when the defect is excited.

1. Introduction

Natural brown type IIa diamonds owe their colour to multi-vacancy defects introduced by plastic deformation [1, 2, 3]. Positron annihilation studies show that clusters of perhaps 50 vacancies exist and are optically active [4]. The clusters have also been seen by TEM experiments [5]. These clusters are not found in diamonds which have been treated to remove the colour or in naturally colourless diamonds. Although CVD diamonds are often brown, positron annihilation experiments do not reveal the presence of large vacancy clusters. In CVD brown diamonds, the brown colour is lost following anneals at temperatures lower than that needed for natural brown diamonds. It may be that the colour arises from vacancy disks [3] or graphitic forms of carbon.

In type IIa diamond, dislocations are often found on {111} planes and lie along ⟨110⟩ direction. This is not the case in CVD diamonds where they lie in bundles close to the (001) growth axis. Previous modelling [6] has shown that the cores of individual dislocations can be reconstructed and the straight dislocation along (001) has a competitive core energy with dislocations with zig-zag segments lying along [101] and [011]. Many of these grown-in dislocations occur in bundles, and such is their strain that they are easily seen in birefringence. Remarkably, the birefringence exhibits an approximate four fold rotational symmetry which implies that the same numbers of dislocations are found in each of the four (110) quadrants. In section 2 we review possible arrangement of the dislocations in the bundle.

The absorption spectrum for a brown CVD diamond not only shows the continuous absorption seen in natural browns but also broad absorption bands around 1.4, 2.38, 3.44 and 4.5 eV. The last is due to neutral substitutional nitrogen centres [1, 7]. Of particular interest is the band at 2.4 eV which can result

Figure 1. Birefringence seen in crossed-polarisers of two 4-petalled defects in CVD diamond. The plane of view is (001). The horizontal and vertical axes lie along ⟨100⟩ and the polarisers lie along ⟨110⟩. Figure provided by P. Martineau.

in a pink colouration of the diamond. We discuss in section 3, the assignment of this absorption band and why their spectra are so broad.

2. Birefringence of dislocation bundles in CVD diamond

Homoexpitaxial CVD diamond contains bundles of edge and mixed dislocations of density about $\sim 10^5$ cm^{-2}, lying approximately along the [001] growth direction [8]. Although the dislocations lie close to the (001) directions, they fan out at an angle of a few degrees in ⟨110⟩ directions. Detailed studies using X-ray topography show that, in many cases, the dislocations are observed in groups of four or more lines diverging from a defect on or near the substrate. Where only four lines are observed to emanate from a given point, the dislocations deviate from [001] in such a way that their projections onto the (001) plane lie along ⟨110⟩ Often, those with projections along [110] or $[\bar{1}\bar{1}0]$ have $[1\bar{1}0]$ Burgers vectors, and those with projections along $[1\bar{1}0]$ or $[\bar{1}10]$ have [110] Burgers vectors. These dislocations are predominantly of edge type [9]. Diamonds containing such bundles exhibit a birefringence with a few strikingly bright regions with a characteristic pattern consisting of four approximately equally bright petals as shown in figure 1. There is a close connection between these four petalled defects and dislocations. X-ray topographs reveal that the regions where birefringence is seen, coincides with dense bundles of dislocations [9].

The four petalled pattern shown in figure 1 is not invariant under a rotation of the crystal about [001], *i.e.*, the pattern changes with the directions of the crossed polarisers (figure 2). Thus, the strain field is linked to the underlying crystallography as expected from a dislocation model and is not the case for a spherical inclusion. The additional lobes seen when the polarisers are parallel to ⟨100⟩ are weaker than the bright lobes seen for the case when the polarisers lie along ⟨110⟩.

The spatial extent of the birefringence in the (001) plane is about 13% of the crystal size or about 300-400 μm, given that the crystal is about 3 mm in size. Clearly, the birefringence of a single dislocation does not extend over this distance, and the resultant strain due to a bundle of many dislocations must be considered.

We note that the dislocations arise only in the epilayer and not the underlying substrate. Since dislocations cannot terminate within the crystal, the bundle must be composed of dislocation dipoles with zero net Burgers vector. We have found [10] that the four petalled pattern arises from a compact defect on the substrate surface with dimensions less than 20 μm. We discuss here possible arrangements

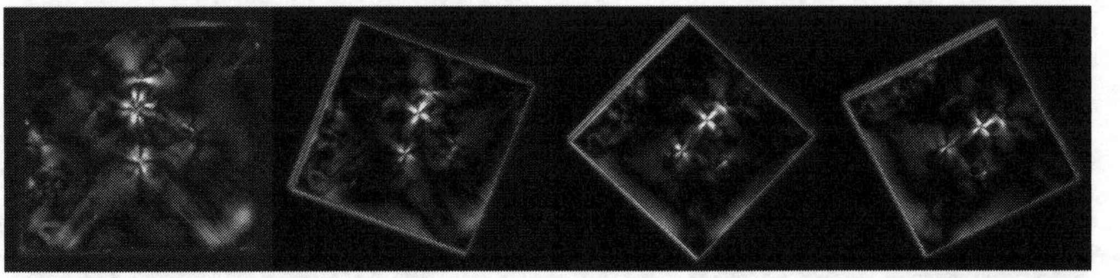

Figure 2. Birefringence of a petalled defect when the crystal rotated with respect to the cross-polarisers by 0, 22, 45, and 67° from ⟨100⟩. Note 8 dark arms are seen when a polariser is parallel to ⟨100⟩. Exposure times are 1.9, 0.5, 0.23, and 0.37 s revealing a weaker birefringence in the case when the polarisers are along ⟨100⟩. Figure provided by P. Martineau.

for the dislocations and their bundles to account for the experimentally observed birefringence.

2.1. Birefringence of a single edge dislocation

The strain fields of a bundle of dislocations can be found using isotropic elasticity theory by summing the strains due to individual edge dislocations. The principal values of the strain tensor and their directions are then calculated. The phase difference between the two polarised components of the light ray at the exit of the crystal is

$$\delta = 2\pi(n_1 - n_2)t/\lambda, \tag{1}$$

where n_1 and n_2 are the refractive indices of the crystal along the principal directions of strain, λ is the wavelength (550 nm), and t is the optical thickness of the crystal taken to be 2.52 mm. The refractive indices are related to the strain through the strain-optic coefficients p_{ij}. For stressed cubic materials,

$$n_1 - n_2 = -n^3(p_{11} - p_{12})(\epsilon'_{11} - \epsilon'_{22})/2, \tag{2}$$

where ϵ'_{ii} are the principal values of the strain tensor and n the refractive index of diamond.

We have previously calculated $p_{11} - p_{12}$ to be −0.3 [11], and this compares favourably with experiment [12].

Thus, light polarised along the principal direction associated with ϵ'_{11} moves with the slower speed if $\epsilon'_{11} > \epsilon'_{22}$. This direction corresponds with tensile strain. We note that δ is simply related to the shear strain $(\epsilon'_{11} - \epsilon'_{22})/2$ and hence a map of δ gives the shear strain field in the (001) plane. The intensity of birefringence is given by $I = E^2 \sin^2(2\phi)\sin^2(\delta/2)$ where ϕ is angle between a principal axis of strain and a polariser [13].

For an edge dislocation lying along z, with the Burgers vector of magnitude b, taken to be $a/\sqrt{2}$, and parallel to x, the strain tensor ϵ_{ij} and its principal values ϵ'_{ii} are well known and lead to a shear strain of [14]

$$\epsilon'_{11} - \epsilon'_{22} = -2Ax/r^2, \qquad \theta = \frac{1}{2}\tan^{-1}\{(y^2 - x^2)/(2xy)\}. \tag{3}$$

Here $A = b/(4\pi(1 - \nu))$, ν is Poisson's ratio, $r^2 = x^2 + y^2$ and θ the angle between the Burgers vector and a principal direction of strain. We note that $n_1 - n_2$ decreases with distance from the core as $1/r$.

The strain field, proportional to δ, and birefringence of a single edge dislocation with Burgers vector along [100] are shown in figure 3 for the cases when the polariser is along [100] and [110]. The birefringence reveals a characteristic dipole pattern which turns into a four petalled pattern when the crossed polarisers are turned through 45°. The patterns can be simply understood from the principal strains and their directions given above. However, the shear strain map and birefringence of a single edge dislocation cannot describe the four petalled defect and alternative models must be considered.

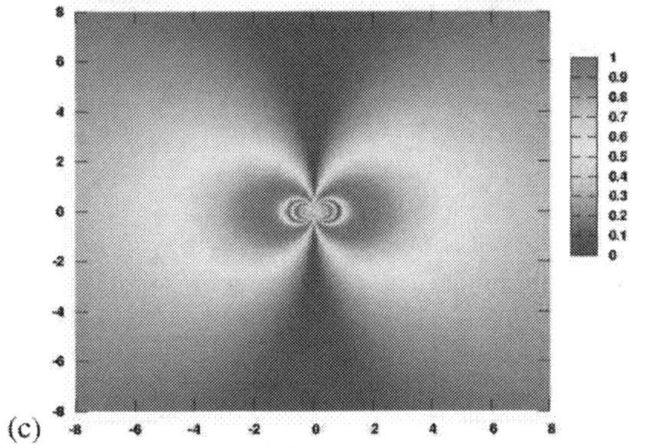

Figure 3. (a, b) Birefringence and (c) map of $|\sin(\delta)|$ in the (001) plane for a single edge dislocation lying along z with Burgers vector along x. In (a) the polariser lies along x and at 45° to x in (b). Note the dipole character in (a) and four petalled form in (b). Dark arms lie along y for both cases. (c) δ-map showing dipole character and dark arms along y. The horizontal and vertical axes are in μm. Colour version online.

2.2. *The four petal defect*

X-ray topography shows that the petalled defect is associated with groups of edge dislocations emanating from a defect on or near the substrate surface and fanning out along $\langle 110\rangle$ directions. Many of these dislocations have $[\overline{1}10]$ Burgers vectors and lie about 3-5° off [001] in the (110) plane, while those with $[01\overline{1}]$ Burgers vectors lie at a similar angle to [001] but in the $(01\overline{1})$ plane [9]. Here we shall investigate the strain field of the simplest arrangement having this structure consisting of a linear array of dislocations in each quadrant. We note that this arrangement has a four fold rotational axis and such a symmetry is necessary to explain the strain field of the four petalled defect shown in the lower part of figure 4. The other four petalled defect lying in the upper part of figure 4 possesses two fold rotational symmetric strain field which has been discussed earlier [10].

We placed 20 edge dislocations in the first quadrant all lying along $[\sin\chi, \sin\chi, \cos\chi]$ and intersecting the (001) plane at equidistant points along [110] as indicated schematically in figure 5(a). The dislocations all have the same $a/2[1\overline{1}0]$ Burgers vector. The separation of individual dislocations in each arm is 0.1 μm. Dislocations in the other three quadrants are found by rotating those in the first quadrant.

The dislocations introduce additional half-planes outside the core of the defect, which produce a net compression along the tangential direction and a radial tensile field. This is necessary to reproduce the radial slow axis shown in figure 4.

Figure 5 shows the calculated strain field and birefringence. The strain map, figure 5(b), reveals a four petalled pattern with dark arms corresponding with zero shear strain along $\langle 110\rangle$ as indeed observed. Since δ is zero along the four $\langle 110\rangle$ directions, it follows that the birefringence vanishes along these directions for all directions of the polariser. This accounts for the four fold pattern seen in birefringence for polarisers lying other than along $\langle 100\rangle$. For points along say [100], we find that the principal axes

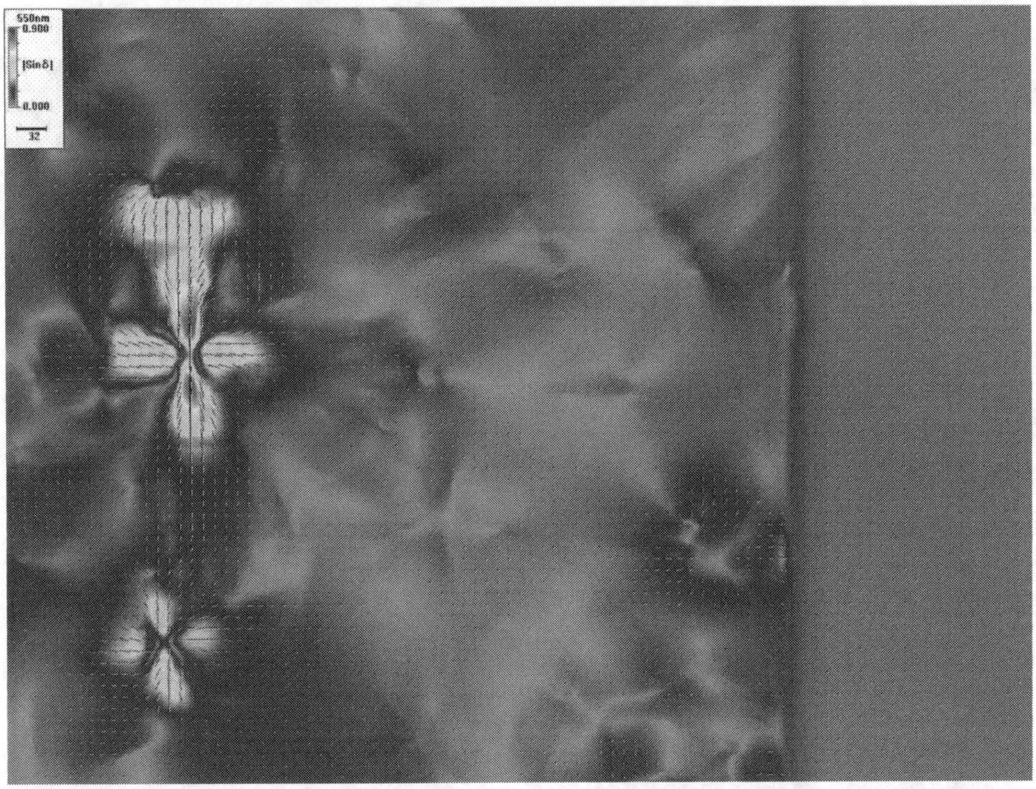

Figure 4. Phase difference $|\sin(\delta)|$ between the two polarised waves at the exit of the specimen of thickness 2.52 mm showing the four petalled defect. The dashed lines parallel to the [100] and [010] edges of the specimen indicate the principal directions of strain for the slow direction of light propagation. Note that for points along [100], the direction of the slow axis is also along this direction. Figure provided by P. Martineau. Colour version online.

of strain also lie along $\langle 100\rangle$ directions and in fact the tensile axis lies along this [100]. This is also in agreement with the experiment as shown in figure 4.

Consider now the birefringence for polarisers lying along $\langle 100\rangle$. In this case, the four $\langle 100\rangle$ directions are also dark as these directions coincide with the principal axes of strain. Figure 5 shows the birefringence map for this case and we find 8 arms in agreement with observation. The long exposure time required to reveal this structure is also in agreement with the theory as shown when figure 2 and figure 5 are compared. Finally, we note that the extent of the birefringence is around 300 μm and consistent with the observations.

2.3. Summary

We have analysed the strain field and birefringence of several linear arrays of dislocations which could be responsible for a number of petalled defects seen in single crystal CVD diamonds. The strain field of a single edge dislocation as well as its birefringence cannot explain the magnitude of the birefringence associated with the four petalled defect. The birefringence of four linear arrays of edge dislocations located on four arms lying along $\langle 110\rangle$ directions with $\langle\bar{1}10\rangle$ Burgers vectors, leads to a strain field with four bright petals along $\langle 100\rangle$ as observed. The birefringence associated with the defect is sensitive to

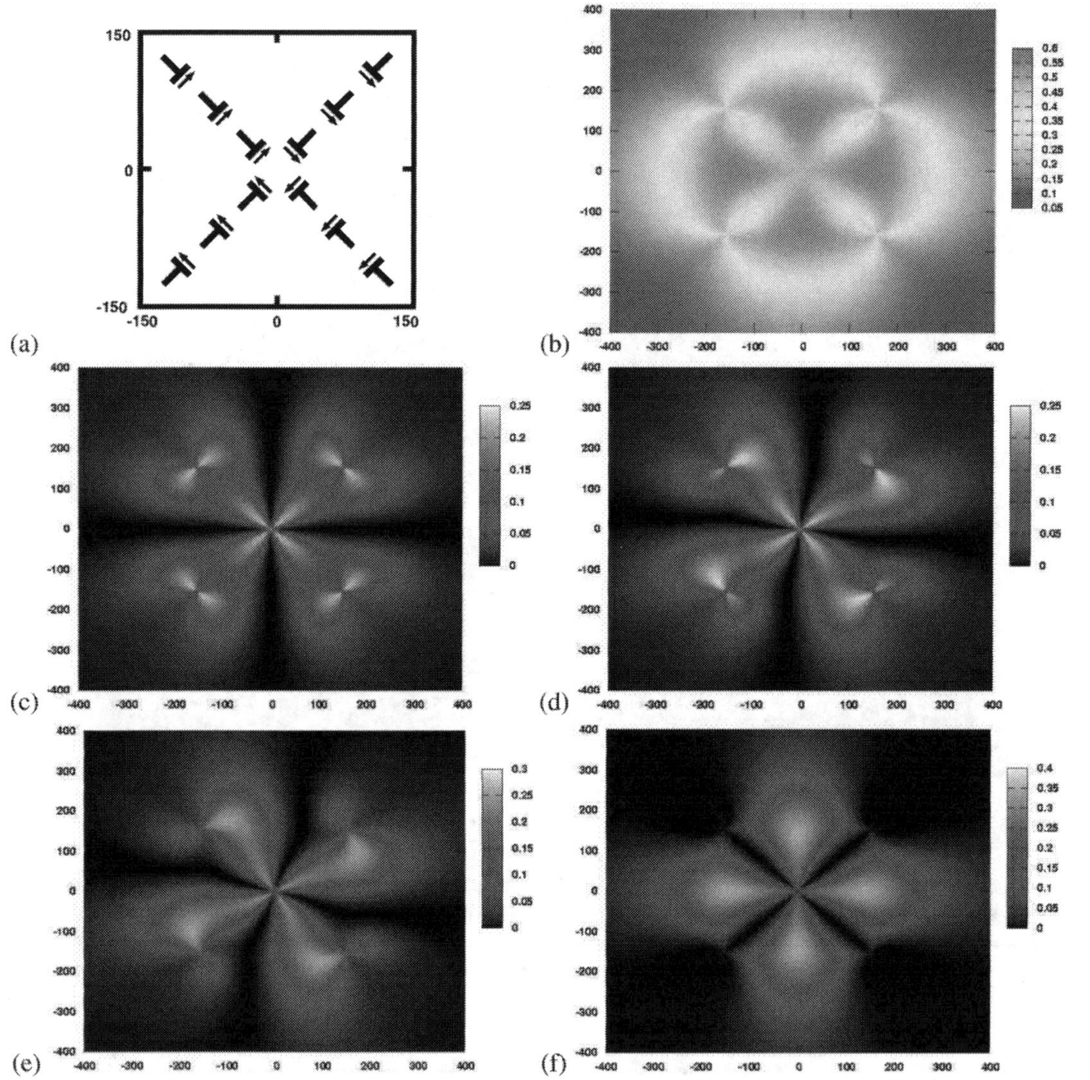

Figure 5. The calculated strain field and birefringence of the four arm defect. (a) Projection of dislocation lines onto (001). The dislocations are inclined at angle χ = 3.5° to [001] and are separated by 0.1 μm. The projection shows the four fold symmetry of the bundle and the directions of the Burgers vectors. (b) Delta or strain map of the bundle, (c-f) birefringence for polarisers at angles of 0, 15, 30, and 45° to [100]. Note the extent of the birefringence is about 300 μm and that there is a weaker birefringence when the polarisers are parallel to [100] than at an angle of 45° with [100]. The birefringence shows 8 bright lobes in the first case and 4 in the second. Scale along x and y is in μm.

the directions of the polarisers. If these are along ⟨110⟩, then four bright lobes are seen also along ⟨100⟩. if the polarisers are parallel to ⟨100⟩, then eight weak lobes are seen. Such patterns reflect the four fold symmetry of the defect. With the arrangement of dislocations and Burgers vectors shown in figure 5(a), with the extra half planes of atoms in each dislocation lying outside the core, the principal axis of strain associated with the slower polarised ray is radial, as observed. The birefringence is seen to extend to

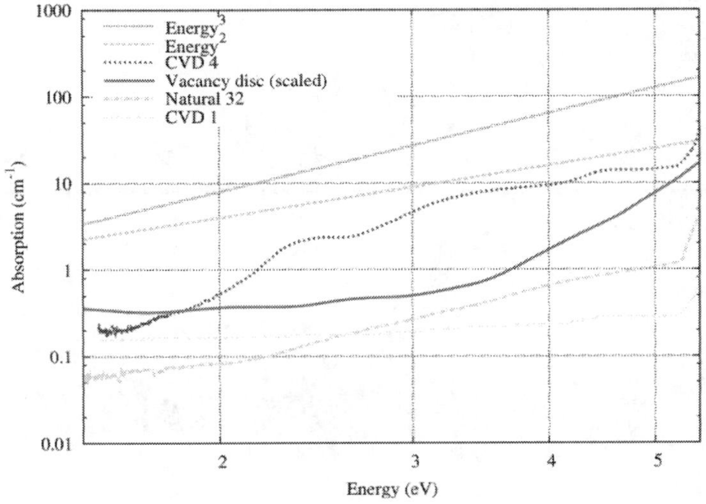

Figure 6. Observed absorption spectra (log-log scale) of two CVD brown diamonds and vacancy disk thought to relate to natural brown diamonds [3]. Note additional point defect related broad absorption peaks at 2.38, 3.44 and 4.5 eV in the browner CVD sample. Colour online.

about 300 μm and this requires about 20 dislocations in each arm or 80 dislocations in the bundle with a separation about 0.1 μm. This the diameter of the the defect at its core is about 4 μ m.

The shear strain, averaged over the epilayer thickness, introduced by these dislocation arrays and along the cube axes is about 1.2×10^{-5} leading to a stress of about 7 MPa at 180 μm from the core and which decays as $1/r$ where r is the distance to the core. The model of the arrays of dislocations, in particular the four fold symmetry, have implications for the defect possibly lying on the substrate which act as a source for the arrays.

3. Absorption peaks in CVD diamond

Figure 6 shows the absorption spectrum of a brown CVD diamond. In addition to the continuum responsible for the brown colour in natural IIa diamonds, broad absorption peaks are detected at 2.38, 3.44 and 4.5 eV.

The peak at 4.5 eV is due to neutral substitutional nitrogen [15] while the 3.44 eV peak has been related to negatively charged nitrogen-hydrogen vacancy centres ($V\mathrm{NH}^-$) [16]. This centre is after N_s^0, the most dominant one detected by electron paramagnetic resonance [17]. It exhibits photo- and thermo-chromism [16] as an anneal around 800 K leads to the loss of neutral nitrogen donors as well as the intensities of the 2.38 eV (520 nm) and 3.44 eV (360 nm) bands. Following this anneal, the concentration of $V\mathrm{NH}^-$ as measured by EPR [17] and the intensity of a broad band centred at 1.37 eV (900 nm) increases. We review here the calculations of the optical properties of these defects carried out previously [18], and we investigate the origin of the lack of a zero phonon line and the breadth of the 3.44 eV and 1.37 eV bands.

Figure 7 shows the structure of the $V\mathrm{NH}$ defect along with that of $V\mathrm{N}_2$. It is clear that the two defects are very similar. Further, the band structures of $V\mathrm{NH}^-$ and $V\mathrm{N}_2^-$, shown in figure 8 are strikingly similar with each defect having three gap levels.

Although density functional theory is unable to describe optical transition energies accurately, we have used the errors in the transitions for $V\mathrm{N}_2$ to correct the transitions for $V\mathrm{NH}$. In a similar way, we can correct the acceptor level of $V\mathrm{NH}$ from the error we make in the same level for $V\mathrm{N}_2$.

We found [18] the spin 0 state is more stable than spin 1 for $V\mathrm{N}_2$ and $V\mathrm{NH}$. The vibrational modes of neutral ($S = 0$) and negative $V\mathrm{NH}$ defects are listed in table 1. The value for $V\mathrm{NH}^0$ in the $S = 0$ configuration is in good agreement with one observed at 3123 cm^{-1} [16, 19] associated with $V\mathrm{NH}^0$ and

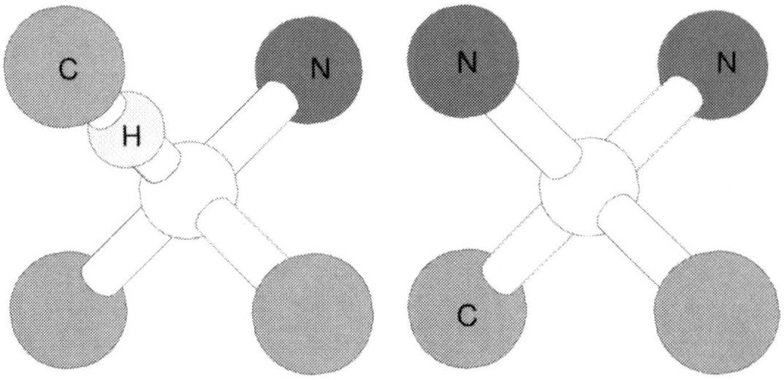

Figure 7. VNH and VN_2 defects. Note that if the H atom was moved to lie on top of its carbon neighbour, the defect would have the same structure and symmetry and be isoelectronic with VN_2.

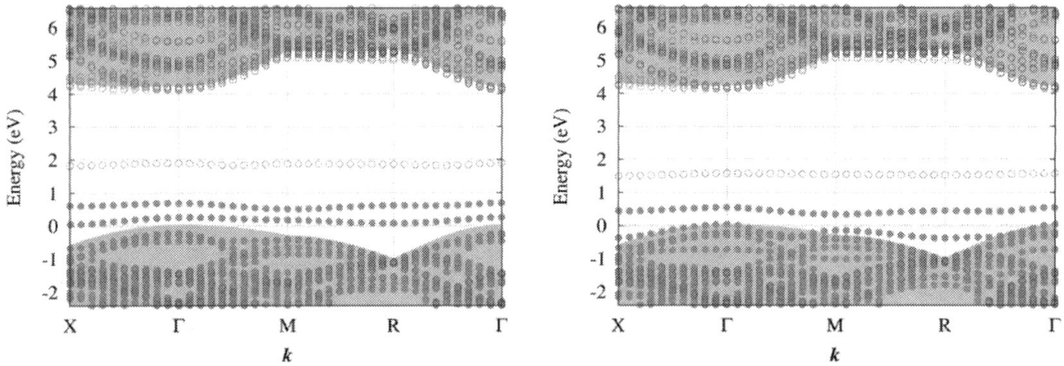

Figure 8. Band structures along high-symmetry paths in the Brillouin-zone and in the vicinity of the band gap with E_v aligned to zero for the spin singlet configurations of VN_2^0 (left), and $V\mathrm{NH}^0$ (right). Filled and empty symbols indicate filled and empty bands, respectively, with the underlying shaded area indicating the bands in a defect-free supercell of the same size. Note there are three localised gap levels.

Table 1. C–H stretch mode frequencies (cm^{-1}) calculated for VNH in the neutral and negative charge states. These values are obtained using the GGA.

Defect	$q = 0$, $S = 0$	$q = 0$, $S = 1$	$q = -1$
VNH	3190	2858	2941

supports a spin-singlet ground state of VNH.

For VN_2, which has C_{2v} symmetry, the three gap levels have in ascending energy, b_1, a_1 and b_2 symmetry. Although, VNH does not have C_{2v} symmetry, we shall refer to the gap levels as b_1, a_1 and b_2. The two highest in all three defects, a_1 and b_2, originate from bonding and anti-bonding orbitals made up primarily of dangling bonds on the two carbon radicals, The lowest b_1 level is localised on the N atoms while the level arising from C-H falls deeper in the valence band.

Both defects support internal $a_1 \rightarrow b_2$ optical transitions in the neutral (process I) and negative (process II) charge states as well as charge transfer transitions. The $b_1 \rightarrow b_2$ transition is dipole forbidden. Thus, each defect is expected to possess at least four optical transitions as charge transfer can take place by either by exciting a valence electron to a b_2 level for the neutral defect (process III),

Table 2. Calculated and experimental transition energies (see figure 9).

Defect	E_1	E_2	E_3
Calculated values, eV			
VN_2	1.297	0.969	2.7
VNH	1.129	0.84	2.3
Corrected value, eV			
VN_2	2.46	1.26	3.4
VNH	2.3	1.1	3.0
Experimental values, eV (nm)			
VN_2	2.46 (504)	1.26 (984)	3.4 (365)
VNH	2.38 (520)	1.37 (900)	

or exciting a electron from the b_2 level to the conduction band for the negatively charged defect (process IV). These four lowest energy optical transitions are schematically shown in figure 9.

The internal transition energies for neutral (E_1) and negatively charged (E_2) VN_2 are found by evaluating the total energies for the two single particle determinantal wave functions. These yield 0.97 eV for VN_2^- and 1.3 eV for neutral VN_2. The former is in reasonable agreement with experimental value of 1.26 eV for the H2 optical centre [20] but the latter is in poor agreement with the experimental value of 2.46 eV [21]. We suppose this is because of errors in the correlation energy. As discussed above, we correct the energies for VNH by adding the error in the energies for VN_2. Table 2 gives the corrected transition energies.

Previously we used density functional theory to place the acceptor level of the VN_2 defect at $E_v+2.7$ eV [22]. This can be compared with an experimental value of $E_v+3.4$ eV based on a study of the photochromacity of VN_2 [23]. This energy is close to a weak absorption band of VN_2^0 labelled H13 [24]. Hence, the H13 absorption band is due to a charge state change involving the transfer of a electron from the valence band to the defect as in $VN_2^0+\nu \rightarrow VN_2^- + h$. The acceptor level of VNH can also be evaluated and is given in table 2. Similarly we used the error found for VN_2 to correct the acceptor level of VNH, table 2.

4. Summary

In summary, VNH has four optical transitions shown schematically in figure 9. Each charge state has an internal transition and we label these transitions as processes I and II. There are also two transitions involving the valence or conduction band and a change in the charge state either by excitation of a valence electron creating VNH^- (process III) or by ionisation of VNH^- with electron emission to the conduction band (process IV).

The acceptor level of the defect establishes the minimum energy required for a charge state change from neutral to negative leading to a free hole. For a photochromic effect, the hole must be trapped by a nearby different defect. The defect then remains in the negative charge state and the internal absorption due to the neutral one is bleached. Thus, the acceptor level referenced to the valence band top is the bleaching energy for the neutral defect. Recovery can take place by the reverse process, or by exposure to light enabling electron emission from the negative defect to the conduction band. In the latter case, the minimum energy for recovery of the neutral defect is the distance of the acceptor level to the conduction band. These energies are known for VN_2 and are given in table 2. These experiments have shown that H2 is bleached with light of energy greater than 2.06eV (600nm) (process IV), and recovers with intense light of energy 3.4eV (365 nm, process III) [23]. The sum of the energies being the band gap

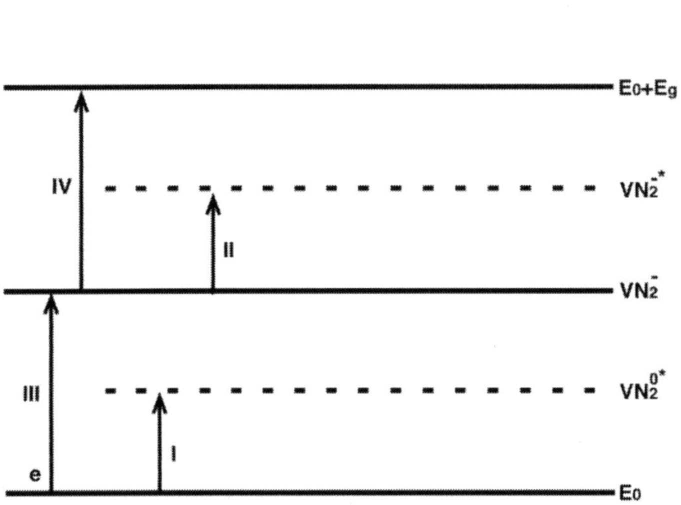

Figure 9. Configuration energies of VN_2 and VNH defects showing the five states involved in the visible and UV optical transitions. The lowest line denotes the ground state energy $E = E_0$ of the crystal with a neutral defect. The next line denotes the energy of the ground state crystal and an internally excited neutral defect VN_2^{0*}. The third is the energy of the crystal with a free hole and the charged defect VN_2^- in its lowest energy state. The fourth represents the energy of the crystal with a hole and the internally excited charged defect VN_2^{-*}. The top line is the energy of a free electron in the conduction band and a hole in the valence band with the defect neutral and in its lowest energy state, *i.e.* the band gap energy $E_0 + E_g$. The energies of processes I, II and III are denoted by E_1, E_2 and E_3. E_3 is just the acceptor level of the defect referred to the valence band top. Assuming weak lattice relaxation, process IV has energy $E_g - E_3$.

confirms that the photochromic effect is due to a charge state change with an acceptor level at $E_v + 3.4$eV. Importantly, the effect is not limited by ionisation of the trap which is probably nitrogen. In principle, observations of a photochromic effect in VNH could yield its acceptor level but, from the similarity of its properties with VN_2, we expect such a level to be close to that of VN_2.

Table 3. Vertical absorption and emission energies as well as zero phonon lines (ZPL) calculated for VN^-, VN_2^- and VNH^-.

	Vertical absorption (eV)	Vertical emission (eV)	ZPL (eV)
VN^-	1.95	1.55	1.74
VN_2^-	1.36	0.87	1.08
VNH^-	1.12	0.52	0.80

It is clear from figure 6 that the absorption bands due to VNH are very broad. This is reminiscent of colour centres in alkali halides where it is related to a large structural change when the defect is excited and shown schematically in figure 10.

We have examined the effect of relaxation of the excited state on the vertical absorption and emission energies of VNH^-, VN^- and VN_2^- found by calculating the total energy differences between the ground and excited states when the structures are frozen or when relaxed. These are given in Table 3. For VN^-

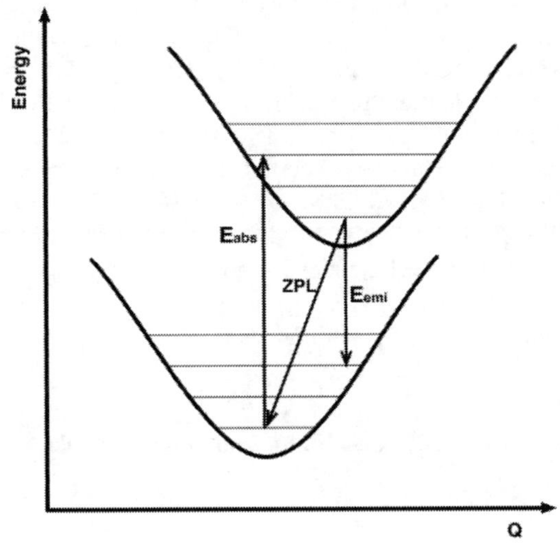

Figure 10. Configuration energy diagram showing schematically the ground and excited states of *V*NH. The peaks in absorption and emission are given the vertical transitions E_{abs} and E_{emi}. The zero-phonon line is the transition energy between the two minima. Its intensity decreases and the spectra broaden as the two minima move further apart.

and VN_2^- these energies differ by 0.4 and 0.5 eV which account for a weak zero-phonon line and broad vibronic sidebands. In VN^-, the integrated intensity of the ZPL is only 3% of the overall absorption. For VNH^-, the effect is larger as the difference between the absorption and emission energies is now 0.6 eV. This we expect a weaker zero-phonon line which could explain why it is not seen. Sharp line luminescence is not detected from VNH^- and our theory suggests a broad spectrum starting around 0.5 eV.

5. Discussion

We find VN_2 and *V*NH posses three localised gap levels. The upper pair are related to bonding and antibonding orbitals on the two C radicals and the lower one comes from the N lone pair and the C-H unit. The positions of the highest occupied and lowest unoccupied levels of VN_2 and *V*NH are very similar and it follows that the optical activity of the three defects is also similar and derived from the properties of the two carbon atoms having dangling bonds. We have 4 electrons to distribute among these levels for the neutral defect and 5 electrons for the negatively charged defects. The ground and excited states of VN_2 have symmetry $^1A_1$ and $^1B_1$ respectively but their energy difference is calculated to be far too small. This reflects the fragility of the theory in treating excited states. In the negatively charged defect, the ground and excited states have $^2B_1$ and $^2A_1$ symmetry and their calculated energy difference is in fair agreement with the observed H2 transition.

We expect VN_2 and *V*NH to possess similar optical properties to VN_2, and in particular similar internal optical transitions in the two charge states and similar acceptor levels. With this proviso we can discuss the absorption bands seen in CVD diamond. Table 2 shows that the two broad bands at 2.38 eV (520 nm) and 1.37 eV (900 nm) are internal transitions in *V*NH and VNH^-. This is supported by the bleaching and recovery data, which show these bands to be anti-correlated [16].

The 360 nm band is more contentious. It lies close to that required for a charge state change of VNH^0 (process III) and should be analogous with the H13 band in VN_2^0. However, it seems that this band does not anneal out at the same time as the 2.38 eV (520 nm) band indicating that a distinct defect is involved. Nevertheless, we recall that the intensity of the H13 band of VN_2 at 364 nm is weak, and we suppose that this is also true in *V*NH so that the broad 360 nm band hides the weak absorption where VNH^0 changes charge state. We remark that the existence of VNH^- shows the defect has an acceptor level, and must absorb light with an energy close to this level. The weakness of the transitions may be because the dipole matrix element is between an extended state in the valence band, and the localised a_1 state in the gap and this is smaller than the overlap between two localised gap states, as is involved in an internal transition. In the case of VN_2, the charge state change has been investigated optically with intense light of energy 600 nm and 364 nm (2.07 and 3.4 eV). Similar energies should effect photochromic behaviour in VNH.

Finally, we note that the optical bands of VNH are anomalously broad. It appears that there a larger structural relaxation when VNH is excited than VN_2^- or VN^-. This would certainty broaden the spectra as the intensity of the ZPL relative to the broad vibronic side band would be $\exp(-\Delta E/\hbar\omega)$. Here ω is a typical phonon frequency taken to be ~ 500 cm^{-1}. This gives the Huang-Rhys factor of 3.7 required to give the ZPL intensity of VN^-. ΔE is the difference between the vertical absorption energy and the ZPL. Thus, the fractional intensities of the ZPLs are 3%, 1% and 0.4% for VN^-, VN_2^- and VNH^-, respectively. Hence, broad absorption is to be expected for VNH^- with a weak zero phonon line.

6. Acknowledgements

L Hounsome, N Fujita, P M Martineau, M Gaukroger, D Fisher, R Khan and U Bangert are thanked for many valuable discussions.

References

[1] Jones R, 2009 *Diamond Relat. Mater.* **18** 820–826
[2] Fisher D, Evans DJF, Glover C, Kelly CJ, Sheehy MJ and Summerton GC, 2006 *Diamond Relat. Mater.* **15** 1636–1642
[3] Hounsome LS, Jones R, Martineau PM, Fisher D, Shaw MJ, Briddon PR and Öberg S, 2006 *Phys. Rev. B* **73** 125203
[4] Mäki JM, Tuomisto F, Kelly CJ, Fisher D and Martineau PM, 2009 *J. Phys. Cond. Matter* **21** 364216
[5] Bangert U, Barnes R, Gass MH, Bleloch AL and Godfrey IS, 2009 *J. Phys. Cond. Matter* **21** 364208
[6] Fujita N, Blumenau AT, Jones R. amd Öberg S and Briddon PR, 2007 *Phys. Status Solidi A* **204** 2211–2215
[7] Nazaré MH and Neves AJT, 1987 *J. Phys. C* **20** 2713–2722
[8] Martineau PM, Lawson SC, Taylor AJ, Quinn SJ, Evans DJF and Crowder MJ, 2004 *Gems Gemol.* **40** 2
[9] Gaukroger MP, Martineau PM, Crowder MJ, Friel I, Williams SD and Twitchen DJ, 2008 *Diamond and Related Materials* **17** 262–269
[10] Pinto H and R. J, 2009 *J. Phys. Cond. Matter* **21** 364220
[11] Hounsome LS, Jones R, Shaw MJ and Briddon PR, 2006 *Phys. Status Solidi A* **203** 3088–3093
[12] Lang AR, 2009 *Diamond and Related Materials* **18** 1–5
[13] Born M and Wolf E, 2001 *Principles of Optics*, 7th edition, p 824 , Cambridge
[14] Bullough R, 1958 *Phys. Rev.* **110** 620–623
[15] Jones R, Goss JP and Briddon PR, 2009 *Phys. Rev. B* **80** 033205
[16] Khan RUA, Martineau PM, Cann BL, Newton ME and Twitchen DJ, 2009 *J. Phys. Cond. Matter* **21** 36
[17] Glover C, Newton ME, Martineau P, Twitchen DJ and Baker JM, 2003 *Phys. Rev. Lett.* **90** 185507
[18] Goss JP, Briddon PR, Pinho H and Jones R, 2010 *Phys. Status Solidi A* **207** 2049–2053
[19] Fuchs F, Wild C, Schwarz K and Koidl P, 1995 *Diamond Relat. Mater.* **4** 652–656
[20] Lawson SC, Davies G, Collins AT and Mainwood A, 1992 *J. Phys. Cond. Matter* **4** 3439–3452
[21] Davies G, Nazaré MH and Hamer MF, 1976 *Proc. R. Soc. London, Ser. A* **351** 245–265
[22] Goss JP, Briddon PR, Jones R and Sque S, 2003 *J. Phys. Cond. Matter* **15** S2903–S2911
[23] Mita Y, Nisida Y, Suito K, Onodera A and Yazu S, 1990 *J. Phys. Cond. Matter* **2** 8567–8574
[24] Collins AT, 1983 *J. Phys. C* **16** 6691–6694

Journal of Physics: Conference Series **281** (2011) 012024 doi:10.1088/1742-6596/281/1/012024

An analysis of vacancy clusters and sp2 bonding in natural type IIa diamond using aberration corrected STEM and EELS

I S Godfrey[1,2] and U Bangert[1]
[1]School of Materials, The University of Manchester, Manchester M1 7HS, UK
[2]SuperSTEM Laboratory, STFC Daresbury Laboratory, Warrington, WA4 4AD, UK

Email: Iain.Godfrey@postgrad.manchester.ac.uk

Abstract. The link between brown colour and vacancy structures in natural diamond and has been studied through the techniques of aberration corrected scanning transmission electron microscopy (AC-STEM) and electron energy loss spectroscopy (EELS). Bright field (BF) and high angle annular dark field (HAADF) STEM imaging modes have revealed discrete patches of stronger contrast, of the order of 1nm in size, that are similar to simulated images of vacancy clusters. Core loss spectra show a pre-edge feature for brown diamond corresponding to π^* states. This feature is absent for colourless and heat treated brown diamonds. Additional spectra acquired in regions containing dislocations for both brown and white (colourless) diamonds showed a second pre-edge feature indicating that dislocations have electronic states associated with them and are therefore optically active.

1. Introduction

Type IIa natural diamond is noted for having an extremely low concentration of impurities. In fact the trace amounts of nitrogen (<1ppm) that have been reported in this material are undetectable with IR spectroscopy [1]. Therefore we must look to other crystallographic anomalies to explain the remarkable difference in optical absorption that is observed between brown and colourless diamonds. From transmission electron microscopy (TEM) studies we know that natural diamonds contain many dislocations that have been introduced by geological processes [2]. However it has been demonstrated that these extended defects are not directly responsible for brown colour as they occupy the structure of both brown and colourless diamonds in similar concentrations; although the question still remains as to whether certain types of dislocations are optically active. Further evidence that dislocations are not responsible for brown colour is given by the fact that heat treatment at ~2500°C under a stabilizing pressure of 7GPa removes the brown colour but not the dislocations [3].

In a quest to identify the structural anomaly responsible for the brown colour, density functional theory (DFT) has shown that another extended defect, the spherical vacancy cluster, has the correct electronic properties to allow optical absorption to occur [4]. The motion of jogged dislocations (intersecting dislocations having different slip systems) yields vacancies which coalesce into stable spherical clusters that are lined with pi-bonds. The bonding between the carbon atoms in the localized regions of the vacancy clusters will be sp^2 hybridised and this should be detectable with EELS. It has been demonstrated that the vacancy cluster has a low formation energy and an optical absorption similar to that of brown diamond [5]. In addition, positron annihilation spectroscopy (PAS) studies have shown that the spherical vacancy cluster acts as a deep trap for positrons, which explains why the average positron lifetime is longer for brown diamond than for colourless [6-7].

The aforementioned techniques provide sufficient evidence to support the link between vacancy clusters and brown colouration in type IIa diamonds and negate the possibility that other extended defects composed of carbon interstitials, (such as those observed in nitrogen containing type 1a diamond), are the dominant defect structures. Ab initio density functional theory has shown that extended interstitial defects can arise in diamonds containing sufficient quantities of nitrogen (~0.2 at. %) but not in chemically pure type IIa diamond [8]. Substitutional nitrogen plays an important role in the trapping of vacancies and the growth of platelet defects. Moreover, interstitial defects do not result in the formation of large quantities of dangling bonds and hence the sp^2 bonded structures that are evident in brown diamond.

Aberration corrected STEM and EELS are powerful techniques for studying vacancy defects and their associated electronic states. The excellent spatial resolution (<0.1nm) and small depth of focus (~8nm) of the STEM probe allow vacancy clusters to be imaged via BF and HAADF imaging modes. The signal for the BF image originates from the central diffraction disk. Interpreting the contrast in BF images can be difficult as the observed lattice fringes can undergo a reversal of sign with changing crystal thickness and defocus conditions. Additionally, variations in electron phase at defects can distort the structural information of imperfect crystals. As the name suggests the HAADF signal is derived from electrons that are scattered to high angles. The scattering phenomenon is Rutherford like in nature and is approximately proportional to Z^2, i.e. the HAADF signal increases with atomic number and thickness. Although HAADF images are easier to interpret because of the Rutherford scattering behaviour, unexpected contrast can sometimes be observed due to high-angle scattering contributed by phonons and Bloch wave interference effects [9].

Core-loss EELS is sensitive to electronic states originating from defects and the pre-edge structure can reveal defect states in the band gap for diamond [10-11]. The predominant feature of the diamond k-edge is the ionization that occurs at 290eV due to k-shell electrons being promoted to σ^* states. Features appearing before the ionization edge in the range 282-288eV have been assigned to π^* states [12].

2. Vacancy cluster contrast

Thin electron transparent specimens of untreated brown, annealed brown and colourless type IIa diamond were assessed using the aberration corrected SuperSTEM at Daresbury Laboratory. Electron imaging was carried out under UHV conditions (~$1x10^{-9}$ Torr) and the specimens were free from carbon contamination. The high magnification BF and HAADF images shown in Figure 1 reveal some differences between the crystal structures of the three specimens. It is important to note that the distortions observed in the images of the colourless and heat treated samples are due to micro charging under the highly focused electron probe. This phenomenon is most problematic with insulating materials in the AC-STEM and is indicative of the electronic properties of the material. In the case of colourless diamond the contribution from sp^2 bonding is negligible and the band gap contains fewer electronic states accordingly. The BF images of the brown diamond show localised regions of stronger contrast that are absent from those for the colourless and annealed specimens. The darker patches that are observed for the brown diamond are of the order of 1nm in size, which correlates with the PAS measurements and also agrees with the HR-TEM observations that reveal similar features in a number of natural brown diamonds [13].

Journal of Physics: Conference Series **281** (2011) 012024 doi:10.1088/1742-6596/281/1/012024

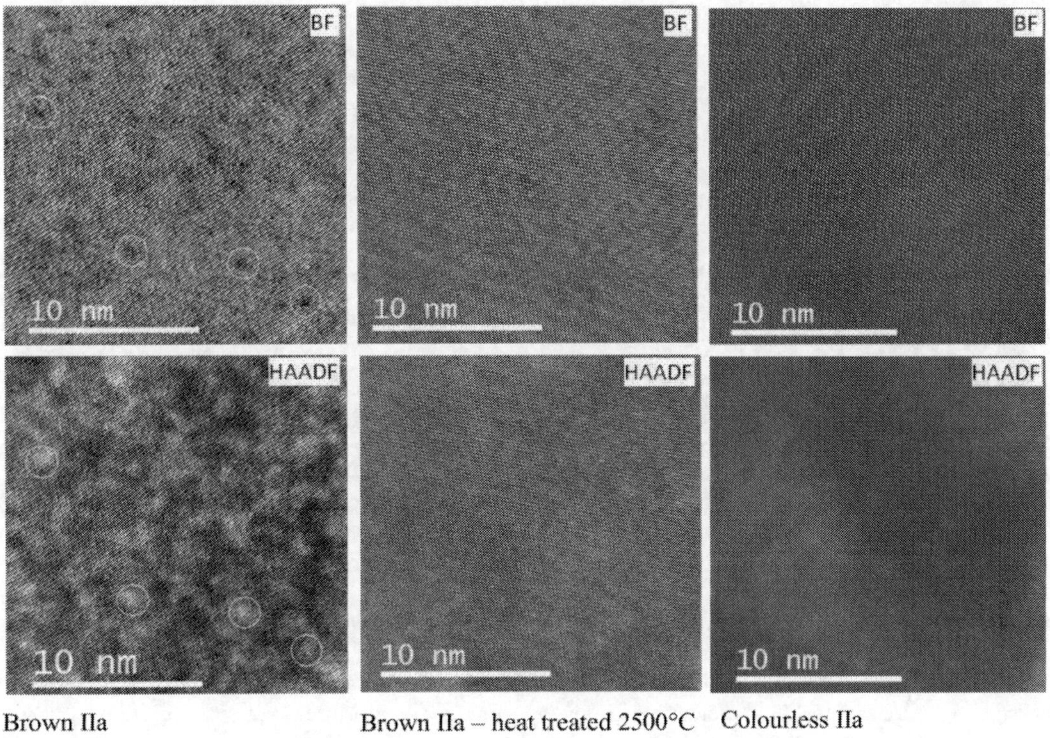

Figure 1. Comparison of BF and HAADF image contrast for brown, heat treated brown and colourless natural type IIa diamonds.

The corresponding features that are observed in the HAADF images have a bright intensity which is unexpected, i.e. one might expect to see weak contrast of opposite sign where there are clusters of vacancies. However, this phenomenon can be explained by considering the effect that a relaxed structure of vacancies might have on the relative excitation of s-type and p-type Bloch states in the diamond crystal. Channelled p-type states that would normally pass through a defect free crystal without scattering interact with the displaced atoms around the vacancy cluster causing s-type states to be excited. The resultant scattered waves add cumulatively giving rise to bright intensity and strong contrast in the HAADF image. A detailed explanation of Bloch wave interference effects in imperfect crystals is given by Perovic et al. [9].

2.1. *Through-focal image series*

By acquiring a through-focal image series in both BF and HAADF it was possible to study the contrast of the vacancy clusters in more detail. This involved acquiring a number of images at different values of defocus, starting at -20nm underfocus and moving to +20nm overfocus, with focus being at the beam entrance surface. The thickness of the specimen in the region of interest was determined from the number of electron mean free paths acquired from EELS measurements and was found to be approximately 80nm. Concentrating on one of the vacancy clusters in the BF series (Figure 2), we observe strong contrast of bright intensity at -20nm underfocus, very little contrast approaching focus and then stronger contrast of opposite sign at +20nm overfocus. This effect is not unexpected as the contrast displayed by a cluster will vary depending on its position relative to the electron probe and therefore will not necessarily be visible when the surface of the specimen is in focus. The fact that the feature is invisible at around -10nm defocus means that it is in-focus at a depth probed 10 nm below the beam entrance surface and hence is positioned at this depth. The feature may therefore be disregarded as a surface asperity.

By comparison the HAADF series shows strong contrast of bright intensity throughout and does not display any reversal of sign.

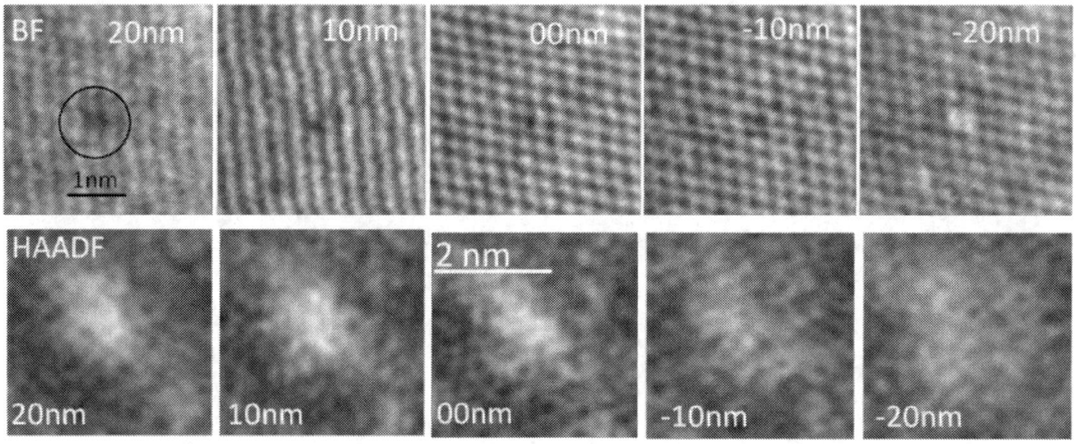

Figure 2. Through-focal image series acquired between -20nm and +20nm, showing how the contrast of a vacancy cluster changes in BF and HAADF.

2.2. STEM Image Simulation

Image simulations were performed to determine the contrast that one would expect from a multi-vacancy structure in experimental images. The simulations employed here use the multi-slice method [14] which transmits and propagates the electron wave function through successive slices of the crystal structure, integrates the electron intensity over the detector area and outputs the proportion of the incident intensity detected as a function of probe position. For STEM image simulation, the multi-slice calculation is performed for each position of the incident electron probe.

Experimental parameters were selected for the 100kV SuperSTEM with Nion aberration corrector. BF and HAADF simulations of a 54 vacancy cluster positioned at an arbitrary depth of 7nm below the beam entrance surface of a 30nm crystal were produced. A defocus range of -20nm to +20nm was selected, in keeping with the experimental conditions. The angular range of the BF and HAADF detector was 0-6 mrad and 70-210 mrad respectively. The output from the multi-slice calculations is shown in **Figure 3** below.

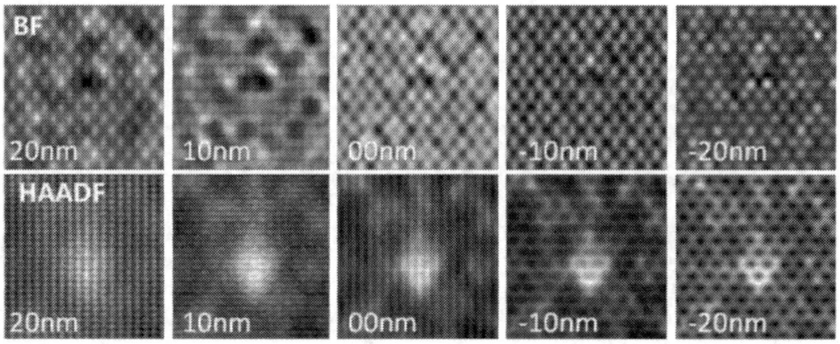

Figure 3. Simulated BF and HAADF image series of v54 cluster positioned 7nm below beam entrant surface. The diamond crystal thickness is 30nm. An rms amplitude of vibration of 0.02Å was applied.

Comparing the small scale features highlighted in the experimental images and the simulated vacancy cluster shown in Figure 3, we observe similar characteristics. Between -20nm underfocus and focus in the BF series the contrast is weak and the cluster is barely visible. However, on moving through focus to +20nm overfocus the contrast becomes stronger and the cluster appears dark.

Although the feature in Figure 3 does not display a complete reversal of contrast it does appear less prominent at underfocus and this agrees with the simulation. Furthermore, if one correlates the defocus condition under which weak contrast is observed in both the experimental and simulated images, then the approximate position of the defect within the volume of the specimen can be determined. For example, one could ascertain from the simulated image series that a cluster positioned 10nm below the surface of the specimen would be largely invisible in the electron image. In comparison to the BF series, the contrast observed in the HAADF series is always of the same sign and is of similar intensity to that seen experimentally.

3. Core-loss EELS pre-edge structure

Core-loss EELS measurements were carried out to investigate sp^2 states in brown, heat treated brown and colourless natural diamond specimens. The diamond k-edge spectra presented here were acquired with a GATAN Enfina spectrometer having a collection angle of up to 19mrad. Spectrum images were acquired with a pixel dwell time of 0.5 s over the energy loss range 248 eV to 382 eV. The energy resolution determined from the FWHM of the zero loss peak was 0.3 eV.

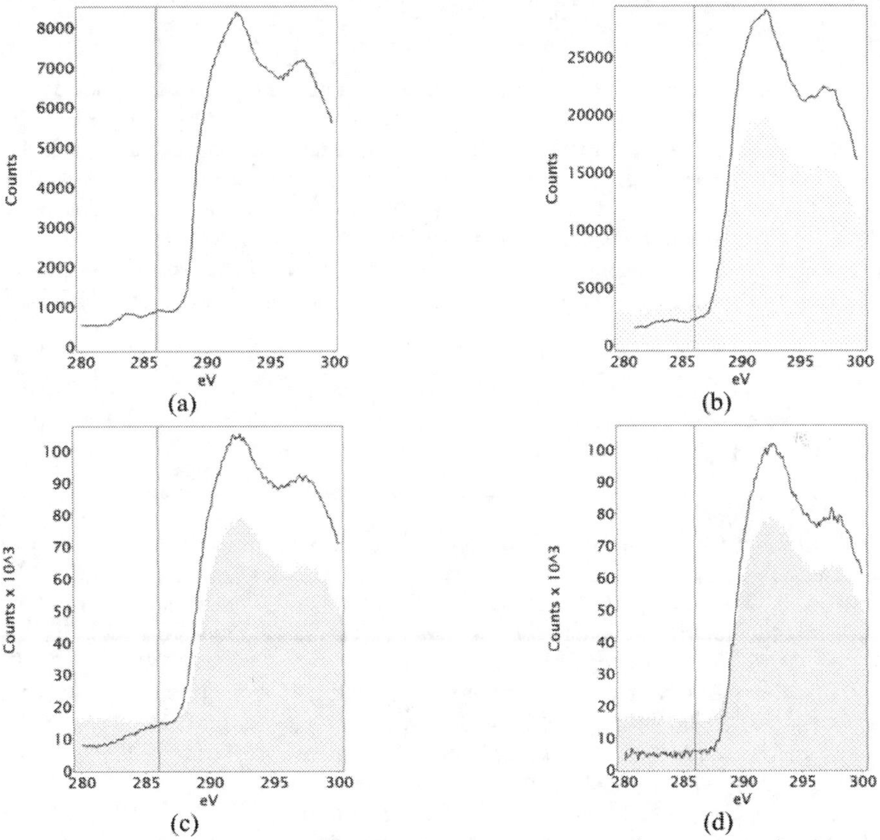

Figure 4. Core-loss spectra showing the difference in the pre-edge structure for (a) untreated type IIa brown diamond near dislocation, (b) untreated type IIa colourless diamond near dislocation (c) type IIa brown diamond annealed at 1900°C and (d) type IIa brown diamond annealed at 2500°C.

Figure 4 shows the difference in the structure immediately before the diamond k-edge for treated and untreated natural diamonds. Interestingly, two peaks are observed for the brown diamond (a), one at ~284eV and another at 286eV. By comparison the 286eV peak is absent for the colourless diamond (b), although a small contribution is seen at 284eV. The double feature before the main ionisation edge has been reported previously in nanocrystalline diamonds by Okada et al [12] and has been attributed to sp^2 states. However, evidence exists in this research and previous studies of low-loss EELS [15] of diamond to suggest that the 284eV peak originates from electronic states at dislocations. The fact that dislocations have electronic states associated with them begs the question as to why they do not contribute to the brown colouration. In the absence of further quantitative information, it is assumed that these optically active dislocations are not present in sufficient concentration to cause brown colour.

Comparing the spectra for the natural brown diamonds heat treated to 1900°C (c) and 2500°C (d), there is still evidence of the sp^2 states in the former but not the latter. This finding agrees with the annealing studies which show that the brown colouration is not removed until temperatures in excess of 2000°C are reached. In fact the brown colouration becomes less and less prominent until it finally disappears at around 2500°C [3]. Furthermore, PAS measurements have demonstrated that the disappearance of the brown colour at the higher treatment temperature coincides with a reduction in the average positron lifetime [6]. One hypothesis is that the spherical vacancy structure grows by the uptake of mobile vacancies at elevated temperature until a critical size is reached where the structure collapses and forms an optically inactive dislocation loop. A similar mechanism for the removal of brown colour has been discussed by Hounsome et al [5].

4. Conclusion

A correlation exists between the contrast observed at vacancy clusters in experimental STEM images of type IIa natural brown diamond and the contrast predicted by simulated images of similar vacancy structures. Phase contrast is the principle mechanism for observing contrast in the BF image, whereas high angle scattering due to Bloch wave interference is the probable cause of contrast of opposite sign to that expected in the HAADF image. Core-loss EELS shows a double pre-edge peak for brown diamond that is annealed out by heat treating to 2500°C. The peak at 286eV originates from sp^2 states associated with the π-bonded lining of the vacancy clusters. However, the 284eV peak can be attributed to electronic states at dislocations in both brown and colourless natural diamonds.

5. Acknowledgements

The author would like to thank Dr. D. Fisher and Dr. P. Martineau of the Diamond Trading Company for the provision of diamond specimens and fellow colleagues at SuperSTEM for their support in STEM training.

References

[1] Fisher D 2009 Brown diamonds and high pressure high temperature treatment *J.Lithos* 2 **112** 619-24

[2] Willems B, Martineau PM, Fisher D, Van Royen J and Van Tendeloo G 2006 Dislocation distributions in brown diamond *Phys. Stat. Sol.* A **203** 12 3076-80

[3] Willems B, 2006 PhD Thesis Antwerp

[4] Fujita N, Jones R, Öberg S and Briddon PR 2009 Large spherical vacancy clusters in diamond – Origin of the brown colouration *Diamond & Related Materials* **18** 843–45

[5] Hounsome L, Jones R, Martineau PM, Fisher D, Shaw MJ, Briddon PR, and Öberg S 2006 Origin of brown coloration in diamond *Phys. Rev.* B **73** 125203

[6] Mäki J-M, Tuomisto F, Kelly C J, Fisher D and Martineau P M 2009 Properties of optically active vacancy clusters in type IIa diamond J. Phys.: Condens. Matter **21** 364216 10pp

[7] Avalos V and Dannefaer S 2003 Vacancy-type defects in brown diamonds investigated by positron annihilation *Physica* B **340-342** 76-79

[8] Goss JP, Coomer BJ, Jones R and Fall CJ 2003 Extended defects in diamond: The interstitial platelet Phys. Rev. B **67** 165208

[9] Perovic DD, Rossouw CJ and Howie A 1993 Imaging elastic strains in high-angle annular dark field scanning transmission electron microscopy Ultramicroscopy **52** 353-359

[10] Muller DA, Tzou Y, Raj R and Silcox J 1993 Mapping sp^2 and sp^3 states of carbon at sub-nanometre spatial resolution Nature **366** 23-30

[11] Egerton RF and Whelan MJ 1974 The electron energy loss spectroscopy and band structure of

diamond Philosophical magazine **30** 4 739-749

[12] Okada K, Kimoto K, Komatsu S and Matsumoto S 2002 Sp^2 bonding distributions in nanocrystalline diamond particles by electron energy loss spectroscopy *J. Appl. Phys.* **93** 3120

[13] Barnes R, Bangert U and Martineau P 2006 HR-TEM imaging and image simulation of vacancy clusters in brown diamond *Phys. Stat. Sol.* A **203** 12 3081–87

[14] Kirkland E J Advanced Computing in Electron Microscopy Plenum New York 1998

[15] Bangert U, Barnes R, Gass M H, Bleloch A L and Godfrey I S 2009 Vacancy clusters, dislocations and brown colouration in diamond J. *Phys. Condens. Matter* **21** 364208 6pp

Dislocation dipole annihilation in diamond and silicon

J Rabier and L Pizzagalli
Institut PPRIMME, Département de Physique et Mécanique des Matériaux - UPR 3346 CNRS, Université de Poitiers, ENSMA - SP2MI, BP 30179, F-86962 Chasseneuil Futuroscope Cedex, France

E-mail : jacques.rabier@univ-poitiers.fr

Abstract. The mechanism of dislocation dipole annihilation has been investigated in C and Si using atomistic calculations with the aim of studying their annihilation by-products. It is shown, in C as well as in Si, that dipole annihilation yields debris that can be depicted as a cluster of vacancies, or alternately by two internal free surfaces. These defects have no strain field and can hardly be seen using usual TEM techniques. This suggests that the brown colouration of diamond could be due to microstructures resulting from deformation mechanisms associated with dipole formation and their annihilation rather than to a climb mechanism and vacancy aggregation. In silicon where a number of dipoles have been evidenced by TEM when dislocation trails are found, such debris could be the missing link responsible for the observation of strong chemical reactivity and electrical activity in the wake of moving dislocations.

1. Introduction

Point defects or association of point defects have been assigned to some changes in physical properties in plastically deformed semiconductors. In diamond, it has been recognized that the brown colour of diamond is concomitant with the presence of dislocations but not directly related to dislocations. Rather this colour change has been attributed to point defect clusters [1, 2], the presence of which are related to some specific plastic deformation conditions. Those defects are usually thought to be associated with climb mechanisms leading to vacancy aggregation [3]. However, the relationship between deformation conditions and the actual formation mechanism of these vacancy clusters is not clearly understood yet. As far as silicon is concerned, a puzzling question remains about the presence of points defects in the wake of moving dislocation in the lower temperature range of plasticity evidenced through EPR measurements (see for example [4]). Chemical etching in the same range of deformation temperature has also evidenced trails lying in crystallographic directions behind dislocations in their glide plane [5]. Those trails also appear to have an electrical signature [6].

Some effort has been devoted to understand the relevant features in the deformation microstructure of diamond [7, 8]. In brown diamond many paired perfect dislocations and small dislocation loops are found building the microstructure suggesting a link to the current brown colour mechanism involving vacancy clusters or disks. However no significant change in the microstructure before and after high pressure high temperature (HPHT)-treatment was evidenced although this treatment alters the brown colouration. The authors suggest that changes should occur at dislocation core.

Very few observations were devoted to the characterisation of deformation microstructures of silicon in the regime where trails are evidenced. Recent observations [9] put forward the large density of dislocation dipoles found in deformation microstructures obtained in those conditions. Faint contrasts were also observed in the neighbourhood of dislocations dipoles. The structural objects associated to these faint contrasts have not been determined. It is thought that they could correspond to dipole annihilation by-products in agreement with their geometrical features. However these TEM observations are difficult to relate unambiguously to the fingerprints of trails.

It comes out from these TEM studies that a common feature of these deformation microstructures in diamond and silicon is the presence of dislocation dipoles and small dislocation loops. Unidentified defects present in these microstructures should be associated to physical signatures. The question then arises as to whether specific defects such as clusters of point defects may be associated with such populations of dipoles.

Indeed, several computational works dealing with dipole annihilation in various materials have put forward that the classical dipole annihilation issues have to be revisited [10-12]. Dipole annihilation by-products, beside those relevant to screw orientations which restore the perfect crystal [13] have been found to consist of debris of almost atomistic dimensions which are difficult to evidence by TEM. For example in alkali halides it was shown that voids result from vacancy dipole ''annihilation'' which is consistent with the experimental results of plastic deformation tests and TEM observations [14] as well as with atomistic computations [10].

In this context the present paper investigates the possible transformation of small height vacancy dipoles in diamond and silicon into vacancy clusters using atomistic simulations.

2. Simulation techniques

There are several issues which are of importance in the numerical simulations of small dislocation dipoles.

(i) The choice of the interatomic potential. It has to describe correctly the initial configuration i.e. dislocation core structure and any defect resulting from dislocation dipole annihilation. Since no potential has proved to be superior to all others, several ones were employed.

(ii) The nature of the core of the dislocation building the dipole. For the sake of simplicity and also since the microstructure associated to dipoles contains perfect dislocations (diamond) or were found in the low temperature deformation range (silicon), we chose to study the dipole annihilation of two perfect dislocations. Since the most common core issued from calculations is the 60° S_1 core (see [15]), this core configuration has been chosen.

(iii) The size and the limit conditions for the computational cells. Perpendicular to the dislocation line, computations were performed using either periodic boundary conditions, or fixed surfaces conditions; the computation cell is periodic along the dislocation line so that energies are computed per unit line of dislocation.

Calculations have been performed using systems of size 96x192x3, i.e. 55296 atoms. Initial configurations have been built by inserting a vacancy dipole of 60° dislocations with the S_1 core, with opposite Burgers vectors, with a variable distance, *h*. Since stable vacancy clusters were already found lying on (111) planes in diamond, the dipole plane was chosen as to be a (111) plane. The geometry of the system is depicted figure 1. After relaxation of the dipole, a second calculation is done, with a manual procedure as to open a vacancy cluster.

Different potentials were used for each material: Tersoff [16], EDIPA and EDIPB [17] potentials for diamond, Tersoff [16], EDIP [17], Stillinger-Weber [18] potentials for silicon.

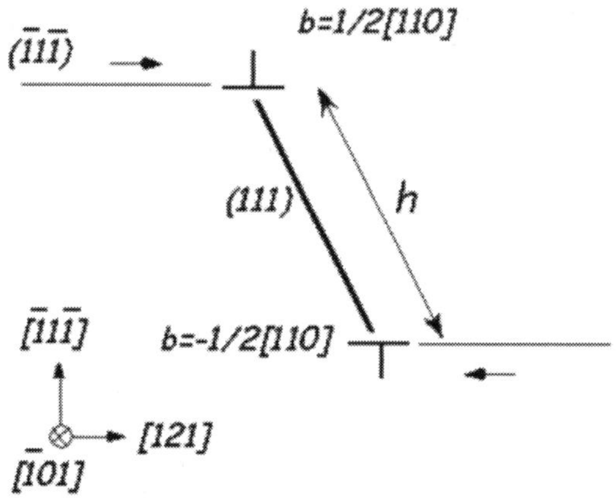

Figure 1. Geometrical features of a vacancy dislocation dipole.

3. Results

Owing to the close agreement between calculations performed using periodic boundary conditions, and fixed surfaces conditions, only results using periodic boundary conditions are reported in this paper.

For various dipole heights, the energy of a vacancy dipole first modelled by elasticity and then relaxed atomistically was computed. Allowing these vacancy dipole configurations to open on part of the dipole height yields a relaxed configuration which is open all along the dipole height. The resulting defect can be described as a vacancy cluster aligned along (111) or as two (111) internal free surfaces. No strain field is found associated to such configuration: the parent dislocations building the dipole do not exit anymore so that this defect can be considered as the result of dipole annihilation i.e. "debris". These two types of configurations are shown in figure 2 for the same dipole height.

Figure 2a. Vacancy dislocation dipole, h=3.9nm, diamond, Tersoff potential [16].

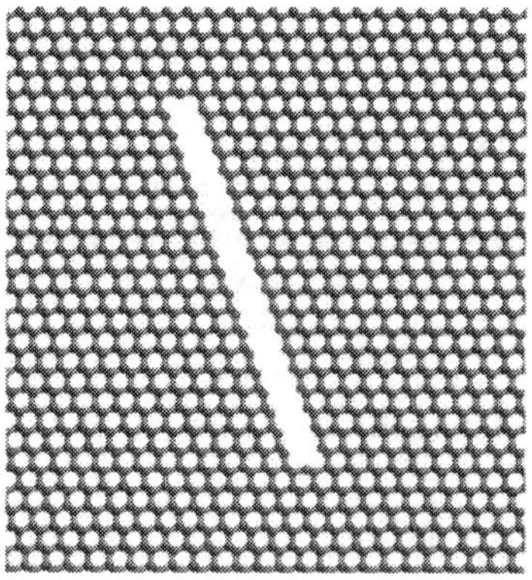

Figure 2b. Vacancy cluster resulting from dislocation annihilation, is as figure 2a.

The energy of the relaxed configuration after opening the dipole and the energy of the starting dipole configuration were compared for different dipole height, *h*. Depending on *h*, it is found that the stable configuration is either a vacancy dipole or debris. The dipole critical heights h_c for which a vacancy dipole converts into debris have been calculated with the different potentials. The results are given in tables 1 and 2 respectively for diamond and silicon. It is found that h_c depends quite strongly on the interatomic potential, however it is shown that debris are stable at low dipole heights for any potential both for diamond and silicon.

From an elastic point of view this calculation compares the elastic energy of a dislocation dipole and that of two internal free surfaces. To achieve reliable results, the interatomic potentials have to describe correctly the dislocation core energetics as well as the surface energy. The surface energy we can extract from our computations is in good agreement with the (111) surface energy computed with the same potential taking the energy of the debris as $E=2\gamma h$. However the present calculations do not take into account (111) surface reconstructions; such reconstructions contribute to lower the surface energy so that the critical heights in this paper are under estimated.

Table 1. Vacancy dipole critical heights in diamond depending on the interatomic potential.

C	Critical height h_c (nm)	Potential
	4.9	Tersoff [16]
	13.4	EDIP A [17]
	11.3	EDIP B [18]

Table 2. Vacancy dipole critical heights in silicon depending on the interatomic potential.

Si	Critical height h_c (nm)	Potential
	2.4	Tersoff [16]
	3.1	EDIP [18]
	1.2	SW [19]

4. Discussion

The geometry of the defect resulting from vacancy dipole annihilation is similar to that conjectured by Jones et al [1, 2]. Previous interpretation of vacancy cluster formation in diamond relies upon some climb mechanism associated to trails of high concentration of vacancies which aggregates to form cluster and aggregation of vacancies [2, 3]. Such circular disks of vacancies were found to be unstable as compared to the energy of a perfect vacancy loop, calculated according to elastic theory, for a loop radius of 1.2 nm containing 200 vacancies.

Discrepancy in the critical size of the cluster between the results of Jones et al [2] and this paper can come not only from the different geometry of the problem but also from the way the calculation was handled: ab initio methods in Jones et al [2], interatomic potentials in this work. Furthermore infinite dipoles are used in the present simulations; this does not take into account the energy of dislocation segments which necessarily close a defect of finite dimensions.

It is shown here that similar defects can result from the annihilation of a vacancy dipole in both diamond and silicon, although the critical height for the stability of this debris is quite different in the two materials. The possible occurrence of such defects in these two materials suggests that the analysis of deformation microstructures has to be revisited focusing upon dipoles and dipole annihilation by-products. However the formation mechanism of vacancy defect cluster related to dipole annihilation fits quite well the observations that clusters are likely to be located in the wake of moving dislocations. Such processes obviously hold for jogged dislocations and dipoles: extension clearly relates to dislocation movement.

Associated with dipole annihilation, several stages can lead to the formation of debris:

(i) large vacancy dipoles can decompose into vacancy loops that, when shrinking can transform into debris,

(ii) small dipoles can "annihilate" directly leaving debris.

The two pictures are somehow different as far as the deformation temperature is concerned. The first process requires point defect diffusion although the second one is an athermal process. In this second case it means that brown diamonds could be associated with a relatively low deformation temperature.

The formation of the debris has been found in this work using perfect dislocations as parent dislocations. However dislocations in semiconductors have been found to occur under several core structures, particularly they are found dissociated in quite a wide range of temperature. Is the formation of the debris liable to occur only if the deformation is controlled by perfect dislocations? Dipoles of dissociated dislocations in diamond or zinc blende crystals give rise to dissociated dipole S, or Z dipoles as in usual FCC materials. Is the final product of the annihilation of such extended dipoles in silicon and diamond debris like those depicted in this paper? Such calculations about the annihilation of extended dipoles have been done for FCC metals [11, 12] but the comparison with semiconductors cannot be followed up. Indeed different from what was found here, vacancy clusters were found to lie along (110) joining the two parent dislocations in FCC metals. It is clear that the structure and properties of the by-products resulting from the annihilation of extended dipoles should be investigated for semiconductors.

Another point that has to be addressed about dipole formation in silicon and diamond is the following. Do the deformation microstructures of semiconductors generate dislocation dipoles in the usual way as other materials do, or are dipoles the signature of a specific mechanism relevant to materials with diamond structure? Indeed those materials have been found to possess dislocations with multiple core structures associated to very different mobility [15]. Perfect dislocations and dissociated dislocation have been evidenced in these two materials which are likely to be relevant to the shuffle and glide set respectively. In silicon a perfect ''shuffle core'' was evidenced at high stress and low temperature and a dissociated ''glide core'' at medium temperature and low stress [20, 21]. Those two kinds of dislocation cores have been found to behave quite differently: perfect ''shuffle core'' dislocations are impossible to move at temperatures below the BDT [22] once they have been stopped. This seems to be corroborated by recent computations [23] showing that their lower energy configurations are sessile.

In that context one has to consider the possible occurrence of sessile core structures during plastic deformation at moderate temperatures. Then the nucleation of glissile dislocations can be facilitated in the stress field of pre-existing sessile dislocations in a {111} glide plane a few atomic distances from that of the parent dislocation. Interaction of these new dislocations, located in a different atomic glide plane, with the parent ones will give rise to the formation of dislocation dipoles. Annihilation mechanisms will result in by-products such as vacancy loops and debris. These by-products can provide nucleation sites of jog formation on mobile dislocations, interacting with them, feeding the formation of new dipoles and in fine the possible formation of debris in the wake of moving dislocations.

5. Conclusion

Calculations performed in this paper show that dipole annihilation can be considered as a possible mechanism of formation of vacancy clusters involved the brown coloration of diamond. Such defects could also exist in silicon under smaller dimension and could be related to the formation of dislocation trails. This has to be proved experimentally. However, due to the lack of strain fields associated with these defects, by-products of dipole annihilation will be difficult to image using TEM techniques. Besides looking at experimental evidence of such mechanisms, several points need to be addressed: the mechanism associated to the transformation between a vacancy dipole and debris and the relevance of this annihilation by-products for extended dipoles.

References

[1] Jones R 2009 *Diamond and Related Materials* **18** 820
[2] Hounsome LS, Jones R, Martineau PM, Fisher D, Shaw MJ, Briddon PR and Öberg S 2006 *Phys. Rev.* B **73** 125203
[3] Fisher D 2009 *Lithos* **112** 619
[4] Jones R 2000 *Mat. Sci. Eng.* B **71** 24
[5] Eremenko V Demenet JL and Rabier J 2009 *Phys. Stat. Sol.* (c) **6** 1801
[6] Eremenko V and Yakimov E 2004 *Eur. Phys. J., Appl. Phys.* **27** 349
[7] Willems B Martineau PM, Fisher D, Van Royen J and Van Tendeloo G 2006 *Phys. Stat. Sol. (a)* **203** 3076
[8] Willems B, 2006. Structural defects and colour-treated diamond: a transmission electron microscopy study. *Unpublished PhD thesis*, University of Antwerp.
[9] Eyidi D, Eremenko V, Demenet JL and Rabier J 2009 *Physica* B **404** 4634
[10] Rabier J and Puls MP 1989 *Phil. Mag.* A **59** 821
[11] Aslanides A and Pontikis V 2000 *Phil. Mag.* A **80** 2337
[12] Wang H, Xu DS, Yang R and Veyssière P 2008 *Acta Materialia* **56** 4608
[13] Wang H, Xu DS, Yang R and Veyssière P 2009 *Acta Materialia* **57** 3725
[14] Strunk H 1984 *Proceedings of the Ninth Yamada Conference on Dislocations in Solids*, Tokyo (Utrecht: VNU science Press) 483
[15] Rabier J, Pizzagalli L and Demenet JL 2010 *Dislocations in Solids*, ed JP Hirth and L Kubin vol 93 (North-Holland, Amsterdam Elsevier) p 47
[16] Tersoff J 1989 *Phys. Rev.* B **39** 5566
[17] Lucas G, Bertolus M and Pizzagalli L 2010 *J. Phys: Condensed Matter* **22** 035802
[18] Justo JF, Bazant MZ, Kaxiras E, Bulatov V and Yip S 1998 *Phys. Rev.* B **58** 2539
[19] Stillinger FH and Weber TA 1985 *Phys. Rev.* B **31** 5262
[20] Rabier J, Cordier P, Demenet JL and Garem H 2001 *Mat. Sci. Eng.* A **309–310** 74
[21] Rabier J and Demenet JL 2001 *Scripta Materialia* **45** 1259
[22] Rabier J and Demenet JL 2005 *Phys. Stat. Sol.* (a) **202** 944
[23] Pizzagalli L, Godet J and Brochard S 2009 *Phys. Rev. Lett.* **103** 065505

International Conference on Extended Defects in Semiconductors (EDS 2010) IOP Publishing
Journal of Physics: Conference Series **281** (2011) 012026 doi:10.1088/1742-6596/281/1/012026

Effect of off-axis growth on dislocations in CVD diamond grown on {001} substrates

Nick Davies[1,3], Riz Khan[1], Philip Martineau[1], Mike Gaukroger[1], Daniel Twitchen[2] and Harpreet Dhillon[2]

[1]DTC Research Centre, Belmont Road, Maidenhead, Berkshire, SL6 6JW, UK
[2]Element Six, King's Ride Park, Ascot, Berkshire, SL5 8BP, UK
[3]Author to whom any correspondence should be addressed.

Email: nick.davies@dtc.com

Abstract. Single crystal CVD synthetic diamond samples grown on substrates close to (001) have been studied using X-ray topography, photoluminescence imaging and Nomarski microscopy. The substrates used for this study were polished up to 15 degrees from (001). Nomarski images of the final growth surface have been compared with X-ray section topographs and photoluminescence images, both sampling a plane close to the surface of a (010) cross-sectional slice. The photoluminescence images provide evidence of the direction of step flow growth which was compared with the surface morphology. Samples grown on substrates polished off-(001) about [010] by greater than 10 degrees exhibit regions possessing both [001] and [101] dislocations as a result of on-axis and off-axis step flow growth respectively. Our previously published model suggests relatively low angle surface inclinations arising from risers are needed to switch dislocation direction from [001] to [101] line direction, and core energies per unit length suggest a minimum riser angle of 10 degrees is required. Samples grown on substrates polished greater than 10 degrees from (001) exhibit dislocations with a [101] line direction consistent with the model. Below this angle the dislocations do not switch line direction. Possible mechanisms influencing dislocation line direction via off-axis growth are discussed.

1. Introduction

There is currently much interest in the use of single crystal chemical vapour deposition (CVD) synthetic diamond in applications which make use of its exceptional optical, thermal, electronic and/or mechanical properties. While point defects in single crystal CVD diamond have been studied for many years [1], relatively little has been reported on the extended defect content of this material and how these may be influenced via the control of growth parameters. Extended defects, such as dislocations and stacking faults, are likely to influence the suitability for applications such as electronic devices. Extended defects are also known to affect the birefringence of diamond material [2] as a result of the associated strain; this is likely to be significant in optical applications such as laser windows, lenses and Raman lasers.

Comparing different diamond types, natural type IIa diamond contains relatively high dislocation densities (up to 10^9 cm^{-2}) [3] resulting from plastic deformation. These are arranged either in slip bands or in networks resulting from polygonisation [2, 4]. In contrast, dislocations in single crystal CVD diamond material tend to nucleate at or near the interface with the substrate on which it is

Published under licence by IOP Publishing Ltd

homoepitaxially grown and possess line directions close to perpendicular to the local growth surface. Birefringence and X-ray topography can be used to study these dislocation configurations and their variation from sample to sample [2]. Gaukroger *et al* [5] used X-ray projection topography to analyse the Burgers vectors of dislocations in samples of single crystal CVD diamond grown on (001) substrates. Two types of dislocation with line directions close to the [001] growth direction were identified. Mixed 45° [001] dislocations were found to be associated with damage at the substrate surface. Clusters of four or more edge [001] dislocations emanating from points at the substrate surface were also identified. Fujita *et al.* [6, 7] have presented the results of *ab initio* modelling of both edge and mixed <001> dislocations, and Blumenau *et al.* [8, 9] have reported the results of *ab initio* calculations for 60° and screw dislocations and an investigation of their dissociation. The core energy per unit length for these different dislocation types has been used by Martineau *et al.* [10] to calculate a minimum surface inclination angle (e.g. riser angle) with respect to dislocation line direction above which the dislocation line direction changes from [001] to [101]. In this paper we further test this hypothesis and report growth on substrates miscut from the {001} direction.

2. Experimental details

2.1. CVD diamond samples

The homoepitaxial diamond samples studied here were grown by Element Six Ltd. using microwave plasma CVD on {001} type Ib CVD diamond substrates possessing either (010) or (110) side faces processed into parallel-sided plates. The samples investigated were all standard CVD synthetic diamond similar to those in previous publications [2, 10]. The substrates were polished between 4° and 15° from [001] about [010] in order to investigate the effect of off-axis growth on the dislocations. Table 1 lists details of the samples discussed in this paper.

Table 1. Sample details.

Sample	Substrate preparation
A	Polished 4° from [001] about [010]
B	Polished 15° from [001] about [010]
C	Polished 10° from [001] about [010]

2.2. Luminescence imaging

Photoluminescence images were recorded using a DiamondView instrument that exposes samples to above band gap UV radiation to excite surface luminescence and can give information about spatial variations in the intensity of luminescence emanating from point defects or dislocations. After X-ray topography analysis {001} polished cross-sections were processed from sample C and images recorded.

2.3. X-ray topography

X-ray topographs were recorded using a Lang camera fitted to a rotating anode X-ray generator having a molybdenum target, zirconium filter and 0.2 mm x 0.2 mm apparent source size. The operating conditions of 40 kV, 40 mA produced a bright source of molybdenum Kα_1 X-rays at 17.4 keV. A 250 μm slit collimator was used at a distance of 0.7 m from the source. Preliminary X-ray topographs were generated rapidly on dental film before high quality topographs, requiring longer exposure times, were recorded on nuclear plates (Ilford L4) and processed using standard techniques.

Section topographs were recorded using a {533} reflection. With molybdenum Kα_1 radiation, use of a {533} reflection allows samples to be set up in such a way that the plane sampled by the X-ray beam is within 2° of a {100} plane. It also allows topographs to be recorded with relatively little projection distortion because, at 81.37°, 2θ for the Bragg condition is fairly close to 90°. Exposure times were typically one hour.

3. Results and discussion

Figure 1 shows a {533} X-ray section topograph sampling a region close to the centre of sample A. The topograph indicates the dislocations are propagating very close to the [001] direction, which is parallel to the growth direction. The dislocation structure of sample A is typical of CVD diamond material grown upon a [001] orientated substrate, and demonstrates that growing on a substrate polished at a low off-axis angle does not significantly influence the dislocation propagation direction.

Figure 1. {533} X-ray section topograph sampling central region of sample A (substrate polished 4° from [001] about [010]).

Figure 2 shows a {533} X-ray section topograph sampling a region close to the centre of sample B. This sample has been grown on a [001] orientated substrate which has been polished 15° from [001] about [010]. Here we observe two distinct growth pseudo-sectors characterised by different dislocation propagation directions. There is an 'on-axis' growth pseudo-sector towards the left of the sample where the dislocations propagate in the [001] direction. To the right of the sample is an 'off-axis' growth pseudo-sector, where the dislocations propagate close to the [101] direction. From the analysis of this sample we suggest that the dislocation direction of propagation has been "switched" from [001] to close to [101] via growth upon substrates with their major face polished off-[001], and the required off-axis angle is relatively small compared to the change in dislocation propagation direction. A 10° off-axis change has resulted in dislocations propagating 25° off the [001] direction. A growth pseudo-sector boundary is created between the 'on-axis' and 'off-axis' regions, which is likely to be defined by the difference in the relative growth rates for each pseudo-sector.

The surface morphology of sample B has been investigated using Nomarski microscopy and in particular the differences between the regions of 'on-axis' and 'off-axis' growth have been studied. The left hand image in figure 3 is a Nomarski micrograph of the surface of the 'on-axis' region of sample B showing regular, low angle steps that were flowing across the surface in a direction close to <100> when growth was terminated. In contrast, the right hand image in figure 3 is a Nomarski image of the surface of the 'off-axis' region of sample B showing coarse and irregular steps. Both images were recorded under the same magnification conditions. The surface morphology of sample B suggests the mode of growth is significantly different for the 'on-axis' and 'off-axis' regions, with the 'off-axis' growth having a higher proportion of riser growth.

Figure 2 {533} X-ray section topograph sampling central region of sample B (substrate polished 15° from [001] about [010]).

Figure 3 (a) Nomarski micrograph of the 'on-axis' growth surface of sample B. (b) Nomarski micrograph of the 'off-axis' growth surface of sample B. The width of the image in both cases is 144μm.

Figure 4 shows a {533} X-ray section topograph sampling a region close to the centre of sample C. This sample has been grown on a [001] orientated substrate which has been polished 10° from [001] about [010]. This sample exhibits 'on-axis' and 'off-axis' growth pseudo-sectors, with the dislocation line direction being [101] in the 'off-axis' region. This sample demonstrates the dislocation direction of propagation can be changed from [001] to [101] by growth on substrates polished off-axis by a relatively small angle.

Figure 5 shows a luminescence image of a polished {001} cross-section of sample C (the sample holder can be seen as a circular feature within image). The cross-section was produced to correlate closely with the region sampled by the {533} X-ray section topograph. The orange luminescence is due to nitrogen-vacancy defects and the diagonal striations are caused by variations of concentration due to the differential incorporation of such point defects in the regions of the crystal where growth has been on the terraces and risers of steps [2]. The degree of orange luminescence appears to differ between the 'on-axis' and 'off-axis' regions, with the 'off-axis' region appearing brighter. The difference in apparent brightness may be due to the 'bright' striations becoming brighter in the 'off-axis' region or the 'off-axis' region possessing a higher proportion of 'bright' striations, suggesting differences in the proportion of terrace and riser growth in the two pseudo-sectors. Blue dislocation luminescence can also be observed to correlate with those dislocations imaged in the {533} X-ray section topograph in figure 4. Growth horizons running parallel to the substrate surface can also be observed shortly after the start of growth and then in the early stages of growth.

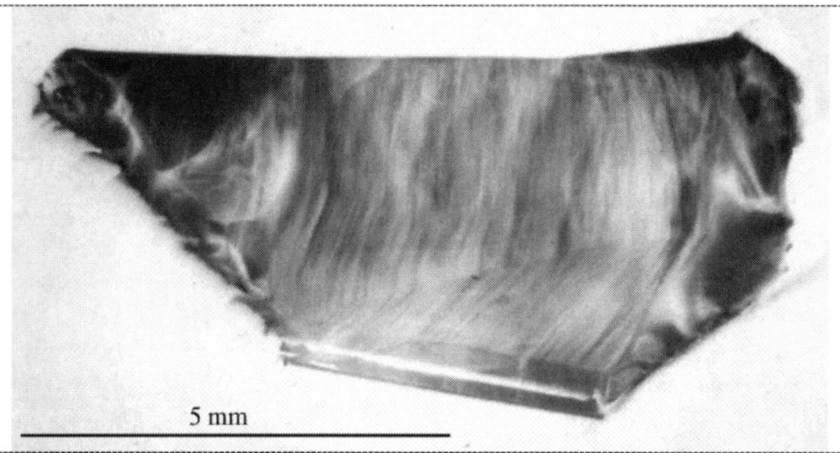

Figure 4 {533} X-ray section topograph sampling central region of sample C (substrate polished 10° from [001] about [010]).

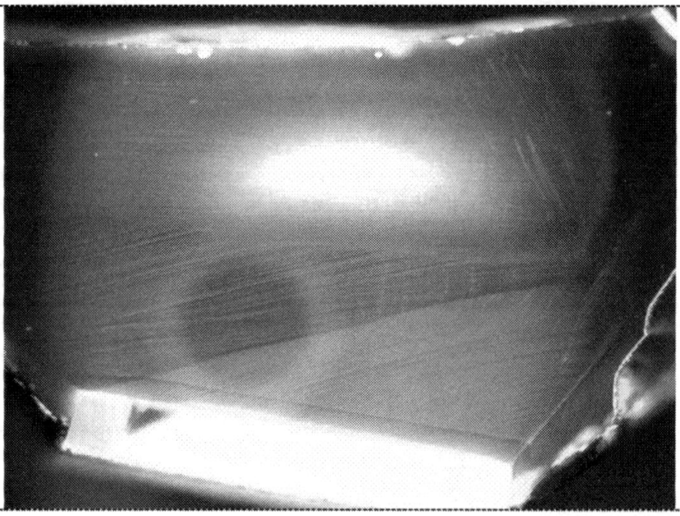

Figure 5 UV-excited photoluminescence image of a cross-sectional slice of sample C.

The surface morphology and diagonal striations seen in the DiamondView image confirm step-flow growth [10]. Martineau *et al.* [10] proposed a model for the step flow growth and dislocation interaction in which the dislocation direction of propagation was parallel to the growth direction within terrace growth ([001] line direction) and at 45° to the growth direction ([101] line direction) within riser growth. The risers in the study were inclined only by approximately 20° (the dislocations did not propagate perpendicular to the riser surfaces). It could be seen in TEM images that the dislocations propagated in the [101] direction during riser growth. Geometric arguments coupled with theoretical values for dislocation core energies per unit length [6 – 8], led to the suggestion that the energy of the [101] dislocation is less than that of a dislocation with a line direction perpendicular to the riser surface, and if a riser were incumbent on the dislocation it would change direction from [001] to [101]. A critical riser angle was proposed and calculated using dislocation core energies per unit length from *ab initio* calculations [6 - 8], above which [001] dislocations switch to [101] dislocations.

In the present study we have observed the switching of dislocation line direction from [001] to close to [101], or [101] in some cases, but the relatively low spatial resolution achievable with X-ray topography has made it difficult to determine whether the dislocations are straight or made up of segments with different directions. Figure 6(a) illustrates two mechanisms by which the angle of the growth horizon with respect to [001] and terrace/riser growth may cause dislocations to propagate in the [101] direction. In (a), due to the fact that the growth horizon is parallel to the off-axis substrate both terraces and risers are inclined at an angle above the critical angle needed to switch dislocation line direction from [001] to [101]. In figure 5 the growth horizons in the 'off-axis' region appear to be parallel to the substrate, which is consistent with the terraces being inclined at an angle possibly greater than that required to convert [001] dislocations to [101]. Figure 6(b) illustrates how terrace and riser growth could cause dislocations to propagate at a direction close to [101]. The surface morphology of sample B shows obvious differences in the step flow growth for the 'on-axis' and 'off-axis' pseudo-sectors. The 'on-axis' region shows regular low angle steps, and terrace growth dominates over riser growth. In contrast, the 'off-axis' region shows very coarse irregular steps, where neither terrace or riser growth dominates. The increase in riser content for the 'off-axis' region will lead to a deflection of the dislocations toward the [101] direction, because during riser growth the dislocation line direction will be switched to [101]. For this case, an average dislocation line direction of 25° away from [001] is not inconsistent with an approximately equal riser and terrace content with the riser growth resulting in [101] propagation and the terrace growth results in [001] propagation.

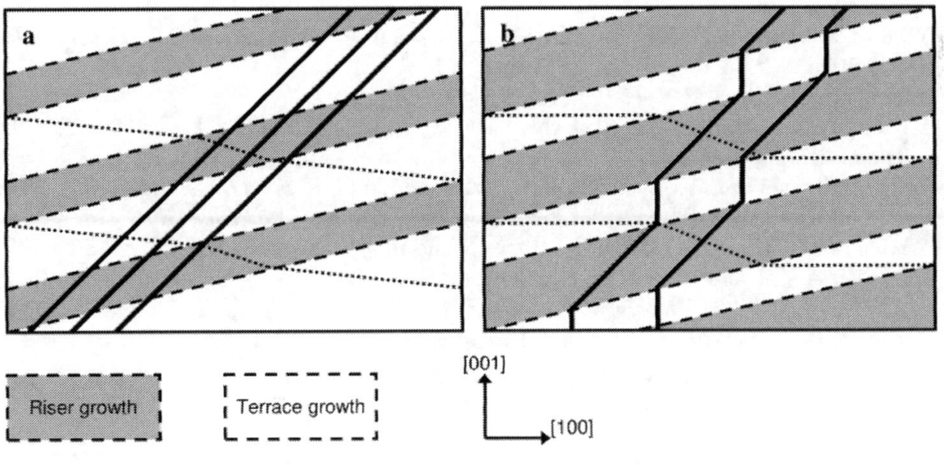

Figure 6 Schematic cross-sectional view of the effect of step flow growth on dislocations during growth on off-axis substrates. Bold lines represent dislocations, dashed lines boundaries between terrace and riser growth and dotted lines growth horizons. (a) Both terrace and riser angles are above critical value to switch dislocation line direction to [101]. (b) Riser angles are above critical value to switch dislocation line direction to [101].

4. Conclusion

Single crystal CVD diamond samples grown on (001) substrates polished off-axis by between 10° and 15° from (001) have been studied using X-ray topography, Nomarski microscopy and luminescence imaging. Results clearly show dislocations propagating in, or close to, the [101] direction in some regions of the sample rather than the [001] direction which is more usual for CVD diamond samples grown on (001) substrates polished on-axis. The off-axis angle required to deflect the dislocations is relatively small compared to the resulting angle of deflection of the dislocations (25° - 45°). The model relating the dislocation line direction with surface inclination as a result of step-flow growth as proposed by Martineau *et al.* [10] has been used to explain the deflection of dislocations during growth on off-axis substrates.

Two mechanisms are proposed for the deflection of dislocations during growth on off-axis substrates. It is possible that either the orientation of the growth horizon with respect to (001) or differences in the riser and terrace content between on-axis and off-axis growth pseudo-sectors, or a combination of the two, play a role in diverting dislocations from [001] to [101] or close to [101]. Our work here suggests that either model may dominate in different scenarios.

Further work, such as TEM studies, to investigate the case where the dislocations propagate close to the [101] direction would provide a better understanding of the above mechanisms. The high resolution provided by TEM could enable determination of the extent to which dislocations are straight or made up of alternating segments. Also, stopping the growth run before the 'on-axis' growth dominates would allow the surface morphology of the 'off-axis' region to be studied in greater detail and the relative proportions of terrace and riser growth to be accurately determined.

Acknowledgements
The authors would like to thank John Freeth and Hugh Leach for sample preparation.

References

[1] Mainwood A 1999 *Phys. Stat. Sol. A* **172** 25
[2] Martineau P M, Lawson S C, Taylor A J, Quinn S J, Evans D J F and Crowder M J 2004 *Gems and Gemology* **40** 2
[3] Willems B, Martineau P M, Fisher D, Van Royen J and Van Tendeloo G 2006 *Phys. Stat. Sol A.* **203** 3076
[4] Sumida N and Lang A R 1981 *Phil. Mag. A* **43** 1277
[5] Gaukroger M P, Martineau P M, Crowder M J, Friel I, Williams S D and Twitchen D J 2008 *Diam. Relat. Mater.* **17** 262
[6] Fujita N, Blumenau A T, Jones R, Oberg S and Briddon P R 2006 *Phys. Stat. Sol. A* **203** 3070
[7] Fujita N, Blumenau A T, Jones R, Oberg S and Briddon P R 2007 *Phys. Stat. Sol. A* **204** 2211
[8] Blumenau A T, Heggie M I, Fall C J, Jones R and Fraunenheim T 2002 *Phys. Rev. B* **65** 205205
[9] Blumenau A T, Jones R, Frauenheim T, Willems B, Lebedev O I, Van Tendeloo G, Fisher D and Martineau P M 2003 *Phys. Rev. B* **68** 014115
[10] Martineau P M, Gaukroger M P, Khan R, Evans D J F 2009 *Phys. Stat. Sol. C* **6** 1953

Journal of Physics: Conference Series **281** (2011) 012027 doi:10.1088/1742-6596/281/1/012027

Surface-phosphorus interaction in Si nanocrystals

A. Carvalho,[1] B. Celikkol,[2] J. Coutinho[1] and P.R. Briddon[3]

[1] I3N, Physics Department, University of Aveiro, 3810-193 Aveiro, Portugal
[2] Ceramics Laboratory, École Polytechnique Fédérale de Lausanne (EPFL), CH-1015 Lausanne, Switzerland
[3] School of Electrical, Electronic and Computer Engineering, University of Newcastle upon Tyne, Newcastle upon Tyne NE1 7RU, United Kingdom

E-mail: aicarvalho@ua.pt

Abstract. The properties of phosphorus-related defects in freestanding silicon nanocrystals with different surface compositions are investigated using density functional theory models. Both hydrogen (H-) and hydroxyl (OH-) covered surfaces are considered. The dependence of the formation energy of substitutional phosphorus on the distance to the surface is found to be stronger in hydroxyl-covered crystals. The formation of tricoordinated defects, energetically favourable in hydrogenated nanocrystals, is less likely in nanocrystals with hydroxylated surfaces. Further, the ionisation energy level of P-doped nanocrystals is found to be deeper in OH-covered nanocrystals than in H-covered nanocrystals.

1. Introduction

Freestanding silicon nanocrystals (Si-NCs) offer exceptional potential for electronic structure manipulation through the control of confinement effects and the hybridisation of surface and silicon states. Si nanoparticles grown using plasma deposition techniques can be dispersed in a variety of liquids to form inorganic inks, and printing-based film deposition techniques are currently under development[1, 2]. In order to exploit this class of materials, it is necessary to achieve a detailed understanding of the influence of the surface radicals and adsorbed molecules on the properties of Si-NCs.

Especially in small nanocrystals, surface atoms can have an important contribution to the highest occupied (HO) and lowest unoccupied (LU) molecular orbitals, changing the electronic energy levels and the optical absorption spectra[3, 4, 5, 6]. The effect of oxygen on hydrogen-terminated crystals has been explored by previous studies, showing that both surface hydroxylation and silanone groups lead to a redshift of the optical absorption threshold[3, 6], whereas oxidation of the core can either lead to a blueshift or to a redshift[4, 6]. Here we show, using theoretical calculations, that the surface composition also affects the properties of impurities, even at the nanocrystal core, changing the respective potential for segregation and ionisation energy. As a case study, we will consider substitutional phosphorus, an n-type dopant which shows limited doping efficiency in silicon nanocrystals and quantum dots [7, 2].

2. Modelling details

The simulations were carried out within the framework of density functional theory, as implemented in the AIMPRO code[8, 9]. Core electrons were accounted for using the dual space separable pseudopotentials of Hartwigsen, Goedecker and Hutter[10]. A Padé parametrisation of a functional based on the local density approximation was employed for the exchange and correlation energy[11], with spin polarisation for paramagnetic systems. The basis set consists of Cartesian s, p and d-type Gaussian functions centred on each atom. Each atom contributes with a contracted 'C44G*' basis consisting of $(4, 4, 1)$ (s, p, d) Gaussian orbitals, obtained as described in Ref. [12].

Silicon nanocrystal models consisted of spherical clusters of 87 Si atoms (with about 1.3 nm core diameter). The calculations were performed using real-space boundary conditions, thus eliminating spurious cluster-cluster interactions and the need for neutralising background charge. The electron affinity $[I(-/0)]$ and ionisation energy $[I(0/+)]$ were calculated by direct comparison of the total energies of the clusters in different charge states $[E(C^q)]$,

$$I(-/0) = E(C^0) - E(C^-),$$

and

$$I(0/+) = E(C^+) - E(C^0),$$

where C^q represents the energy of the relaxed cluster in charge state q. The origin of the energy scale is pinned to the vacuum level. In bulk silicon, $I(0/+)$ corresponds to the work function.

The calculated ionisation potential of some silanes and small silicon clusters for which experimental data is available, along with the work function and electron affinity of bulk silicon, are given in Table 1. The ionisation energies are in excellent agreement with experiment. However, as expected, the silicon electron affinity is overestimated, reflecting the bandgap underestimation typical of LDA calculations. Still, the systematic error introduced by this approximation should not vary appreciably amongst the calculations presented.

Table 1. Calculated vertical ionisation energies (I_v) of small silanes and silicon clusters, and work function and electron affinity of bulk silicon. The Ref. column refers to the experimental data. All the values are in eV.

Property	system	calc.	Exp.	Ref.
$I_v(0/+)$	SiH_4	12.33	12.36	[13]
$I_v(0/+)$	Si_2H_6	10.76	10.53	[14]
$I_v(0/+)$	Si_4	8.13	8.2±0.1	[15]
$I_v(0/+)$	Si_7	8.18	7.8,7.90	[15, 16]
$I_v(0/+)$	bulk Si[a]	5.32	5.15±0.08	[17]
$I_v(-/0)$	bulk Si[a]	4.64	4.05–4.29	[18, 19]

[a] obtained using total energies from a Si_{1000} supercell calculation. Alignment of the energy scale origin with the vaccuum level was performed using a (100)Si surface.

3. Undoped Si-NCs

The 87-atom Si-NC core can be passivated by hydrogen radicals, reconstituting the four-fold coordination of all surface Si atoms. The resulting cluster shows a clean gap of 2.98 eV between the highest occupied and lowest unoccupied single-particle Kohn-Sham energy levels (HOKS and LUKS). Hydroxyl terminations replacing for the hydrogen terminations change the extension of the highest occupied and lowest unoccupied levels, which have now a partial contribution from the oxygen $2p$ orbitals, increasing the electron affinity and reducing the Kohn-Sham bandgap to 1.22 eV (Fig. 1).

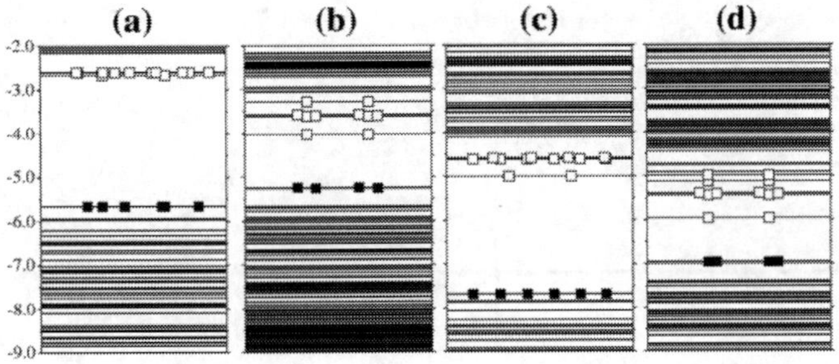

Figure 1. Kohn-Sham one-electron energy levels (eV) of (a) $Si_{87}H_{76}$ cluster, (b) $Si_{87}(OH)_{76}$ cluster, (c) $\{Si_{86}PH_{76}\}^+$ cluster, with the substitutional P^+ at the centre, and (d) $\{Si_{86}P(OH)_{76}\}^+$ cluster, also with the substitutional P^+ at the centre. Filled and empty squares indicate respectively occupied and unoccupied levels.

4. Substitutional Phosphorus

In bulk silicon, P is a shallow donor. However, in nanocrystals the P donor can be regarded as a hydrogen-like atom in a "dielectric box", and the confinement leads to a high electron binding energy[20]. For the P-doped nanocrystals studied [$Si_{86}PH_{76}$ and $Si_{86}P(OH)_{76}$], the HOKS state is partially localised on the central P atom, and the envelope of the wavefunction tail, extended through the nanocrystal, is anti-bonding with respect to the Si-Si bonds, resembling the conduction band states of the bulk crystal. The HOKS state of the OH-covered [$Si_{86}P(OH)_{76}$] cluster, shows a contribution from the oxygen $2p$ states. Figure 1 highlights the similarity between the undoped $Si_{87}H_{76}$ and $Si_{87}(OH)_{76}$ nanocrystals and the isoelectronic crystals where the central Si atom was replaced by P^+.

4.1. Formation energy

The interaction between substitutional phosphorus and the surface leads to a site dependence of the defect formation energy in the cluster. This effect, reported in hydrogen-terminated clusters[7], is much enhanced by the hydroxyl surface (Figure 2). In hydroxylated NCs, P shows a marked preference for the outer Si shell, where it is closer to the surface while preserving four Si neighbours (positions C or D of Figure 2-b). This effect is observed both for neutral and positively charged doped clusters, with quantitative differences: while the formation energy of P^+ at the position E ($d = 5.4$ Å) is 2.1 eV lower than at the centre, in the neutral cluster (Figure 2) it is 1.3 eV lower.

Apparently there is no correlation between the Si-Si bondlengths at site S and the formation energy of P substituting for Si at that position. Hence, strain effects are not the dominant factor. We believe that the attraction between the electropositive P atom and the electronegative oxygen shell is responsible for this effect. This is consistent with the larger formation energy differences in the positive charge state.

4.2. Ionisation level

The ionisation energy [$I(0/+)$] and electron affinity [$I(-/0)$] are useful quantities to predict the direction of charge transfer between doped and undoped crystals. Usually, phosphorus-doped NCs are intended to behave as electron donors. Thus, we define the quantity

$$E(0/+)^{\mathrm{P-bulk}} = I(0/+)^{\mathrm{P}} - I(-/0)^{\mathrm{bulk}},$$

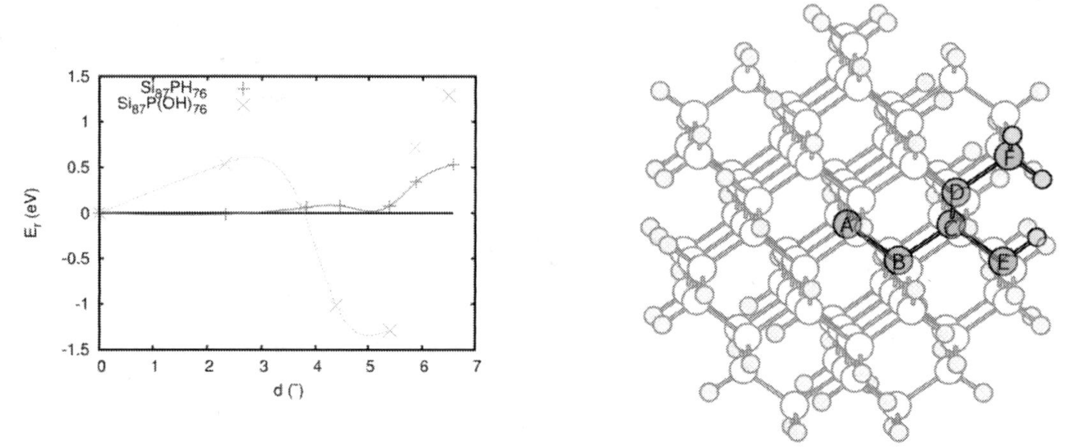

Figure 2. Relative formation energy (E_r) of a four-fold coordinated substitutional P as a function of the distance to the centre (d) (left) both in H-terminated and OH-terminated clusters, and schematic representation of the corresponding positions in a H-terminated cluster (right). The relative energies were calculated in neutral clusters.

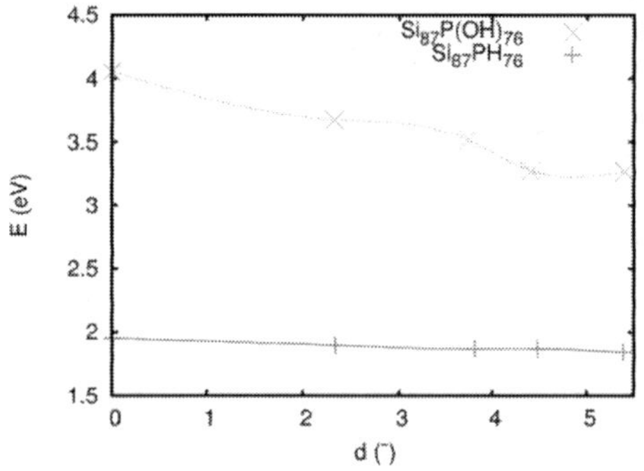

Figure 3. Difference between the ionisation energy of the phosphorus-doped nanocrystal and the electron affinity of the pristine nanocrystal $[E(0/+)^{\text{P-bulk}} = I(0/+)^{\text{P}} - I(-/0)^{\text{bulk}}]$, as a function of the distance to the centre, for nanocrystals with H- and OH-covered surfaces.

which corresponds to the energy necessary to transfer an electron from the n-doped nanocrystal to the undoped crystal, and should ideally be very small or positive.

The calculated electron affinity of undoped $Si_{87}H_{76}$ and $Si_{87}(OH)_{76}$ clusters are respectively 2.13 eV and 1.76 eV. The ionisation energy of phosphorus-doped crystals (with P at the central position) is also about 1.7 eV higher in OH-terminated clusters than in H-terminated clusters with the same number of Si atoms (Figure 3). This is the largest contribution to the difference between the $E(0/+)^{\text{P-bulk}}$ for the hydrogen and hydroxyl-covered nanocrystals. Since the shifts of the Kohn-Sham levels with P substitution is similar in both types of clusters, further research is needed to clarify the origin of the strong binding of the unpaired electron in the $Si_{86}P(OH)_{76}$ cluster.

We now analyse the electron transfer energies $E(0/+)^{\text{P-bulk}}$ (Figure 3). First, we note that the ionisation energy of phosphorus is much larger than in bulk, as found by previous calculations in hydrogenated clusters[7]. In the $Si_{86}PH_{76}$ clusters, it is approximately constant for $d < 6$ Å. In the $Si_{86}P(OH)_{76}$ cluster, $E(0/+)^{\text{P-bulk}}$ is much larger and decreases at a faster rate with the proximity of the surface.

5. Direct interaction between P and surface atoms

Four-fold coordinated P atoms replacing for Si prefer surface sites where they are bound to one surface hydrogen, rather than to the hydroxyl. This is found by comparing the energy of P^+ at different surface sites in a cluster with a mixed surface ($Si_{86}P^+H_{75}OH$). A configuration where P^+ bonds to two hydrogen atoms [$NC - Si_2P(H)_2$] is favoured by 0.55 eV relative to a similar surface site where one of the neighbours is a hydroxyl oxygen.

5.1. Phosphorus-related surface defects

At the surface (positions E and F), P can form tricoordinated defects by releasing one or two surface radicals. The resulting clusters have ionisation energies very close to those of undoped nanocrystals. Si-NCs with a P replacing for atom E and with no hydrogen neighbour [$P_E(0H)$] or a P replacing for atom F and one hydrogen neighbour [$P_F(1H)$], both with coordination three, have gaps of 2.80 eV and 2.94 eV between the HO and LU Kohn-Sham single-electron levels (respectively), close to the 2.98 eV gap of the pristine cluster, and their ionisation energies are only 0.20 eV and 0.04 eV lower than that of the pristine cluster, respectively. Thus, the donor activity is suppressed by the formation of this type of defects. Similar results are found on the hydroxylated clusters.

The relative stability of tricoordinated defects and tetracoordinated substitutional phosphorus, at the surface or at the NC core, depends on the chemical potential of hydrogen, μ_H (μ_{OH} in the case of the OH-terminated clusters), which expresses its availability in thermodynamic equilibrium. In the Si-rich limit, the chemical potential of silicon equals its energy in a bulk silicon reservoir, obtained from a calculation with periodic boundary conditions.[1] To determine μ_H in this limit (μ_H^{Si}), we assume that the source of hydrogen is SiH_4, a typical precursor used for Si-NC growth from the gas phase. The other extreme for μ_{Si} and μ_H, the hydrogen-rich limit, is defined by the conditions $\mu_H^{H_2} = \frac{1}{2}E(H_2)$ and $\mu_{Si}^{H_2} = E(SiH_4) - 4\mu_H^{H_2}$. Similarly, the range of values for the chemical potential of the hydroxyl radical, μ_{OH}, is defined in the Si-rich limit by its energy in silanol (SiH_3OH), and in the O-rich limit by the energies of molecular hydrogen and oxygen [$\mu_{OH}^{H_2,O_2} = \frac{1}{2}E(H_2) + \frac{1}{2}E(O_2)$].

The formation energies of the phosphorus containing defects,

$$E_f(D) = E[C(D)] - \sum_i n_i\mu_i,$$

where $C(D)$ is a cluster containing the defect D, and n_i are the number of atoms of species i that it contains, can be calculated as a function of the hydrogen (or hydroxyl) chemical potential. For simplicity, we have set the chemical potential of phosphorus to its energy at the centre of the cluster. Figure 4 contrasts the results obtained for the clusters with hydrogen and hydroxyl-covered surfaces.

In the $Si_{86}PH_{76}$ clusters, the tricoordinated defects are the most stable form of P in the whole range accessible to μ_H. At the Si-rich limit, the binding energy of hydrogen to P_E is about 1.2 eV. On the other hand the estimated limits for the OH chemical potential, used to calculate the formation energies in $Si_{86}P(OH)_{76}$ clusters, define a wider energy range. Further, in the Si-rich and O-poor limit, the formation energies of tri- and tetracoordinated defects are comparable. However, in the mid-range region, tetra-coordinated defects are most favourable.

6. Conclusion

The chemical composition of the Si-NC surfaces, specifically its coverage by hydrogen or hydroxyl groups, affects the dopant-surface interaction, and consequently the properties of dopant-related

[1] The energy scales of the two calculations were aligned with the aid of an auxiliary calculation performed for a clean [001] silicon surface, where the electrostatic potential of the vacuum was set to zero.

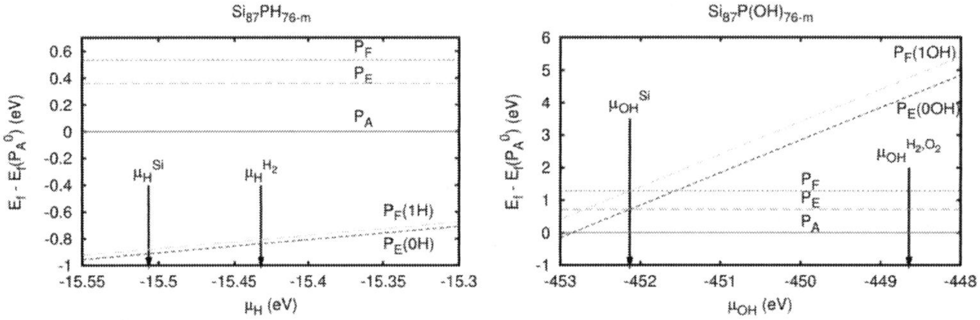

Figure 4. Formation energies of neutral tri- and tetracoordinated P defects in H-covered (left) and OH-covered (right) Si-NCs, as a function of the chemical potentials. $P_S(n\mathrm{H})$ labels a P placed at site S and attached to n hydrogen radicals. The chemical potential of phosphorus is set to be its energy at the centre of the cluster. Estimated limits for the chemical potentials are indicated with arrows.

defects. In the case of P-doped small nanocrystals (87 atoms), we find that hydroxylation suppresses the formation of tricoordinated surface defects. However, it enhances the segregation of P to an outer shell close to the surface, but not directly linked to H or OH. Further, the energy necessary to transfer an electron from a P-doped to an undoped NC $[E(0/+)^{\text{P-bulk}}]$ is higher in OH-terminated crystals than in H-terminated crystals.

Acknowledgments
AC acknowledges the Portuguese Foundation FCT for financial support.

[1] Lechner R, Wiggers H, Ebbers A, Steiger J, Brandt MS, Stutzmann M 2007 *phys. stat. sol. - Rapid Research Lett.* **1** 262
[2] Lechner R, Stegner AR, Pereira RN, Dietmueller R, Brandt MS, Ebbers A, Trocha M, Wiggers H and Stutzmann M 2008 *J. Appl. Phys.* **104** 053701
[3] Puzder A, Williamson AJ, Grossman JC and Galli G 2002 *J . Chem. Phys.* **117** 6721
[4] Zhou Z, Brus L, Friesner R 2003 *Nano Lett.* **3**, 163
[5] Ramos LE, Furthmüller J and Bechstedt F 2005 *Appl. Phys. Lett.* **87** 143113
[6] Eyre RJ, Goss JP, Briddon PR 2008 *Phys. Rev. B* **77** 245407
[7] Melnikov DV and Chelikowsky JR 2004 *Phys. Rev. Lett.* **92** 046802.
[8] Briddon P R and Jones R 2000 *phys. stat. sol.* (b) **217** 131
[9] Rayson MJ and Briddon PR 2008 *Computer Phys. Commun.* **178** 128
[10] Hartwigsen C, Goedecker S and Hutter J 1998 *Phys. Rev.* B **58** 3641
[11] Goedecker S, Teter M and Hutter J 1996 *Phys. Rev. B* **54**, 1703
[12] Goss JP, Shaw MJ and Briddon PR 2007 *Topics in Appl. Phys.* **104** 69
[13] Pullen BP, Carlson TA, Moddeman WE, Schweitzer GK, Bull WE and Grimm FA 1970 *J. Chem. Phys.* **53** 768
[14] Bock H, Ensslin W, Feher F, Freund R 1976 *J. Am. Chem. Soc. 98*, 668
[15] Kostko O, Leone SR, Duncan MA and Ahmed M 2010 *J. Phys. Chem. A* **114** 3176
[16] Fuke K, Tsukamoto K, Misaizu F, Sanekata MJ 1993 *J. Chem. Phys.* **99** 7807
[17] Allen FG and Gobeli GW 1962 *Phys. Rev.* **127**, 150
[18] *New Semiconductor Materials: Characteristics and Properties*, Ioffe Institute, St Petersbourg, available at http://www.ioffe.ru/SVA/NSM/, accessed November 18th 2010
[19] Michaelson H B 1977 *J. Appl. Phys.* **48**, 4729

[20] Chan TL, Tiago ML, Kaxiras E and Chelikowsky JR 2008 *Nano Lett.* **8**, 596

Nanoindentation of semiconductors: experiment and atomistic simulations

D. Chrobak[1,2] and R. Nowak[1]

[1] Nordic Hysitron Laboratory, Faculty of Chemistry and Materials Sciences, Aalto University, Vuorimiehentie 2A, Helsinki-Espoo, FI-02015, Finland

[2] Institute of Materials Science, Faculty of Informatics and Materials Science, University of Silesia, Bankowa 12, 40-007 Katowice, Poland

E-mail: dariusz.chrobak@us.edu.pl

Abstract. With recent developments in advanced materials and nanostructures, particularly those designed for electronics, it is evident that their successful application will depend not only on their electrical properties, but also on their mechanical characteristics. Nanoindentation is a unique method for examining nanostructured materials, as it requires a small volume of the solid and probes the surface layers of particular interest. Nanoindentation of bulk semiconductor crystal structure has been frequently used to study the onset of irreversible deformation - incipient plasticity. Here we present recent experiments supported by molecular dynamics simulations that allow determination of the origin of incipient plasticity in GaAs crystals. It will be demonstrated that, as in case of silicon, plastic deformation of GaAs starts from a pressure-induced structural phase transformation.

1. Introduction

There has been significant recent progress in our understanding of incipient plasticity, coming from nanoindentation experiments [1] supported by atomistic simulations [2]. The onset of plasticity is usually marked by characteristic singularities on indentation load-displacement (*P-h*) curves. There are so called "pop-in events", where indenter suddenly penetrates material at the constant load. Results of nanoindentation experiments by Corcoran *et al* [3] and Gouldstone *et al* [4] indicate that the pop-in events in metals are caused by homogeneous dislocations nucleation under the indenter tip. This view has recently been completed due to high temperature nanoindentation experiments [1] that linked the onset of plasticity in metals with heterogeneous nucleation of dislocations at preexisting vacancy clusters.

Semiconductors subjected to the stress caused by the indenter develop deformasion trough dislocation nucleation or/and the structural phase transformation. This issue comes to light during nanoindentation of silicon crystals. Indeed, the loaded Si-I (cubic diamond - *cd*) structure gradually transforms to metallic Si-II (β-tin) when the pressure of ~11 GPa is attained under the indenter tip. The *in-situ* conductivity measurements combined with nanoindentation experiments [5] reveal that plastic deformation in silicon is initiated by the phase transformation, while the dislocation activity takes place within previously nucleated Si-II phase [6]. During unloading one observes the set of phase transformations that include transition to a mixture of cubic Si-III (bc8) and its rhombohedral modification Si-XII (r8) phases or to amorphous silicon (*a*-Si) [6]. The transformation to Si-III/Si-XII phase is discontinuous resulting in discrete "pop-out event", while transformation to amorphous modification of silicon is gradual and is evidenced by "elbow" on unloading part of *P-h* curve.

In contrast to silicon, for which the origin of incipient plasticity is resolved, the case of GaAs is under continuing debate. Nanoindentation experiments exhibit the pop-in event (the end of elastic deformation) during loading and no singularities during unloading of GaAs crystals [7]. The fact that dislocations and slip traces can be found in cross-sectional TEM observations of residual impressions accompanied by an absence of any trace of non-zinc-blende phase of GaAs made the previous authors to attribute the onset of plastic deformation dislocations nucleation [8]. This hypothesis was supported by continuum theory of dislocations that predicts the theoretical shear stress (τ_c) for homogeneous dislocation nucleation to be a tenth of the shear modulus (G). The comparison with maximum shear stress (τ_T) at the pop-in provides a method capable to predict dislocations nucleation [9]. The maximum shear stress can be calculated from the formula τ_T=0.465p_m, where p_m is the mean contact pressure. Measurements by Bradby *et al* [10] demonstrated that the mean contact pressure at the pop-in events are sufficient to initiate homogeneous dislocation nucleation reaching approximately 11 GPa. On the other hand, Benson *et al* [11] reported that the zinc-blende-to-orthorhombic transition of GaAs starts at 12.5±1.5 GPa and ends at pressure of 17.5 GPa. Therefore, the phase transition based explanation of the incipient plasticity in GaAs should be carefully verified.

The present review addresses the nanoindentation experiments accompanied by simultaneous conductivity measurements used to study the onset of plastic deformation in GaAs [12]. We demonstrated that zinc-blende phase of GaAs (GaAs-I) undergoes metalization and subsequent transformation to orthorhombic rocksalt-like structure (GaAs-II) during loading part of nanoindentation. Phase transformation occurs at the instant of pop-in event being the origin of the GaAs incipient plasticity.

2. Methods

Two 1μm epitaxial GaAs layers with varying Si dopant concentrations were grown by molecular beam epitaxy (MBE) on 350 μm thick GaAs (100) epi-ready substrates. The doping levels of the MBE-grown GaAs layers were N_{Si} = 10^{16} cm^{-3}, and 10^{18} cm^{-3}. Their electro-mechanical properties were studied by nanoindentation coupled with conductivity measurements [12]. Conclusions derived from nanoindentation experiments were supported by theoretical consideration based on molecular dynamics [13] and first-principle calculations [12].

2.1. Nanoelectrical measurements

In order to study the structural changes in GaAs during nanoscale deformation, we used a conductive nanoindentation system with a conductive boron-doped (N_d=10^{21} cm^{-3}) diamond indenter and voltage source (V) to obtain a time correlation of force, displacement, and electric current. The constant voltage is applied to the sample while the magnitude of current running through the tip/sample contact was monitored. The noise floor of the system is below 10 pA, allowing the investigation of pressure-induced microstructural changes.

The nanoindentations were carried out with increasing load over a 20 s period then holding at the maximum load for 10 seconds and finally gradual unloading over 20 s while applying a reverse bias of 3V. The majority of nanoindentations was performed with maximum load of 8 mN.

2.2. First-principle calculations

To study the equilibrium and contact properties of GaAs zinc-blende and rocksalt structures density functional theory (DFT) based calculations were carried out employing the QuantumEspresso code [14]. A pseudopotential method was applied within a local density approximation (LDA) using Perdew-Zunger parametrization of the exchange and correlation functional. The calculations used a supercell based on the face-centered cubic structure and filled with 48 Ga and As atoms, forming 24 consecutive (100) layers of zinc-blende and the subsequent rocksalt phase. The supercell was relaxed along the z-axis and the selfconsistent DFT calculations were applied to relax the structure. The convergence of the total ground state energy was found with plane wave-energy cutoff of 30 Ry and 8x8x2 Monkhorst-Pack mesh in the ***k***-point space. The equilirium properties of zinc-blende and rocksalt phases of GaAs were investigated by calculating their total ground state energy versus unit cell volume curves and subsequently searching common tangent slope that marks equilibrium hydrostatic pressure.

2.3. Molecular dynamics

In order simulate nanoindentation deformation of the GaAs crystal the molecular dynamics (MD) code PARCAS by Nordlund *et al* [15] was used. The interactions between GaAs set of atoms were modeled by three-body Tersoff type potential [16] developed to be able describe both the zinc-blende and rocksalt phases of GaAs. The validity of the selected three-body potential was confirmed by calculations of the many physical properties of both considered structures. The interaction between carbon atoms of cubic diamond structure of indenter with GaAs was realized with Ziegler-Biersack-Littmark purely repulsive potential [17].

MD simulations were restricted to the cuboidal cell (size of 316x316x158 $Å^3$) formed by zinc-blende structure (a=5.635 Å). Four bottom GaAs layers were immobilized while the upper (001) plane was indented by diamond (a=3.567 Å) cube (size of 28x28x28 $Å^3$). The time step of numerical integration of the equation of motion was established to be 0.6 fs assuring the algorithm stability. Each indentation step was composed of indenter shift along [001] direction and then structural relaxation over 20000 time steps.

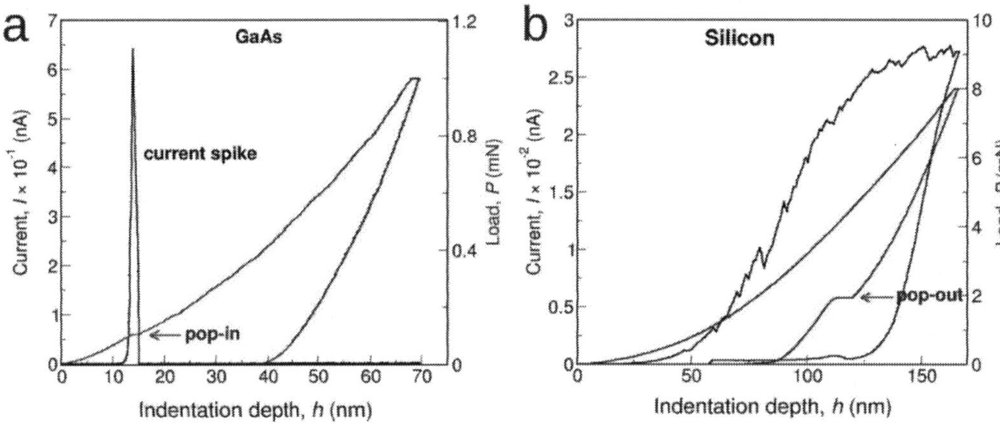

Figure 1. Complex electro-mechanical response to nanoindentation of the low-doped GaAs (a) and Si (b). The sudden current drop ("current spike") recorded for GaAs marks the onset of plastic deformation.

3. Results and discussion

The typical electric response to nanoindentation of low-doped GaAs under reverse bias of 3 V is displayed in the figure 1a. Initially perfect Shottky barrier leaks during loading as a consequence of band gap decreasing with applied pressure [12]. Subsequently, the gradual increase of the current flow is halted when rectifying contact is restored. Sudden drop of the electric current at the moment mechanical pop-in shows the unique correlation between two different phenomena.

The *I-h* characteristic registered for low-doped GaAs sample were compared with that obtained for silicon (figure 1). While "current-spike" is observed for GaAs, the electrical response of silicon to nanoindentation is completely different exhibiting gradual increase of the electric current without any significant singularity (figure 1b). This result agrees with data collected from electrical measurements performed by Ruffell *et al* [5]. They reported similar *I-h* relationship for silicon proving that complex contact composed of metallic indenter and Si-II phase under the tip surrounded by Si-I phase has Ohmic character.

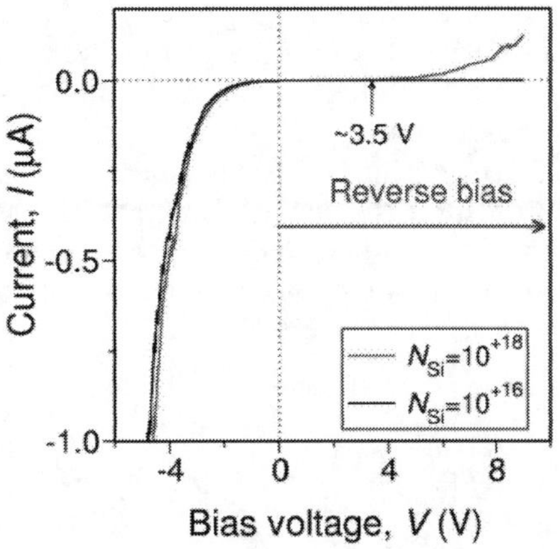

Figure 2. *I-V* characteristics collected for GaAs samples at load of 1 mN (after pop-in) prove that physical phenomenon responsible for pop-in event simultaneously restores Schottky barrier halting the current flow in the system.

The conducting indenter in contact with GaAs surface produces Shottky barrier as seen from *I-V* characteristics (figure 2) collected at load 1 mN (after the pop-in). The applied reverse bias (positive voltage) halts current flow trough the contact for the variously doped GaAs specimens. However, the junction with silicon concentration of $N_{Si}=10^{18}$ cm^{-3} leaks for reverse bias exceeding ~3.5 V. The recorded effect agrees with common expectations based on classical theory of metal-semiconductor junction [18].

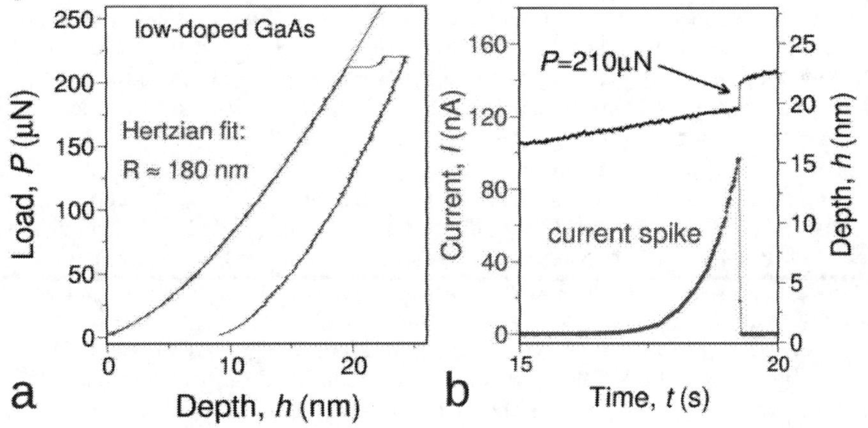

Figure 3. Nanoindentation of low-doped GaAs with maximum load of 220 μN. (a) The *P-h* curve used for evaluation of the indenter tip radius. (b) Detailed observation of the correlation between electric "current spike" and mechanical pop-in event. The current is halted exactly at the moment of pop-in.

The detailed investigation of correlation between "current-spike" phenomenon and pop-in event was accomplished using lower maximum load of 220 μN sufficient to initiate plastic deformation in low-doped GaAs crystal (figure 3a). The initial elastic part of the *P-h* relationship fulfills Hertz equation $P=(4/3)E_r\sqrt{(R)}h^{3/2}$ (E_r – reduced Young's modulus) that presents classical theory of elastic contact between the sphere of radius *R* and isotropic half-space [19]. This made possible to estimate the indenter tip radius to R=180 nm, a parameter necessary for mean contact pressure calculations. The pop-in event occurs at ~18 GPa while indenter penetration depth reaches the value of h=19 nm. Furthermore, the mean contact pressure release Δp_m=2.6 GPa is caused by pop-in. What comes as a revelation is that current increase initiated before the pop-in event drops down immediately at the instant of pop-in (figure 3b) *i.e.* the origin of GaAs incipient plasticity is also a cause of unexpected restoration of the perfect Shottky barrier.

The pop-in events were observed in almost all nanoindentations even though a few *P-h* curves without any singularities were recorded. It is explained [12] that the absence of pop-in event results in no current flow trough the contact due to indentation in the certain place affected by preexisting dislocations. In this case the plastic deformation is expected to occur mainly by dislocation activity.

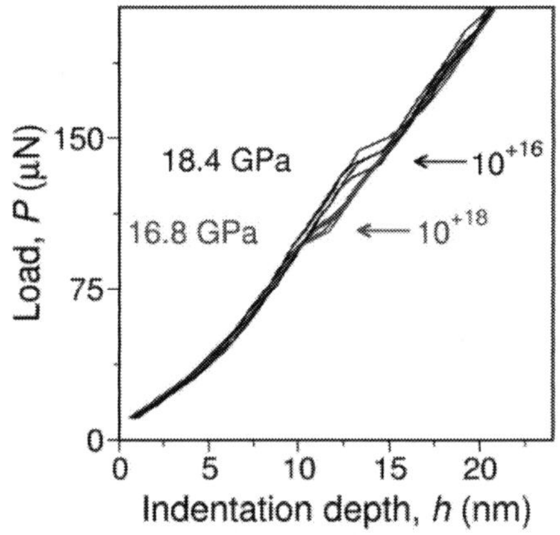

Figure 4. Influence of the silicon concentration on the mean contact pressure at the GaAs pop-in.

The decrease of the mean contact pressure at the pop-in caused by increased doping (figure 4) agrees with the DFT calculated decrease of the hydrostatic pressure at which GaAs-I and GaAs-II phases can exist in equilibrium (figure 5). This suggests a non-dislocation origin of the pop-in events, a conclusion supported by experiments showing that silicon doping of GaAs single crystals increase their critical resolved shear stress [20]. However, the influence of Si doping on the mechanical properties of GaAs is still unclear. Indeed, Yonenaga *et al* [21] showed that the critical stress for dislocation formation in silicon doped GaAs continuously decreases with temperature, being also greater than the critical stress measured for undoped GaAs. On the other hand, there are also literature reports showing that silicon doping makes GaAs softer at low temperatures [22,23]

In consequence, an explanation of the electro-mechanical coupling at pop-in can be derived from an assumption that the GaAs-I to GaAs-II phase transformation marks plastic deformation in GaAs. Findings by Benson *et al* [11] show that phase transformation from GaAs-I to GaAs-II starts at hydrostatic pressure of 12.5±1.5 GPa. The transition is continued until the pressure of ~18 GPa is reached. The rocksalt-like GaAs phase is thermodynamically unstable during pressure releasing and transforms into initial zinc-blende phase or composition of zinc-blende and amorphous phases depending on experimental conditions [11]. Comparison of the mean contact pressures achieved during our nanoindentations with pressure range of GaAs-I → GaAs-II transition makes the phase transformation scenario a favorable explanation of the onset of gallium arsenide plastic deformation. Furthermore, Li *et al* [24] also considered the possibility of phase transformation during Vickers indentation of (110) surface of GaAs because of direct observation of amorphous phase in residual impression.

In order to support our main idea that the onset of plastic deformation (pop-in) in GaAs crystal is caused by GaAs-I → GaAs-II phase transformation rather than dislocation nucleation we studied nanoindentation of GaAs by molecular dynamics simulations [13]. The interaction between the indenter and GaAs crystal revealed a singularity in the *P-h* curve corresponding to a pop-in event.

Analysis of the GaAs structure in the small volume right under the indenter showed a new peak at 90° angle while the prior-to-indentation peak at angle of 110° (characteristic for zinc-blende structure) completely disappeared. Confirmation of nanoindentation induced GaAs-I → GaAs-II phase transition comes from direct visualization of the transformed structure. The six-coordinated rocksalt-like phase was detected in the obtained visualization of the stressed GaAs structure simulated by MD method [13].

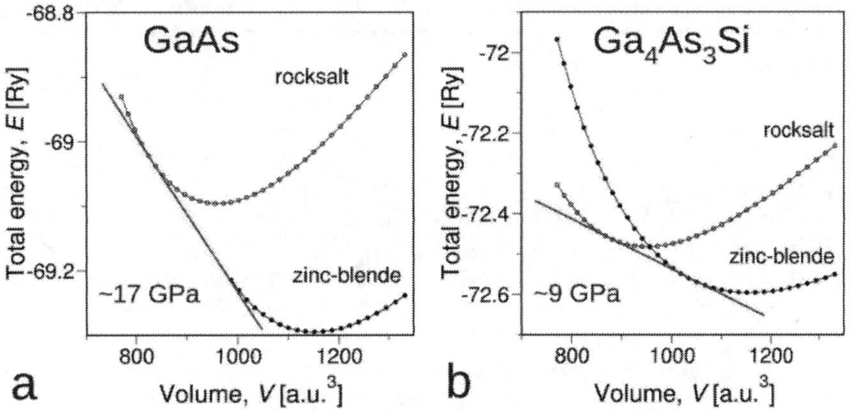

Figure 5. Results of the DFT calculations showing decrease of the equilibrium hydrostatic pressure of the GaAs zinc-blende/rocksalt phases due to silicon doping.

Further argument supporting phase transformation origin of GaAs incipient plasticity comes from first-principle calculations of energy bands in the vicinity of the interface between semiconducting zinc-blende and metallic rocksalt structures. The calculated spacial distribution of energy levels exhibit band bending characteristic for Shottky barrier of metal-semiconductor junction [12]. In this way one can understand why the current flow initiated in the elastic part of nanoindentation drops down at the instant of pop-in event (figure 3b). Namely, if the mean contact pressure is sufficient to start GaAs-I → GaAs-II transformation, the nucleated rocksalt-like phase restores Shottky barrier between small volume of metalic GaAs-II phase and surrounding semiconducting GaAs-I.

4. Conclusion

In summary, the onset of plastic deformation in GaAs crystal witnessed by displacement burst on nanoindentation load-displacement curve that corresponds to sudden halting of the current running through the indenter/GaAs contact is explained in terms of zinc-blende to rocksalt phase transformation. This results contradicts commonly accepted dislocation-based scenario and contributes to our understanding of nanoscale plasticity in III-V semiconductor compounds.

Acknowledgements
The authors acknowledge support from the Academy of Finland (FINNANO project). The calculations were performed at the CSC-IT Center for Science, Finland.

References

[1] Schuh C A, Mason J K, and Lund A C 2005 *Nature Materials* **4** 617
[2] Szlufarska I, Nakano A, and Vashista P A 2005 *Science* **309** 911
[3] Corcoran S G, Colton R J, Lilleodden E T, and Gerberich W W 1997 *Phys. Rev. B* **55** R16057
[4] Gouldstone A, Van Vliet K J, and Suresh S 2001 *Nature* **411** 656
[5] Ruffell S, Bradby J E, Fujisawa N, and Williams J S 2007 *Journal of Applied Physics* **101** 083531

[6] Ge D, Minor A M, Stach E A, and Morris J W 2006 *Philosophical Magazine* **86** 4069
[7] Bradby J E, Williams J S, Wong-Leung J, Swain M V, and Munroe P 2001 *Applied Physics Letters* **78** 3235
[8] Leipner H S, Lorentz D, Zecker A, Lei H, and Grau P 2001 *Physica B* **308-310** 446
[9] Lorentz D, Zeckzer A, Hilpert U, Grau P, Johansen H, and Leipner H S 2003 *Phys. Rev. B* **67** 172101
[10] Bradby J E, Williams J S, and Swain M V 2004 *Journal of Materials Research* **19** 380
[11] Benson J M, Itie J P, Polian A, Weill G, Mansot J L, and Gonzalez J 1991 *Phys. Rev. B* **44** 4214
[12] Nowak R, Chrobak D, Nagao S, Vodnick D, Berg M, Tukiainen A, and Pessa M 2009 *Nature Nanotechnology* **4** 287
[13] Chrobak D, Nordlund K, and Nowak R 2007 *Phys. Rev. Lett.* **98** 045502
[14] http://www.quantum-espresso.org/
[15] Nordlund K, Ghaly M, Averback R S, Caturia M, Diaz de la Rubia T, and Tarus J 1998 *Phys. Rev. B* **57**, 7556
[16] Albe K, Nordlund K, Nord J, and Kuronen A 2002 *Phys. Rev. B* **66** 035205
[17] Ziegler J F, Biersack J P, and Littmark U 1985 *The Stopping and Range of Ions in Matter* (New York: Pergamon Press)
[18] Sze S M 1981 *Physics of Semiconductor Devices* (New York: Willey-Interscience Publisher)
[19] Hertz H 1882 *J. Reine Angew. Math.* **92** 156
[20] Bourret E D, Tabache M G, Beeman J W, Elliot A G, and Scott M 1987 *Journal of Crystal Growth* **85** 275
[21] Yonenaga Y and Sumino K 1989 *Journal of Applied Physics* **65** 85
[22] Rabier J, Garem H, Demenet J L, and Veyssiere P 1985 *Philosophical Magazine A* **51** L67
[23] Boivin P, Rabier J, and Garem H 1990 *Philosophical Magazine A* **61** 647
[24] Li Z C, Liu L, Wu X, He L L, and Xu Y B 2002 *Materials Letters* **55** 200

H-initiated extended defects from plasma treatment: Comparison between c-Si and mc-Si

Heidi Nordmark[1,2], Alexander G. Ulyashin[3], John C. Walmsley[1,2] and Randi Holmestad[1]

[1]SINTEF Materials and Chemistry, NO-7465 Trondheim, Norway
[2]Department of Physics, NTNU, 7491 Trondheim, Norway
[3]SINTEF Materials and Chemistry, P.O. Box 124 Blindern NO-0314 Oslo, Norway

Heidi.Nordmark@sintef.no

Abstract. In this study n- and p-type polished Czochralski-grown Si (Cz-Si) and p-type polished and as-cut multi-crystalline (mc) Si wafers have been directly H plasma treated in a plasma enhanced chemical vapour system in order to study H subsurface defect formation. Raman spectroscopy, secondary ion mass spectroscopy, scanning electron microscopy and transmission electron microscopy have been used to characterise the samples. In polished Cz-Si wafers, H induced defects were only observed up to 1 µm below the surface [1,2], while in similarly treated mc samples H induced defects were observed on grain boundaries and dislocations up to several µm below the surface [3]. It is also established that the distribution of H in the subsurface regions of the Cz-Si substrates after hydrogenation as well as the formation of structural defects depend on the type of doping. Evolution of Si-H_x bonds in hydrogenated Cz Si samples starts at 400 °C, while evolution of H initiated structural defects starts at 600 °C, when SiH_x bonds are mostly dissolved.

1. Introduction

Hydrogen plasma treatment is commonly used to passivate dangling bonds and impurities in Si to improve its carrier lifetime. However, the solubility of H in Si is very low [4], so the introduction of high H concentrations might lead to the formation of defects that act as "lifetime killers". H defect formation is well studied and understood in mono-Si [1-17] but less is known about H segregation and interaction with extended defects like grain boundaries and dislocations in multicrystalline (mc) Si that is very widely used in solar cell fabrication.

Hydrogen plasma treatment can also be used to texture wafer surfaces on Cz Si as it etches and redeposits Si on the surface forming cones with {112}/{113}/{115} surfaces, depending on the hydrogenation temperature[5]. Surface texturing is also very dependent on the formation of {111} platelets in the wafers subsurface [6,7]. The surface morphology is found to differ for different grain orientation[8], surfaces being almost flat for {112} oriented grains and covered with large symmetrical cones for surfaces oriented close to {100}[3].

In this article several samples are studied, both Cz Si and mc Si, and different characterization techniques including scanning electron microscopy (SEM), transmission electron microscopy (TEM), secondary ion mass spectroscopy (SIMS) and Raman spectroscopy, are combined to get a deeper understanding of the H induced defects arising after heavily H plasma treatment.

In our prior work we have studied the temperature evolution of Cz Si samples [1,2] and as-cut and polished mc Si samples [3] in detail. In this article we have added two new samples, to study the influence of substrate temperature during H plasma treatment on H defect formation. Unpublished TEM and SEM images together with new Raman and SIMS results, from the samples studied in [1-3] are combined to try to get a better understanding of the H induced defects.

2. Experimental

n- and p-type polished Cz Si, p-type polished and as-cut mc Si wafers have been studied. Table 1 shows the sample processing details. Samples are denoted as A_d_x_y, where; A is the sample (column 1 in Table1), d the doping (column 2 in Table1), x and y are the post-annealing temperature and time respectively. Hydrogen plasma treatments of all wafers were performed in a Plasma Enhanced Chemical Vapor Deposition (PECVD) setup, as described in Table 1. Subsequent annealing was performed on ~ 1×1 cm samples at temperatures between 200 and 1000 °C for 1 and 10 hours in a furnace in air followed by a rapid cooling. U1 and U2 have not been described previously.

Table 1. Processing parameters and H^+ treatment of the samples. The different samples are denoted as A_d_x_y, where; A is the sample (column 1 in Table1), d the doping (column 2 in Table1), x and y are the post-annealing temperature and time respectively. "None" means that the sample is not post annealed, denoted as 0_0 in the text.

Sample	Doping	H^+ plasma treatment	Post annealing	Ref.
U2	p-type 1 Ωcm	1h, 120 °C 110 MHz	None	New
U1	p-type 1 Ωcm	1h, 180 °C 110 MHz	None	New
S1	n-type 1.8-2.6 Ωcm	1h, 250 °C 110 MHz	None, 1h at 200-1000 °C, 100 °C step, 10 h at 200-600 °C, 100 °C step, 20 h 400 °C	[1,2]
S2	n-type 0.005-0.018 Ωcm	1h, 250 °C 110 MHz	None, 1h at 200-1000 °C, 100 °C step, 10 h at 200-600 °C, 100 °C step, 20 h 400 °C	[1,2]
S3	n-type 1.8-2.6 Ωcm	2 h, 250 °C 13.56 MHz	None, 1h at 200-1000 °C, 100 °C step, 10 h at 200-600 °C, 100 °C step, 20 h 400 °C	[1,2]
S4	p-type 1-10 Ωcm	0.5 h, 250 °C 110 MHz	None, 1h at 200-1000 °C, 100 °C step, 10 h at 200-600 °C, 100 °C step, 20 h 400 °C	[1,2]
I.2/I.4	p-type 10 Ωcm	1h, 250 °C 110 MHz	None/2h 450 °C	[3]
S5	p-type 10-20 Ωcm	1h, 250 °C 110 MHz	None, 1h at 200-1000 °C, 100 °C step, 10 h at 200-600 °C, 100 °C step, 20 h 400 °C	[1,2]
mc Si polished	p-type 10 Ωcm	1h, 250 °C 110 MHz	450 °C, 10h	[3]
mc Si as-cut	p-type 10 Ωcm	1h, 250 °C 110 MHz	450 °C, 10h	[3]

Cross-section samples were prepared for TEM study from every annealing condition by mechanical polishing down to a thickness of about 10 μm in the middle of the 3 mm disk and then thinned to electron transparency using a Gatan PIPS ion mill. The TEM-observations were performed using a

Phillips CM30 operating at 300 kV or a JEOL 2010F operating at 200 kV. The cross section plane used for observations was close to (110).

SEM measurements were performed using a secondary electron detector on a Hitachi s-3400se Field emission SEM (FESEM) operating at 5-20 kV and working distance 8 mm. The resolution of the instrument is 1.5 nm at 15kV.

Raman spectra were recorded with a multichannel spectrometer from Horiba (Jobin Yvon) model T 64000, employing a single monochromator with CCD detector at room temperature. The spectra were excited by a Milennia Pro diode-pumped (Nd:YVO_4 crystal) laser from Spectra-Physics (Model J 40) adjusted to give approximately 100 mW of the 532 nm line on the sample. A slit of 200 μm was used and the resolution was ca. 2 cm^{-1}. All spectra were normalised to the maximum of the optical Si phonon line at ~520 cm^{-1}, and the background luminescence was subtracted from the data.

Secondary ion mass spectrometry (SIMS) was used for analysis of the hydrogen/deuterium distributions (CAMECA ims 7f instrument) in negative secondary ion mode with a 15 keV Cs^+ primary beam. The raw SIMS profiles are given as sputter time versus secondary ion intensity. The sputter time is related to depth and the conversion factor is determined by measuring the SIMS craters with a Dektak 8 surface stylus profilometer assuming a linear dependence.

3. Results and discussion

3.1 Hydrogenated Cz Si samples

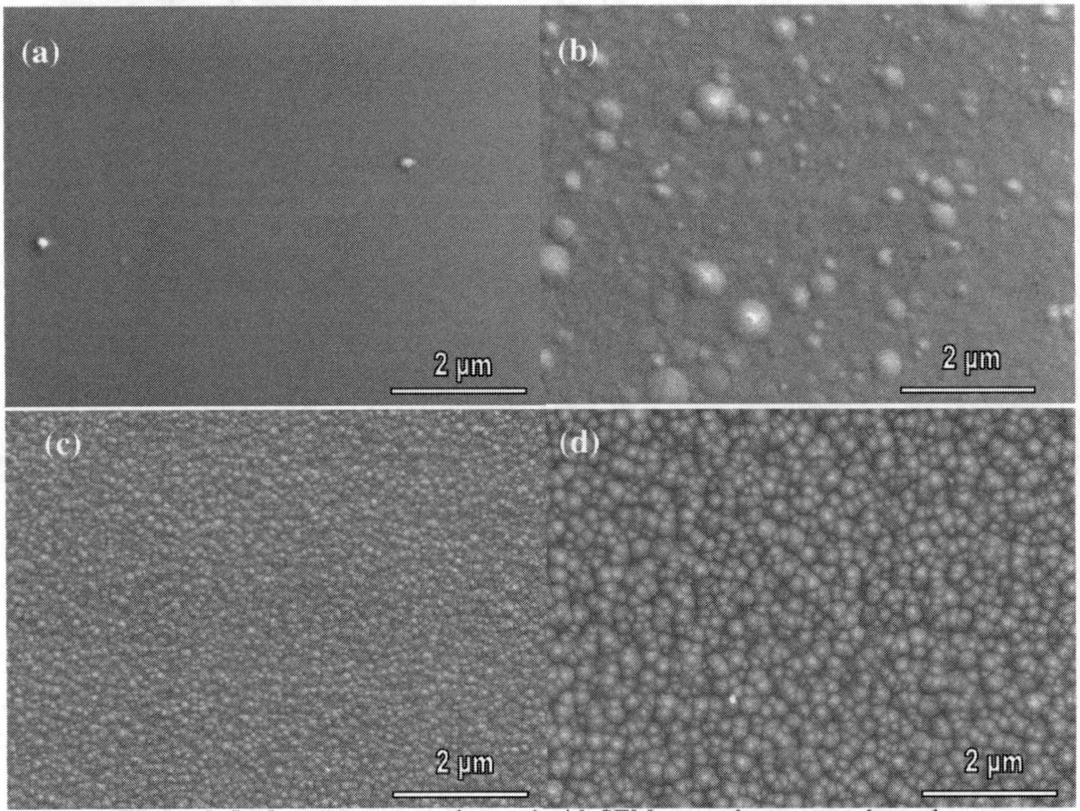

Figure 1. Surface structure as observed with SEM on as plasma treated samples, (a) U2_p-1 Ωcm, (b) U1_p-1 Ωcm, (c) S4_p-1-10 Ωcm, (d) L2_p-10 Ωcm

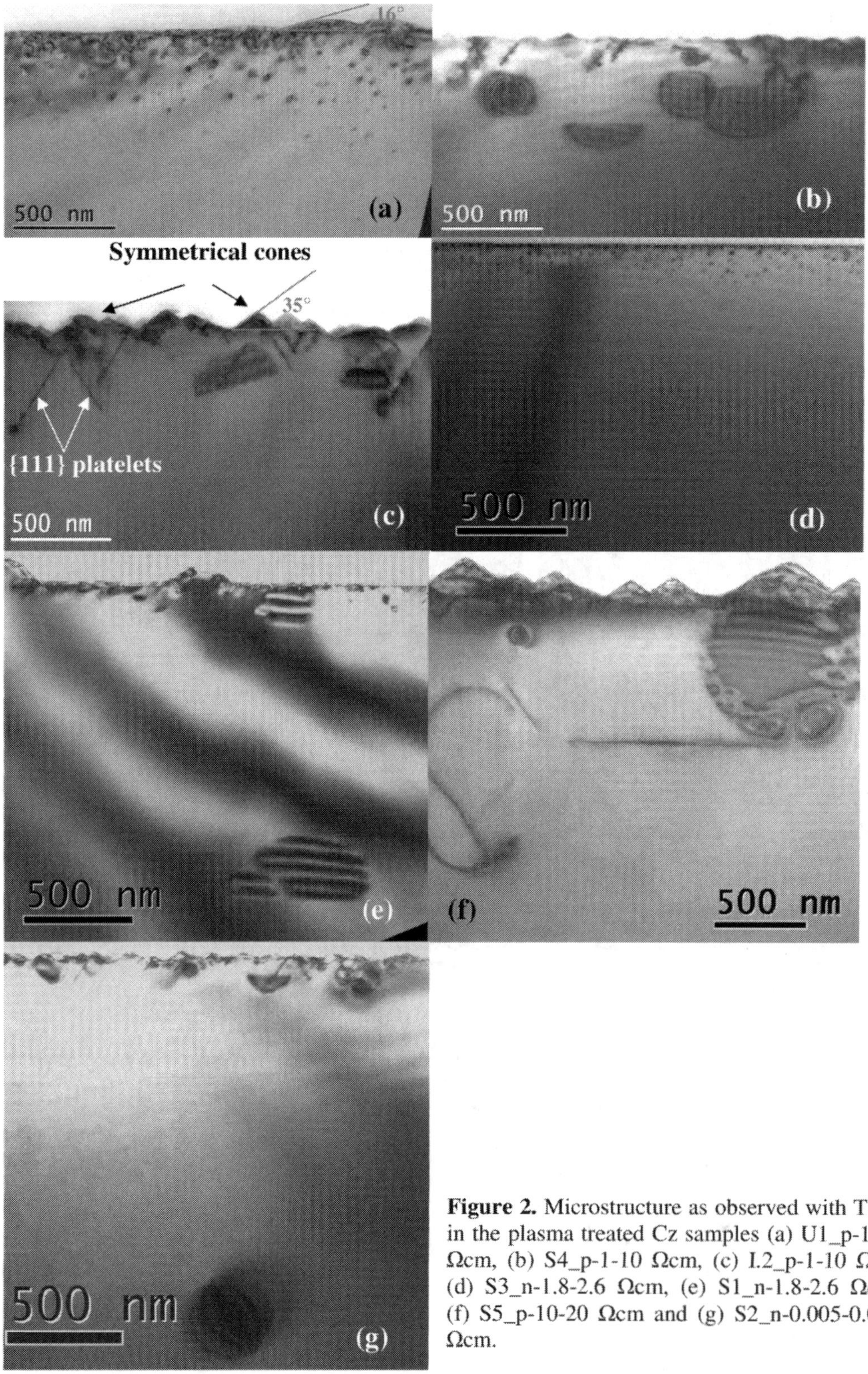

Figure 2. Microstructure as observed with TEM in the plasma treated Cz samples (a) U1_p-1-10 Ωcm, (b) S4_p-1-10 Ωcm, (c) L2_p-1-10 Ωcm (d) S3_n-1.8-2.6 Ωcm, (e) S1_n-1.8-2.6 Ωcm, (f) S5_p-10-20 Ωcm and (g) S2_n-0.005-0.018 Ωcm.

Figure 2 (a)-(d) show TEM images of the Cz Si samples. No image is shown from sample U2_p-1 Ωcm, since no defects were observed in this sample. Due to the low hydrogenation temperature, 120°C, no platelets have formed. Since surface roughness depends on formation of {111} platelets [6,7], the surface is flat. In U1_p-1 Ωcm, plasma treated at 180°C, a high density of small defects, ~20 nm, was observed, formed up to about 1.2 μm below the surface. The formation temperature for platelets is somewhere between 120°C and 180°C. From the literature ~150°C is found to be the critical temperature for hydrogen platelet formation [7,9]. The microstructure is similar in samples S4_p-1-10 Ωcm and I.2_p-1 Ωcm plasma treated at 250 °C. Large platelets, 50-500 nm, are formed up to ~ 0.7 μm below surface. Lower substrate temperature during H plasma treatment gives a higher density of small platelets formed deeper into the bulk.

In Figure 2(d) from sample S3_p-1.8-2.6 Ωcm, very small platelets ~ 20 nm in diameter, can be observed up to 0.5 μm below the sample surface. The low plasma frequency used for 2 hours at 13.56 MHz, compared to 1 hour at 110 MHz for similar doped S1_1.8-2.6 Ωcm, Figure 2(e), led to less surface roughness and formation of smaller platelets [1,2]. Figure 2(e) and (f) show images from samples S1_n-1.8-2.6 and S5_p-10-20 Ωcm, respectively. In both of these samples H induced platelets were observed several μm below the substrate surface. In sample S1 a low density of large platelets, ~0.5 μm, are formed up to 6 μm below the surface [1,2], while in sample S5_p-10-20 Ωcm, a very low density of large platelets were formed up to ~3 μm below the surface. In the high n doped sample S2_n-1.8-2.6 Ωcm, very large platelets, 1-2 μm in diameter, has been observed, formed up to ~ 2 μm below the surface of the sample [1,2]. In low doped samples, platelets are formed deep into the samples. Note the flat cones in the sample H plasma treated at 180 °C make an angle of ~ 16 ° with the surface, Figure 2(a), compared to the other samples H plasma treated at 250 °C make an angle of ~35° with the surface, Figure 2(c). This fits well with the literature [5] and the cones form directions {115}/{112} with the (001) surface respectively.

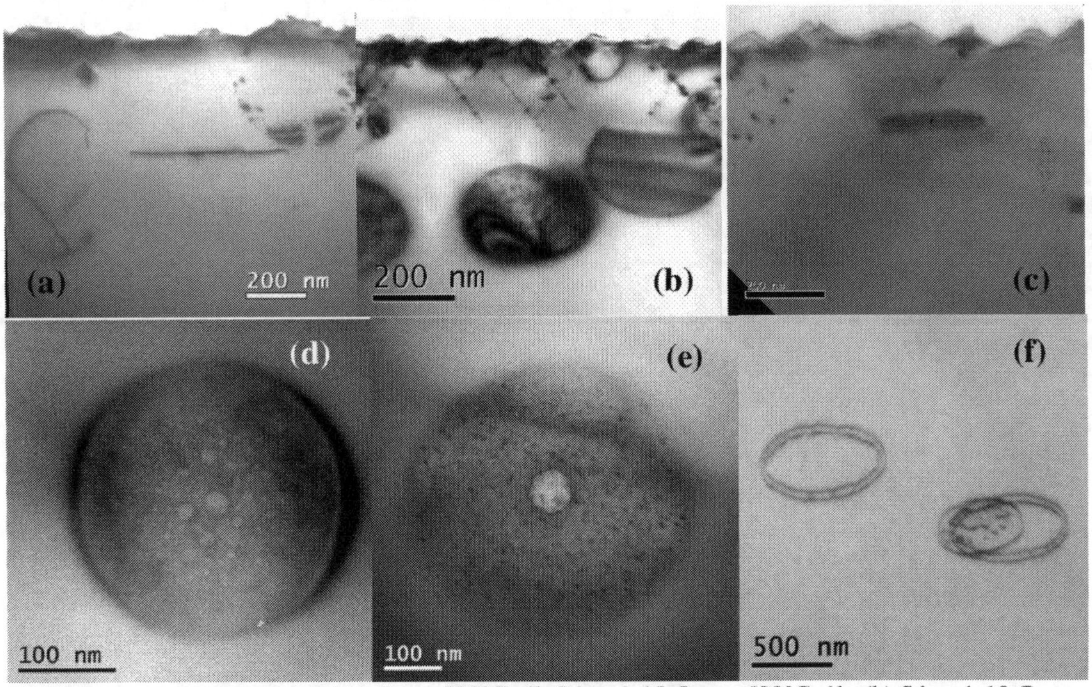

Figure 3. (a) S2_n-0.005-0.018 Ωcm_500°C_4h S4_p-1-10 Ωcm _600°C_1h, (b) S4_p-1-10 Ωcm _600°C_1h, (c) S5_p-10-20 Ωcm _600°C_1h, (d) S1_ n-1.8-2.6 Ωcm_300°C_10h (d) S1_ n-1.8-2.6 Ωcm_400°C_20h (e) dissolved platelets in S1_n-1.8-2.6 Ωcm _600°C_1h.

After annealing at 400 °C not much happened to the microstructure in the samples, except for S3 where the density of platelets is much lower. At the very top defect layer, the defect density decreases and the small defects start to dissolve [1,2]. Larger defects in the subsurface and defects formed deep into the sample are still stable [1.2]. Figure 3(a) shows the same platelets in S2_n-0.005-0.018 Ωcm as in Figure 2(g), after 500 °C annealing for 4 hours in the TEM. The small defects in the cones have dissolved and the large truncated plate in the upper right corner is also partly dissolved.

After 600 °C annealing for 1 hour also platelets formed deep into the samples start to dissolve in n type and low doped p type samples [1,2]. This is seen in Figure 3(b) and (c) where the microstructure after 1 h annealing at 600 °C is shown for sample S4_p-1-10 Ωcm and S5_p-10-20 Ωcm respectively. Dissolution of H induced platelets in n type samples at 600 °C is not shown. While the H induced platelets in S4 are almost unchanged, the platelets in the lower doped S5_p-10-20 Ωcm have started to dissolve. H induced platelets are well known to contain hydrogen bubbles after heavy hydrogenation [8,10,18,19]. Figure 3(c) and (d) show complete H induced platelets formed deep below the surface of sample S1, annealed at T < 600 °C. Figure 3 (d) shows a platelet with bubbles filled with H_2 gas, Figure 3(e) a H induced platelet with central H_2 gas filled cavity and granular contrast. Both types of platelets exist simultaneously in the samples. Figure 3(f) shows partly dissolved H induced platelets annealed for 1 hour at 600 °C. Only dislocations loops and small empty voids are left [1,2]. For annealing at higher temperatures the empty voids will grow in size due to Ostwald ripening [10].

It is important to note that the above-mentioned peculiarities of the H initiated defect formation in Cz-Si samples treated by hydrogen plasma are in a good agreement with those observed in [18-22].

3.2 Polished and as-cut mc Si

In the mc samples the surface is also etched by the H plasma as shown in Figure 4(a) and (b). Figure 4 (a) shows the surface of H plasma treated polished mc Si. The surface roughness and cone shape vary due to different grain orientation [3]. In some orientations the surface is almost flat while in others tilted cones are visible on the surface. The density and size of the cones also varies for the different grain orientations. Due to height differences between the grains, edge peculiarities can be observed at the interface of the grains in the image. Figure 4(b) shows another grain boundary at larger magnification. Fine structures can be observed on the almost flat surface on the right grain.

Figure 4. Surface structure as observed with SEM, (a) Triple grain boundary in polished mc Si_p-10 Ωcm, (b) Grain boundary in polished mc Si_p-10 Ωcm.

The microstructure of a grain with surface normal ≠ {100} is shown in Figure 5. H induced defects in this sample are similar to the H induced defects formed in (100) oriented substrates. The platelet defects still form on {111} and cones still form on the surface, although they are not symmetrical and are more sparsely distributed. Prior investigations have showed that the cones formed in grain orientations other than (100) also have lateral surfaces of {112}, explaining their tilted appearance on the surface [3]. The density of defects in the subsurface and the platelet formation depth of the grain in Figure 5 are similar to the similarly treated Cz sample in Figure 2(c). Grains oriented close to {112} were found to have almost flat surfaces while grains oriented in {100} directions were found to have high density of symmetrical cones on their surfaces [3]. Some of the platelets in Figure 5 have started to dissolve due to post annealing for 10 h at 450 °C.

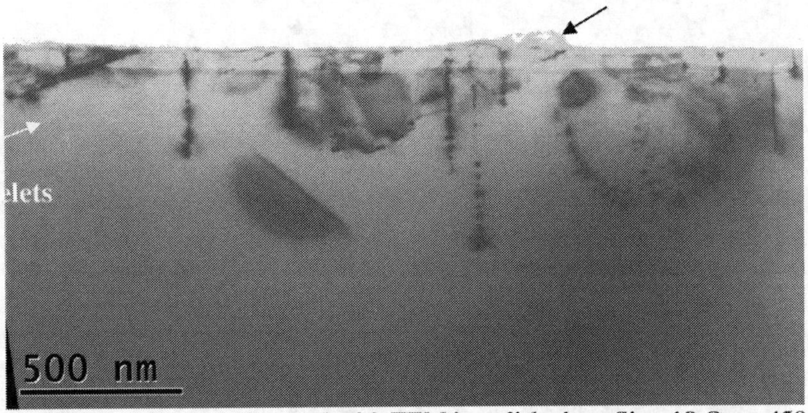

Figure 5. Microstructure as observed with TEM in polished mc Si_p-10 Ωcm_450°_10h.

Figure 6 (a) and (b) show the defect formation in as-cut mc Si samples. A high density of H induced defects can be observed in the subsurface, but also deeper into the sample on defects from sawing and in structural defects like grain boundaries and dislocations [3]. Structural defects act as effective H traps by lowering the platelet formation energy. In similar H plasma treated Cz-Si, Figure 2(c), H induced defect were only observed in the upper 1 μm subsurface of the substrate [3].

3 SIMS and Raman measurements

Figure 7 (a)-(d), show SIMS hydrogen depth profiles from selected Cz-Si samples. The hydrogen detection limit is ~10^{18} cm^{-3}. From Figure 7(a) it is observed that, after hydrogenation, S1_n-1.8-2.6 Ωcm (green curve) and S5_p-10-20 Ωcm (red curve) show similar H concentration profile with a second peak about 3-4 μm into the bulk, and that hydrogen is detected up to about 12-15 μm depth. In I2_p-1 Ωcm _0_0 (black curve), however, the hydrogen concentration drops rapidly below the detection limit at about 0.7 μm depth. Figure 7(b) shows details of the same profiles as 7(a) of the samples in the near-surface area. After 1 hour hydrogenation at 250 °C the samples contain about 1%, or $5x10^{20}$ cm^{-3}, hydrogen in the top layer, decreasing rapidly down to 0.1% at about 0.2-0.5 μm depth, depending on sample doping and type. The hydrogen concentration drops faster in S1_n-1.8-2.6 Ωcm(green curve) and S5_p-10-20 Ωcm (red curve) than in I2_p-1 Ωcm (black curve).

Hydrogen profiles after post annealing are shown for sample S1_n-1.8-2.6 Ωcm and S4_p-1-10 Ωcm in Figure 7(c) and (d), respectively. After annealing at 500 °C for 10 hours in air, hydrogen is still present in the n-type Si sample S1_n-1.8-2.6 Ωcm _0_0, Figure 7(c) (red curve), but the second peak is found closer to the surface, at about 1μm, and the hydrogen level drops below the detection limit at ~10 μm. After 10 hours annealing at 600 °C, almost no hydrogen is detected in S1_n-1.8-2.6 Ωcm (green curve), while in S4_p-1-10 Ωcm a second peak at about 0.2 μm depth can be observed in Figure 7(d) (red curve). In Figure 7(d), Sample S4_p-1-10 Ωcm is similar to I.2_p-10 Ωcm, except from shorter hydrogenation duration, 0.5 h compared to 1h. While the H concentration peak in sample I2_ p-10 Ωcm is very close to the surface of the sample, there is a second H peak in S4_ p-1-10 Ωcm at 0.25μm after 600 °C annealing, suggesting that some H close to the surface has diffused out of the sample.

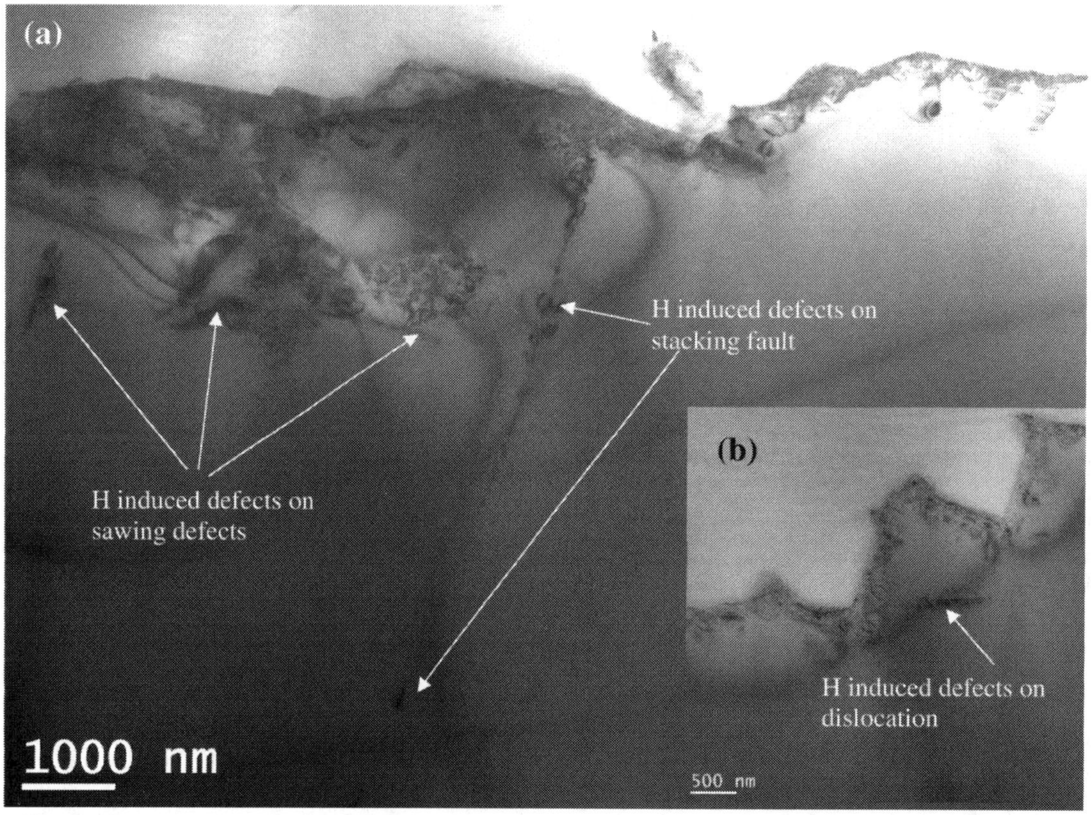

Figure 6. Microstructure observed with TEM in as cut mc Si_p-10 Ωcm. H induced defects form in the subsurface, on sawing defects and stacking faults (a) and on dislocations (b).

The difference in hydrogen profile for p-doped samples I.2_p-10 Ωcm /S4_p-1-10 Ωcm and S5_p-1-10 Ωcm is most likely due to the difference in doping level. Boron is known to trap hydrogen effectively by forming B^{-}-H^{+} complexes [11,12] which are stable up to ~ 250 °C. As a result, hydrogen diffuses deeper into the bulk in a low doped sample such as S5. Phosphorus can also trap hydrogen by forming P^{+}-H^{-} complexes, but this is less probable and these complexes are less stable.

An early SIMS study of deuterium [12] showed the connection between doping and diffusion. In most cases the H concentration of n and p type silicon samples was found to decay exponentially with depth, while p type samples, with resistivity 0.1 and 1 Ωcm showed trap dependent diffusivity [12]. The substrate temperature also changed the profile [12]. Because of the very low solubility of hydrogen [4], high concentrations of in-diffused hydrogen result in hydrogen induced defect formation in the subsurface. Comparing the H concentration profiles of sample S1_n-1.8-2.6 Ωcm, I.2_p-10 Ωcm, S4_p-1-10 Ωcm and S5_p-10-20 Ωcm with the TEM observations in Figure 2 and 3, there is a strong correlation between the hydrogen concentration peaks and the hydrogen platelet formation. In sample S4/I.2, the platelets are formed up to ~ 0.7 μm depth, while in sample S1 and S5, a low density of large platelets is observed several μm below the sample surface[1,2]. The density of platelets formed deep into the sample is higher in S1 than in S5, in accordance with the higher concentration of H observed in S1 than in S5 in this region.

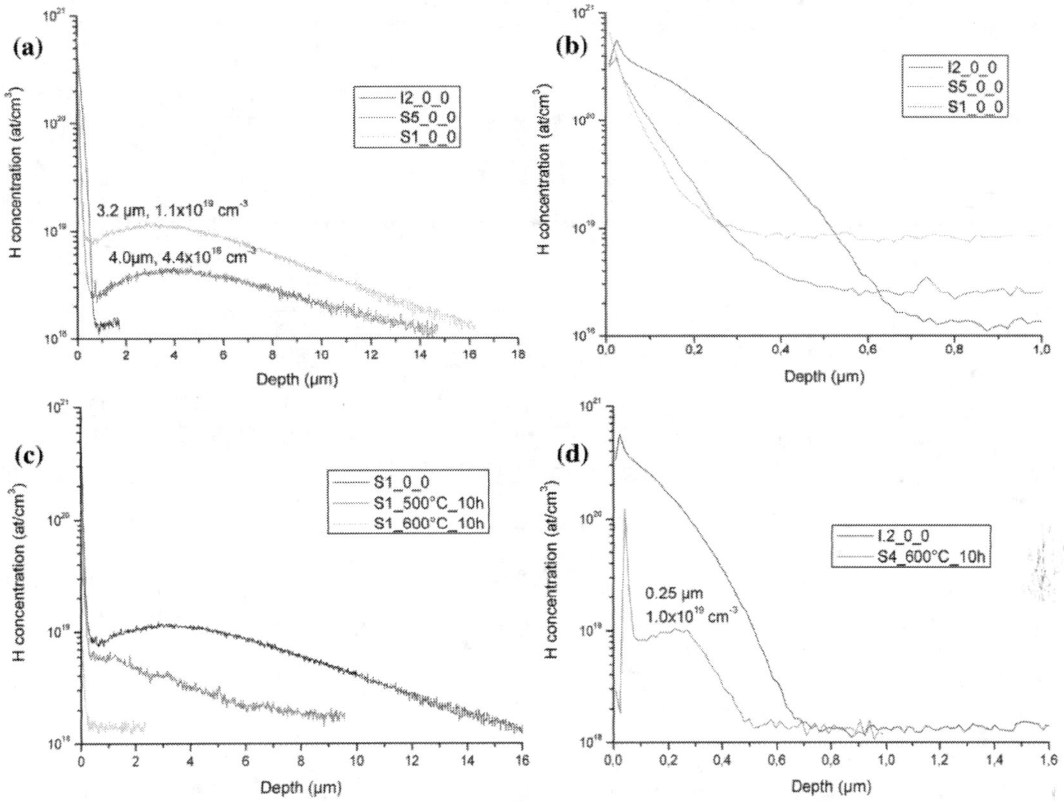

Figure 7. Hydrogen concentrations in hydrogen plasma treated samples measured by SIMS, (a) selected n and p type samples, S1_n-1.8-2.6 Ωcm, S4_p-1-10 Ωcm, S5_p-10-20 Ωcm, before annealing, (b) near-subsurface area, (c) S1_n-1.8-2.6 Ωcm after post annealing (d) I.2_p-10 Ωcm and S4_p-1-10 Ωcm after post annealing

Figure 8 shows Raman spectra from the Cz Si samples. The spectrum from U2_p-1 Ωcm is not shown since no hydrogen was detected, which can be explained by the low-concentration (below the sensitivity limit of the Raman measurements) of hydrogen, incorporated into the Si subsurface at 120 °C. At higher plasma treatment temperatures the Si-H bonds were detected by Raman. The spectra consist of several peaks, as shown by the green lines found by a Lorentzian fit (red) of the curves (black). The different lines are due to different Si-H_x complexes, where x varies from 1-4, and are shifted relatively due to different internal stresses in the supersaturated subsurfaces [8]. The dominant peak in all samples after H plasma treatment is ~ 2095 cm^{-1}. The p type samples, S4_p-1-10 Ωcm_0_0 and S5_ p-10-20 Ωcm_0_0 show the highest signal of Si-H_x bonds in Figure 8(d) and (e) respectively.

After 1 hour post annealing at 300°C the Si-H_x peaks lower slightly, but the shape is similar, indicating that the silicon-hydrogen bonds are stable at this temperature. However, after 1 h post annealing at 400 °C, only ~2095 cm^{-1} bonds in S4_p-1-10 Ωcm are found to be stable, as shown in Figure 9(c). In the other samples, S1_n-1.8-2.6 Ωcm, S2_n-0.005-0.018 Ωcm and S5_p-10-20 Ωcm shown in Figure 9(a), (b) and (d) respectively, the 2095 cm^{-1} bond still dominates but is very weak. No hydrogen related bonds were observed by Raman measurements in S3 under the measurement conditions applied. U1 (and U2) were not post annealed at this temperature. After 20 hours post annealing of S4 at 400 °C, the peak at 2095 cm^{-1} is strongly reduced and the peak at 2106 cm^{-1} dominates, Figure 10(a). After 1 hour annealing at 600°C, SiH_x bonds with peak at ~2106 cm^{-1} are still detected in this sample, as shown in Figure 10(b).

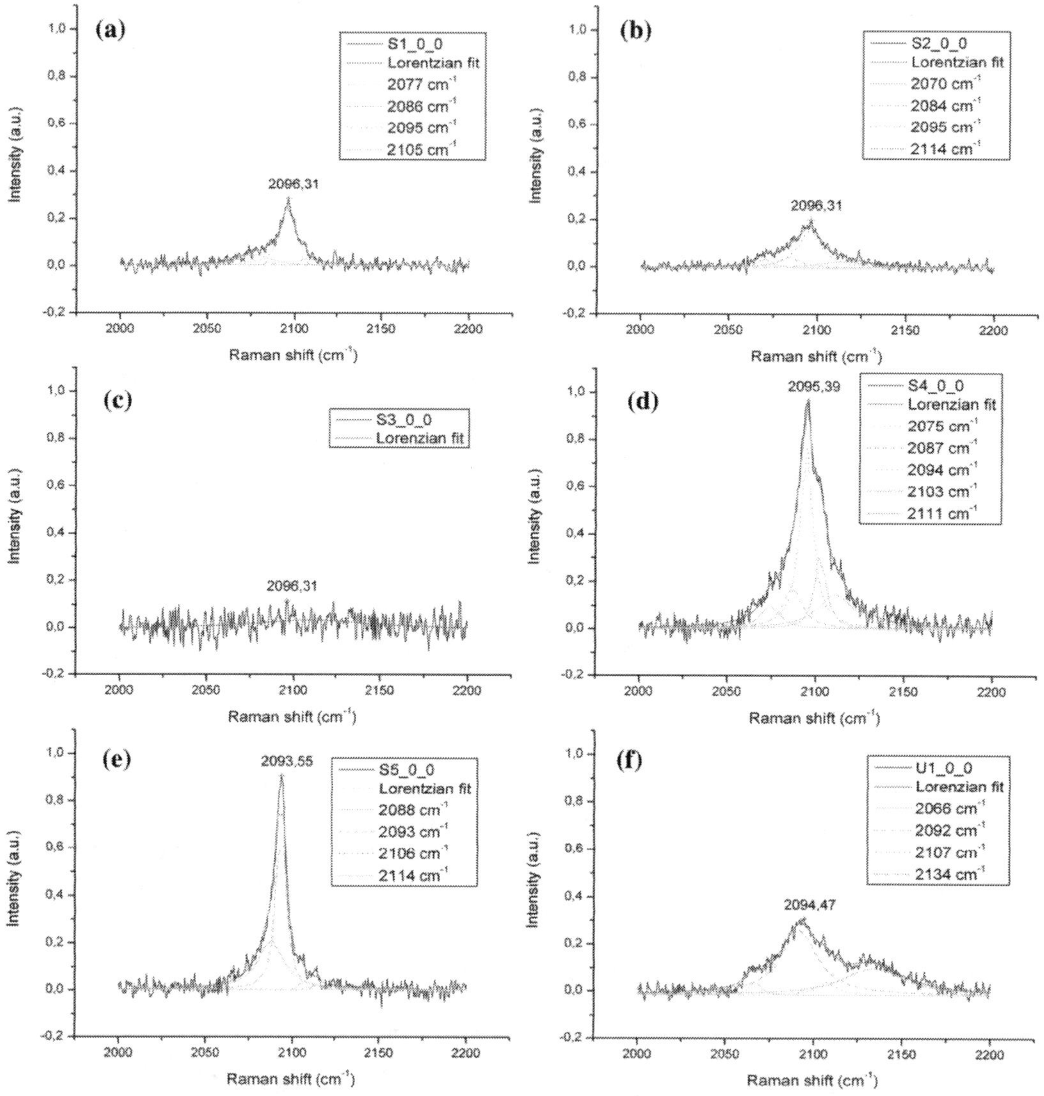

Figure 8. Raman spectra after hydrogenation for the 6 plasma treated samples, (a) S1_n-1.8-2.6 Ωcm_0_0, (b) S2_ n-0.005-0.018 Ωcm _0_0, (c) S3_ n-1.8-2.6 Ωcm _0_0, (d) S4_p-1-10 Ωcm_0_0, (e) S5_p-1-10 Ωcm_0_0 and (f) U1_p-1-10 Ωcm _0_0.

Doping type and level influence the Si-H_x bonds being formed as has also been shown earlier [8,13-17]. Samples plasma treated with a frequency of 13.56 MHz have been found to contain the same amount of Si-H bonds as samples plasma treated with a frequency of 110 MHz [16]. This is not the case in this study, where hardly any Si-H bonds were observed in the low frequency treated sample S3_n-1.8-2.6 Ωcm. In this study several Si-H_x bond peaks were observed as indicated in the Raman spectra. Some of the peaks were very weak and some might be a combination of two or more overlapping peaks that were not resolved. It is hard to draw conclusions about the origin of all the peaks from these room temperature Raman measurements. Low-temperature Raman measurements are required to clarify this question. However, earlier studies of temperature evolution of the different Si-H_x bonds on beveled samples have assigned them to hydrogen induced platelets at different stages of their evolution [16,17,23]. Si-H bonds, with a vibration frequency of about 2095 cm^{-1}, are assigned to the direct plasma damage layer at the top 200 nm of the substrate. Such bonds were found to be stable up to ~350 °C [13,17]. The peak at ~ 2105cm^{-1} has been assigned to Si-H_2 bonds at the platelets inner surface at a final stage, and was found to be the most stable bond above 450 °C [17,24].

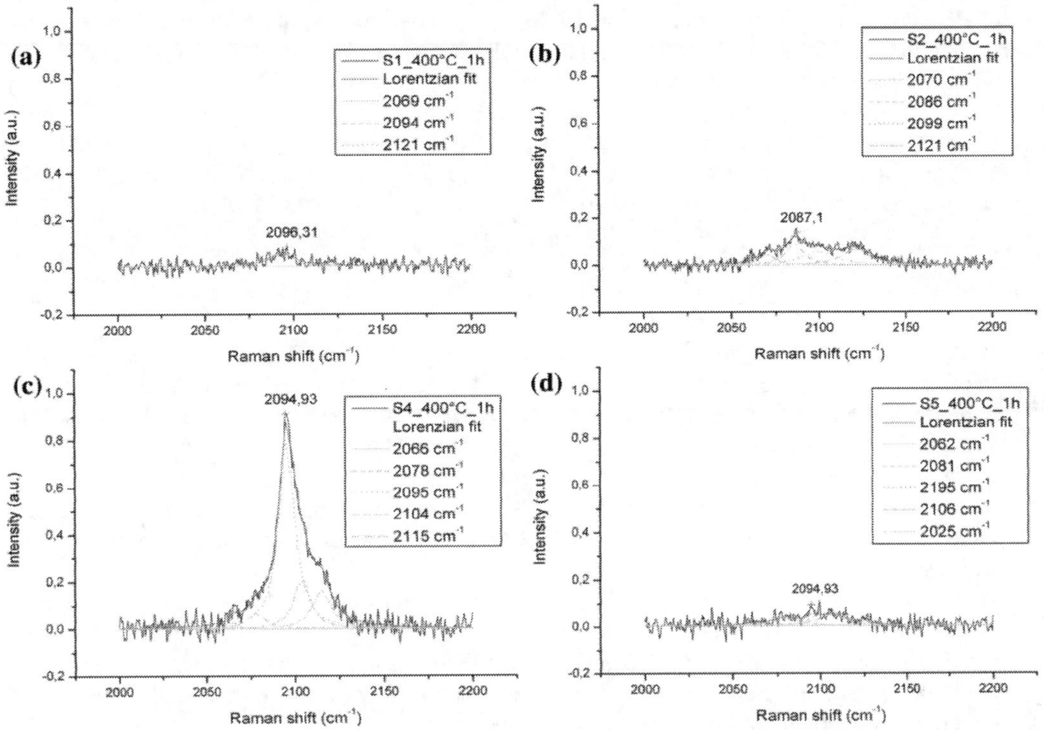

Figure 9. Hydrogen bonds studied by Raman in H plasma treated samples after 1 hour post annealing at 400°C, (a) S1_n-1.8-2.6 Ωcm_400 °C_1h, (b) S2_ n-0.005-0.018 Ωcm_400 °C_1h, (c) S4_p-1-10 Ωcm _400 °C_1h, (a) S5_p-10-20 Ωcm _400 °C_1h

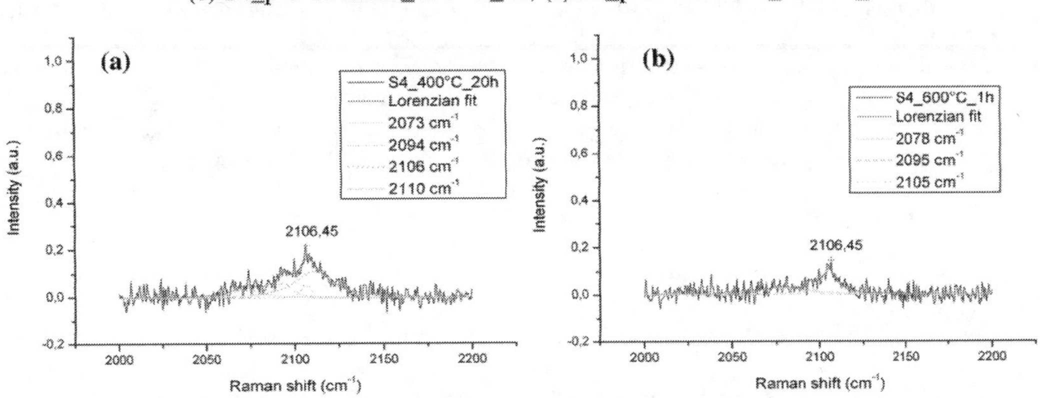

Figure 10. Hydrogen bonds detected by Raman in sample S4_p-1-10 Ωcm after post annealing (a) 20 hours at 400°C and (b) 1 hour at 600°C.

The hydrogen profile observed fits well to the microstructure observed with SEM/TEM. The upper 0.2 μm of the samples consisting of surface cones and numerous defects caused by the plasma is found to be hydrogen rich, both from SIMS and Raman. After annealing at 400 °C for 1 hour, the small defects at the surface are almost annealed out in all samples (observed both in TEM images [1,2] and by the strong reduction of the 2095 cm^{-1} peak in Figure 9), except from S4_p-1-10 Ωcm where no change is observed [1,2]. S4_p-1-10 Ωcm is also the only sample where the hydrogen bonds are stable after this annealing. However, after 20 hours annealing at 400 °C the 2095 cm^{-1} peak is strongly reduced, and the peak at ~ 2106 cm^{-1} dominates also in this sample, Figure 10(a). This peak is also the dominating peak in S4_p-1-10 Ωcm after 1 hour annealing at 600 °C, Figure 10(b). S4_p-1-10 Ωcm, the only sample where the platelets in the subsurface are stable after annealing at 600 °C, still contains a large amount of hydrogen in the subsurface, Figure 7(d), while the hydrogen content is below detection limit in the similar doped n type sample S1_n-1.8-2.6 Ωcm, Figure 7(c). It can be concluded that Si-H_x bonds exhibit a general trend to be dissolved at 600 °C. This result supports similar earlier observations, which were attributed to hydrogen out-diffusion upon annealing at 600 °C [25,26].

4. Conclusions

Our experimental results can be summarised as follows:

- Both surface roughness and defect formation in the sample subsurface after direct H plasma treatment are very dependent on the processing and sample parameters. Lower substrate temperature during H plasma treatment gives a different cone shape and higher density of small platelets formed deeper into the Cz Si bulk than higher substrate temperature.
- In Cz Si samples, H platelet defects form mainly on {111}.
- In mc Si, the density of H induced defects is independent of grain orientation. Some grains with different orientations are more etched than others leading to uneven etching of mc substrates.
- Defects from sawing and structural defects like dislocations and stacking faults in mc Si were found to effectively trap H.
- SIMS and Raman spectroscopy showed that the distribution of H in the subsurface regions of the Cz Si substrate after hydrogenation as well as the formation of structural defects were dependent on the type of doping of this material.
- H induced defects were found to be more stable in p-type samples than in n-type samples with temperature.
- In spite of an evolution of SiH_x bonds at temperatures around 400 °C (Raman), very little evolution of structural defects (TEM) or morphology of the H textured surface (SEM) was observed.
- Evolution of structural defects initiated by hydrogen in the subsurface region of plasma treated Cz-Si samples starts at temperatures 600 °C and above.
- Care has to be taken when exposing both mc and Cz-Si to high concentrations of H during the passivation process used by the Si industry to avoid detrimental H induced defect formation.

References

[1] Nordmark H, Ulyashin A, Walmsley JC, Tøtdal B and Holmestad R 2006 *Nucl. Instrum. Meth. Phys.Res. B* **253** 176
[2] Nordmark H, Ulyashin A, Walmsley JC, Holt A and Holmestad R 2008 *Sol. Stat. Phen.* **131-133** 315
[3] Nordmark H, Ulyashin A, Walmsley JC, and Holmestad R 2009 *J. Appl. Phys.* **105** 033506
[4] Van Wieringen A. and Warmoltz N 1956 *Physica* **22** 849
[5] Ishii M, Nakashima K, Tajima I and Yamamoto M 1992 *Jpn. J. Appl. Phys* **31** 4422
[6] Hwang K-H, Yoon E, Whang K-W and Lee JY 1997 *J. Electrochem. Soc.* **144** 335-339
[7] Hwang K-H, Yoon E, Whang K-W and Lee JY 1995 *Appl. Phys. Lett.* **67** 3590

[8] Ulyashin AG, Job R, Fahrner WR, Richard O, Bender H, Clayes C, Simoen E and Grambole D 2002 *J. Phys.: Condens. Matter.* **14** 13037
[9] Johnson NM, Ponce FA, Street RA and Nemanich RJ 1987 *Phys. Rev. B.* **35** 4166
[10] Grisolia J, Ben Assayag G, Claverie A, Aspar B, Lagahe C and Laanab L 2000 *Appl. Phys. Lett.* **76** 852
[11] Pearton SJ, Corbett JW and Shi TS 1987 *Appl. Phys. A* **43** 153
[12] Corbett JW, Lindstrøm JL, Pearton SJ and Tavendale AJ 1988 *Solar cells* **241** 27
[13] Ma Y, Job R, Huang YL, Fahrner WR, Beufort MF and Barbot JF 2004 *J. Electrochem. Soc.* **151** G627
[14] Düngen W, Job R, Ma Y, Huang YL, Mueller T, Fahrner WR, Keller LO, Horstmann JT and Fiedler H 2006 *J. Appl. Phys.* **100** 034911
[15] Düngen W, Job R, Mueller T, Ma Y, Fahrner WR, Keller LO, Horstmann JT and Fiedler H 2006 *J. Appl. Phys.* **100** 124906
[16] Ma Y, Huang YL, Düngen W, Job R and Fahrner WR 2005 *Phys. Rev. B* **72** 085321
[17] Ma Y, Huang YL, Job R and Fahrner WR 2005 *Phys. Rev. B* **72** 085321
[18] Ghica C, Nistor LC, Bender H, Richard O, Van Tendeloo G and Ulyashin A 2006 *Philosophical Magazine* **86** 5137
[19] Ghica C, Nistor LC, Bender H, Richard O, Van Tendeloo G and Ulyashin A 2007 *J. Phys.D: Applied Physics* **40** 395
[20] Job R, Ma Y and Ulyashin A 2004 *Continuous-nanophase and Nanostructured Materials, MRS Symp.* **788** 571
[21] Job R, Ulyashin AG, Fahrner WR, Beaufort MF and Barbot JF 2002, *Defect and Impurity Engineering Semiconductors and Devices III, MRS Symp. Proc.* **719** 217
[22] Fahrner WR, Job R and Ulyashin AG 2001 *Proc. 2001 1st IEEE Conference on Nanotechnology (IEEE-NANO 2001)* 282-287.
[23] Job R, Ma Y, Huang YL, Ulyashin AG, Fahrner WR, Beaufort MF and Barbot JF 2004 *Solid State Phenom.* **95-96** 141
[24] Job R, Ulyashin AG, Fahrner WR, Beaufort MF and Barbot JF 2003 *Europ. J. Phys.* **23** 25
[25] Ulyashin AG, Job R, Fahrner WR, Grambole D and Herrmann F 2002 *Solid State Phenom.* **82-84** 315
[26] Job R, Ulyashin AG and Fahrner WR 2001 *Mater. Sci. Semicond. Proces.* **4** 257

Journal of Physics: Conference Series **281** (2011) 012030 doi:10.1088/1742-6596/281/1/012030

Microstructures and Magnetism of Silicon co-implanted with Manganese and Carbon ions

L Chow[1], R R Vanfleet[2], M B Huang[3], J LaRose[3], E Del Barco[1], M Arcuri[1] and H Khallaf[1]

[1]Department of Physics, University of Central Florida, Orlando, FL 32816-2385, USA
[2]Department of Physics and Astronomy, Brigham Young University, Provo, UT 84602, USA
[3]College of Nanoscale Science and Engineering, University at Albany-SUNY, Albany, NY 12203, USA

E-mail: chow@mail.ucf.edu

Abstract. Dilute magnetic semiconductors based on transition metal doped silicon have attracted intense interest in recent years due to their compatibility with current silicon technology. Here we present transmission electron microscopy, secondary ion mass spectrometry and ferromagnetic resonance studies of silicon implanted with 1×10^{16} ions/cm^2 of Mn ions and silicon co-implanted with both 1×10^{16} ions/cm^2 of Mn ions and 2×10^{16} ions/cm^2 of carbon ions at a substrate temperature of 350 °C. Afterward, the samples were annealed at temperatures between 800 and 1000°C. The SIMS results show a marked difference between the two specimens while the TEM results show similar features in terms of Mn precipitation and particle evolution. The carbon implanted specimens show additional features that appear to be amorphous silicon pockets within the crystalline implant region. Only one specimen (Mn only implant, unannealed) showed any ferromagnetic properties.

1. Introduction

Magnetic semiconductors have been investigated since the early 1960s [1]. However, interests in magnetic semiconductors gradually waned in the 70s and 80s. In the late 1990s, when spintronic devices were proposed [2,3], magnetic semiconductors become an active research topic again. In particular, spintronic devices based on magnetic semiconducting oxides [4,5] and III-V compound semiconductors [6,7] have been shown with T_c above room temperature. Spintronic devices based on Mn-doped silicon are very desirable due to its compatibility with current silicon technology. In the last several years, both MBE grown Mn_xSi_{1-x} thin films [8] and Mn ion implanted silicon [9] have shown large magnetic moments associated with Mn ions. Since then, there have been many reports that tried to understand the basic mechanism which induced the large magnetic moment in Mn doped silicon [10–16].

Here we compared the microstructures and magnetic behaviour of silicon hosts implanted with only Mn ions and silicon samples co-implanted with Mn and C ions. It was reported recently that ferromagnetism can be obtained by implantation of carbon ions in antiferromagentic Mn_5Si_3 [17]. It is known that the addition of ligands can change the lattice occupation of an impurity atom. For example, co-doped carbon ions can shift the lattice location of ion implanted erbium from the tetrahedral interstitial site to the hexagonal interstitial site in silicon [18, 19]. Here we investigated the effect of carbon co-implantation on the depth profiles and clustering of Mn ions after high temperature annealing.

2. Experimental

In this work, we investigate effects of co-doped carbon impurities on the structural and magnetic properties of Si:Mn, in a hope to identify and tailor the possible microstructures responsible for ferromagnetism in Si. The silicon samples used in this experiment were cut from a float zone (FZ) Si (100) wafer (B-doped with resistivity 1-10 Ohm cm). Carbon atoms at 0.25 at. % are uniformly doped within the 250-nm surface layer of Si (100) using ion implantation and thermal annealing. The carbon ion implantations were conducted at room temperature, with energies of 30, 55 and 87 keV, and doses of 5×10^{15}, 5×10^{15} and 1×10^{16} cm^{-2}, respectively. The carbon ion ranges at these energies are estimated to be 97, 167 and 248 nm, respectively. The C-rich Si and Si control are subsequently implanted at 350°C with two Mn implants, one with an energy of 120 keV and a dose of 3.7×10^{15} cm^{-2}, and the other with an energy of 215 keV and dose of 6.3×10^{15} cm^{-2}. The Mn ion ranges are estimated to be 108 and 188 nm, respectively. The ion range estimates are based on room temperature SRIM calculations and may be shorter than achieved when the substrate is heated during implantation. Implantations were done at an angle of 5-7° from normal to minimize ion channelling. In addition, Mn implantations were conducted following C ion implantations. However, due to radiation induced diffusion [20, 21], particularly with the samples maintained at an elevated temperature of 350°C during Mn implantations, it is possible for Mn ions to stop at greater depths. Following Mn ion implantation, thermal annealing is conducted in the range of 800-1000°C for 30 minutes.

The secondary ions mass spectrometry (SIMS) was carried out at the Material Characterization Facility at the University of Central Florida with a CAMECA IMS-3f magnetic sector analyzer. Details of the SIMS measurements can be found in [22]. Transmission electron microscopy specimens were prepared in a cross sectional geometry by tripod polishing. A Tecnai TF20 in both TEM and STEM modes was used to study the samples. Some of the work was done using instruments at Brigham Young University and some on the microscope at the University of Western Cape, Cape Town, South Africa. Energy Dispersive X-ray spectroscopy (EDX) and electron energy loss spectroscopy (EELS) was used to identify the chemical composition of the precipitates.

Ferromagnetic resonance measurements were conducted at room temperature by employing a broad-band (f = 1~40 GHz) coplanar waveguide (CPW) [23] in the transmission mode as an inductive sensor and a 1.4 T electromagnet able to rotate in a plane. The CPW is made of 300 nm thick gold deposited on an undoped GaAs substrate. The sample is placed upside-down on top of the central sensing area of the CPW to achieve the best coupling.

3. Results and discussion

3.1. Secondary ions mass spectrometry

Figure 1 shows the Mn ion distribution of samples A and B as-implanted and after anneals between 800 and 1000°C, as measured by SIMS. The general features of the manganese ion distributions seem to be quite similar to our previous works [9, 13], namely the Mn as-implanted sample (Sample B) in figure 1(b) shows a featureless broad distribution with a long tail, resulting from multiple implantation and hot implantation. After annealing at temperatures between 800 and 1000°C, the profiles of Mn ions become much narrower and multiple peaks appear. We found that the Mn distributions of the annealed co-implanted samples (Samples A) seem to be narrower than the annealed Mn implanted samples (Samples B) as shown in figure 1. The double-peak profiles of both annealed samples (A and B) may be due to the double Mn implantation.

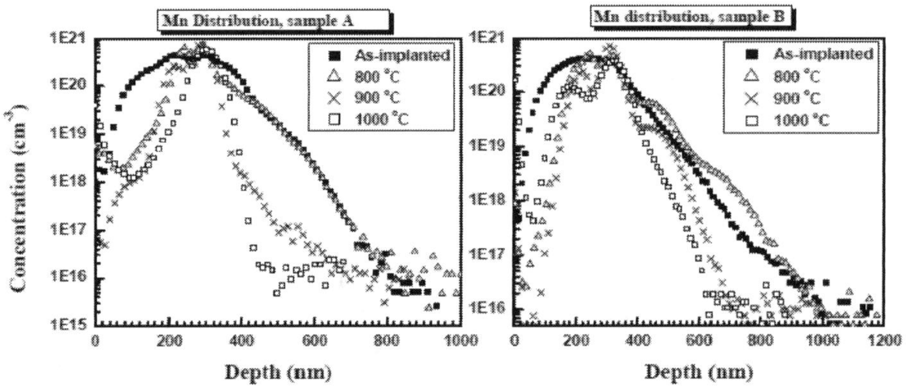

Figure 1. SIMS depth profiles of Mn ions in (a) Mn and Carbon co-implanted silicon, (Sample A), and (b) Mn implanted silicon (Sample B).

3.2. Transmission electron microscopy

The TEM results showed that all samples we investigated were crystalline. No amorphous layer was produced because the implantation was carried out at an elevated temperature (350 °C). Figure 2 shows the as-implanted sample, B-1, has many small precipitates which are confirmed by EDX to be the only regions where Mn is found. This precipitation is in contrast to low temperature implantations where no precipitates are found in the as-implanted specimens. This precipitation indicates that Mn ions are rather mobile at the implantation temperature of 350°C. The precipitate distribution is from a depth of 150 nm to about 400 nm below the surface of the silicon wafer. In a closer view (figure 2(b)), we see that these precipitates have a wide size distribution, ranging from smaller clusters (~ 4 nm) near the surface to larger clusters (up to 14 nm) near the End of Range (EOR) defects. The state, metallic or silicide, of these precipitates is not determined.

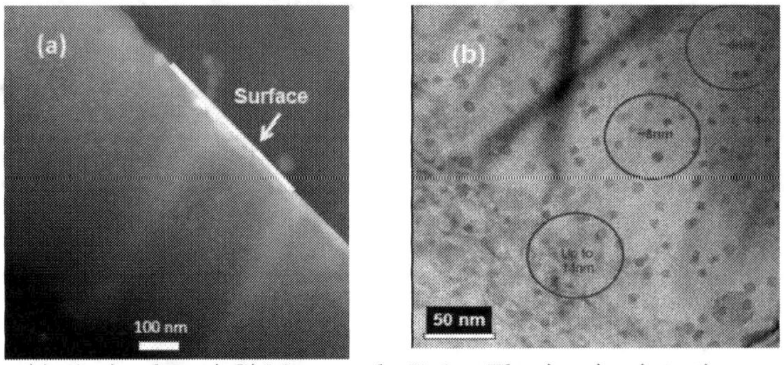

Figure 2. (a) As-implanted Si:Mn sample B-1. The ion implantation was carried at 350°C. Small Mn clusters were formed at depths between 150 to 400 nm. (b) A close-up image showed a distribution of the diameters of these Mn clusters from 4 to 14 nm.

For the sample annealed at 800°C (B-2, figure 3(a)), we noted that the precipitates seem to grow larger and start to transform from a circular shape into a faceted crystallite shape. In a close-up image (Figure 3(b)), fringes are clearly seen. These are Moire fringes that arise from the interference between periodicities in the silicon of the substrate and those in the precipitates.

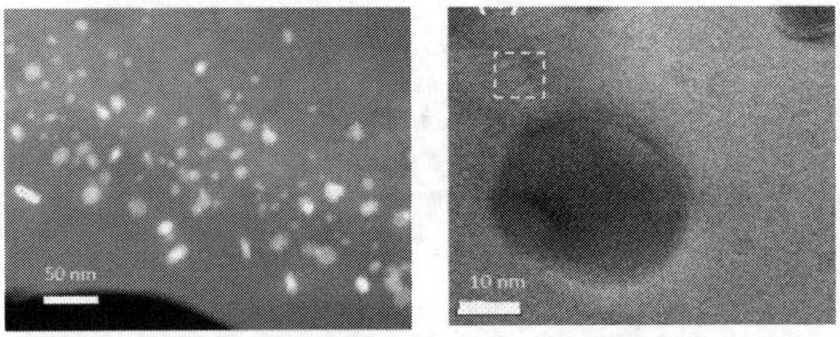

Figure 3. (a) ADF STEM image and (b) Bright Field TEM image of the Mn implanted sample B-2, annealed at 800°C.

For the co-implanted samples, we will detail our findings below for the as-implanted sample (A-1) and for the 800°C annealed sample (A-2). As shown in figure 4(a), the co-implanted sample showed no amorphous regions. Near the EOR there are many crystal defects and in the implant region many precipitates and voids (or amorphous regions). In figure 4(b), an enlarged Annular Dark Field Scanning Transmission Electron Microscopy (ADF STEM) image is shown. In this mode the image is formed by scanning (in a rectangular pattern) a focused probe of electrons across the sample. This allows the ADF imaging mode that generally shows atomic number dependant contrast but can also (if detector angles are somewhat smaller) show some diffraction contrast. In this image we can see that there are two types of defects: those that show brighter intensity or higher average atomic number and are confirmed by EDX to contain Mn and those that appear darker indicating lower average atomic number or lack of crystallinity. No Mn is found in the darker features. Attempts to detect carbon in the dark precipitates by EELS proved negative. Similar spots are also seen after annealing in the A-2 sample (figure 5).

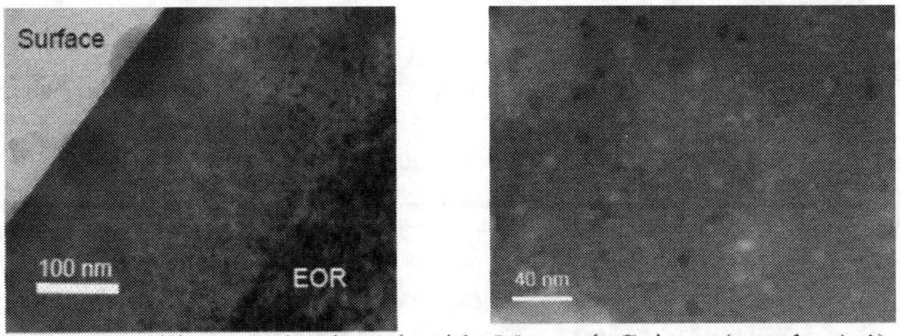

Figure 4. Silicon co-implanted with Mn and C ions (sample A-1). (a) TEM showing implant region with an EOR region full of lattice defects. (b) A higher magnification ADF STEM image showing bright (Mn precipitates) and dark features.

For the co-implanted and 800°C annealed sample (A-2), we noticed certain peculiar features as shown in figure 5(a). We see the typical EOR defects and Mn precipitates (in the 10's of nm ranges) similar to the B-2 sample. We also observed a few very dark spots and many small spots (examples are circled in figure 5(a)). Figure 5(b) is an ADF STEM image. In this view, the very dark spots in the TEM image of figure 5(a) are bright, the Mn precipitates are of medium brightness while many of the small spots show both bright and dark regions. Using EDX, we identify the very dark spots in figure 5(a) and the bright portions of the small spots as Ag precipitates. Silver paint was used to mount the silicon substrate during hot implantation to ensure good thermal contact. We do not expect

Ag to contaminate the silicon substrate during hot implantation and after implantation the silicon substrate was acid washed to remove all silver residues before the high temperature annealing. It appeared that this particular sample (A-2) still had some silver residue when the high temperature anneal was carried out. The many small spots in figures 5(a) and (b) have two types, those that have an associated Ag precipitate and many that do not (usually near the EOR or below). Both types of spots are dark in the STEM image and the dark regions contain no measurable concentration of any element other than silicon. As there is no significant drop in the silicon content (as measured by EELS) at these features, we identify them in both the annealed and as-implanted samples as amorphous silicon pockets associate with carbon co-implantation. The association with carbon implantation would imply some sort of carbon or carbon-silicon precipitation. This would match well with the observed ADF image contrast. However, no excess carbon content was observed by EELS in these dark regions. This would imply a low concentration of carbon that is undetectable in comparison to the background levels of carbon on the specimen or an actual lack of carbon. The exact nature of the formation of this feature is not understood at this time.

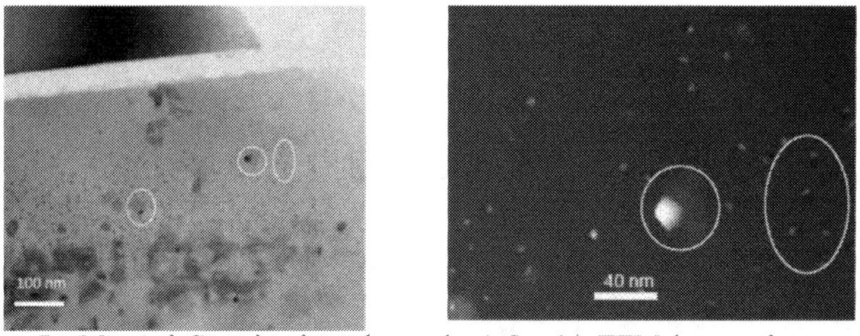

Figure 5. Mn and C co-implanted sample A-2. (a) TEM image shows some very dark spots (two marked with circles). (b) An ADF STEM image. The circled bright spots in (b) correspond to the right most circled dark spots in (a).

3.3. Ferromagnetic resonance spectroscopy

Ferromagnetic resonance measurements were carried out for the eight samples we prepared (A-1 through A-4, and B-1 through B-4). Seven samples showed null results. The only sample that showed a clear FMR signal is the B-1 sample. As shown in figure 6(a), a clear FMR signal is observed in the absorption spectrum at 2.3 kG when a 18 GHz microwave is used. At this frequency, the electron paramagnetic resonance (EPR) peak expected from isolated Mn-ions would be around 6 kG. Therefore the absorption we observed should be ascribed to the dynamics of a rigid magnetic vector resulting from exchange coupling. When the magnetic field is applied in the plane of the substrate, the frequency dependence of the resonance field is expected to follow a straight line when plotted as f^2/H vs H (as observed in figure 6), according to the in-plane FMR resonance condition,

$$\left(\frac{f}{2\pi\gamma}\right)^2 = H_R(H_R + 4\pi M_{eff}) \qquad (1)$$

where the zero field intercept gives $4\pi M_{\mathrm{eff}}$. The slope, which is the gyromagnetic ratio γ, can be used to obtain information of the Lande g-factor.

From our results, we obtained $g = 2.225$ and $M_{\mathrm{eff}} = 1.350$ emu/cm^3. We also carried out FMR measurements as a function of angle with respect to the normal of the substrate. The result is shown in figure 6(b). This result indicates a strong planar magnetostatic energy, likely related to an extended ferromagnetic film embedded inside the silicon substrate.

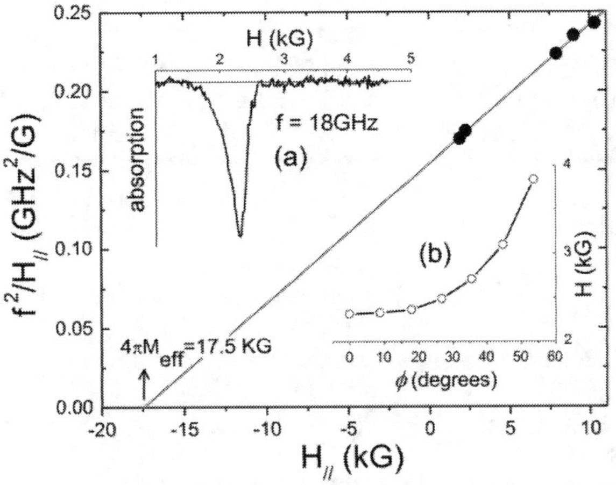

Figure 6. Frequency dependence of the FMR resonance peak with the magnetic field applied in the plan of the substrates. Inserts: (a) FMR spectrum obtained at 18 GHz. (b) Angular dependence of the FMR peak for the same frequency.

4. Conclusion

In summary, we have performed Mn and C ion co-implantation into silicon substrate to compare their physical properties with Mn ion implanted silicon samples. After annealing at different temperatures, we carried out SIMS, TEM and FMR measurements to compare Mn profiles, defect structures and magnetism of these samples. Our SIMS measurements show that the Mn profiles after high temperature annealing seem to be narrower for the co-implanted samples. The TEM measurements show typical EOR defects and Mn precipitates in both Mn implanted and C and Mn co-implanted samples. The co-implanted specimen showed small dark (in ADF STEM) features that appear to be amorphous silicon pockets. These have only been seen in the C implanted specimens. One co-implanted and annealed sample (A-2) showed Ag contamination. After an 800°C anneal, the Ag ions and Mn ions seem to form bi-precipitates. The FMR measurements showed only one sample (B-1) is ferromagnetic at room temperature. It appears that more experiments are needed to resolve the mystery of the magnetism in Si:Mn systems.

References

[1] Tsubokowa I 1960 *J. Phys. Soc. Japan* **15** 1664
[2] Ohno Y, Young DK, Beschoten B, Matsukura F, Ohno H and Awschalom DD 1999 *Nature* **402** 790
[3] Wolf SA, Awschalom DD, Buhrman RA, Daughton JM, von Molnar S, Roukes ML, Chtchelkanova AY and Treger DM 2001 *Science* **294** 1488
[4] Pearton SJ, Heo WH, Ivill M, Norton DP and Steiner T 2004 *Semicon. Sci. Technol.* **19** R59
[5] Chambers SA, Droubay TC, Wang CM, Rosso KM, Heald SM, Schwartz DA, Kittlstved KR and Gamelin DR 2007 *Materials Today* **9** 28
[6] Ohno H, Shen A, Matsukura F, Oiwa A, Endo A, Katsumoto S and Iye Y 1996 *Appl. Phys. Lett.* **69** 363
[7] Beschoten B, Crowell PA, Malajovich I, Awschalom DD, Matsukura F, Shen A and Ohno H 1999 *Phys. Rev. Lett.* **83** 3073
[8] Zhang F M, Liu X C, Gao J, Wu XS, Du YW, Zhu H and Xiao JQ 2004 *Appl. Phys Lett.* **85** 786
[9] Bolduc M, Awo-Affouda C, Stollenwerk A, Huang MB, Ramos FG, Agnello G and LaBella VP 2005 *Phys. Rev. B* **71** 033302

[10] Zhou SQ, Potzger K, Zhang GF, Mucklish A, Eichhorn F, Schell N, Grotzschel R, Schmidt B, Skorupa W, Helm M, Fassbender J and Geiger D 2007 *Phys. Rev. B* **75**, 085203
[11] Lin HT, Huang WJ, Wang SH, Lin HH and Chin HS 2008 *J. Phys: Condens. Mater.* **20** 095004
[12] Zeng L, Helgren E, Rahimi M, Hellman F, Islam R, Wilkens B J, Culbertson RJ and Smith DJ 2008 *Phys. Rev. B* **77** 073306
[13] Chow L, Gonzalez JC, Del barco E, Vanfleet R, Misiuk A, Prujszczyk M, Shunmugavelu A, Chai G and Bak-Misiuk J 2009 *J. Mater. Sci: Mater Electron* **19** S253
[14] Misiuk A, Bak-Misiuk J, Surma B, Osinniy W, Szot M., Story T and Jagielski J 2006 *J. Alloys Comp.* **423** 201
[15] Misiuk A, Barcz A, Bak-Misiuk J, Romanowski P, Chow L and Choi E 2009 *Mater. Sci. Eng. B* **159-160** 361
[16] Shaughnessy M, Fong CY, Snow R, Liu K, Pask JE and Yang LH 2005 *Appl. Phys. Lett.* **95** 022515
[17] Surgers C, Potzger K, Strache T, Moller W, Fischer G, Joshi N and v. Lohneysen H 2008 *Appl. Phys. Lett.* **93**, 062503
[18] Ren XT and Huang MB 2006 *J. Appl. Phys.* **100** 023525
[19] Huang MB and Ren XT 2002 *Appl. Phys. Lett.* **81** 2734.
[20] Pearton SJ, Williams JS, Short KT, Johnson ST, Jacobsen DC, Poate JM, Gibson JM and Boerma DO 1989 *J. Appl. Phys.* **65** 1089
[21] Dzhafarov TD 1989 *Phys. Stat. Sol. (b)* **155** 11
[22] Francois-Saint-Cyr H, Anoshkina E, Stevie F, Chow L, Richardson K and Zhou D 2001 *J. Vac. Sci. Technol.* B **19** 1769
[23] Quddusi HM, Ramsey CM, Gonzalez-Pons JC, Henderson JJ, del Barco E, de Loubens G and Kent AD 2008 *R. Sci. Instrum.* **79** 074703

Journal of Physics: Conference Series **281** (2011) 012031 doi:10.1088/1742-6596/281/1/012031

SIMS and RBS Investigation of Zn Implanted Si

V S Kulikauskas[1], V V Saraykin[2] and V V Privezentsev[3]

[1]Skobel'tsyn Institute of Nuclear Physics, Lomonosov Moscow State University, Leninskie Gory, 119991 Moscow, Russia
[2]Research Institute of Physical Problems, Zelenograd, 124460 Moscow, Russia
[3]Institute of Physics & Technology, Russian Academy of Sciences, Moscow 117218, Russia

E-mail: privezentsev@ftian.ru

Abstract. We present results of secondary ion mass-spectrometry and Rutherford backscattering spectroscopy of the Zn implantation profile in the damaged layer of a n-type Cz-Si(100) substrate, as well as the profile change during annealing. Si wafers were implanted with $^{64}Zn^{+}$ ions at energy of 100keV and ion dose of $2\times10^{14}cm^{-2}$. They were then subsequently annealed at 400°C for 60min and 700°C for 10min. Analysis of experimental data shows that in the as-implanted sample on a silicon surface a damage gauzy amorphous layer with thickness about 80nm was created, located inside the substrate at the ion-implantation depth. After two annealing steps the defects were completely annealed out. From the Zn depth profiles we observe that during annealing the Zn concentration decreases and the Zn peak concentration moves to the wafer surface. The Zn concentration maximum exceeds the solubility limit of Zn in Si, suggesting decomposition of Zn solid solution in Si and Zn precipitation.

1. Introduction

The properties of metal and metal oxide nanoparticles (NPs) in various matrices have been the subject of recent study since such nanostructured materials can be used in modern electronic devices [1]. Among them metallic Zn NPs play an important role. These NPs in a Si substrate can be made in particular by Zn ion implantation with subsequent processing of a Si wafer by various ways. Recrystallization of the damage Si layers implanted by Zn can be induced either thermally [2, 3], by pulsed laser irradiation [4], or by means of an ion beam [5].

Zn implantation of Si with subsequent annealing in neutral (inert) and oxygen atmosphere is the key fabrication stage during Zn NP formation, which subsequently transforms to ZnO NPs. Therefore study of the microstructural properties of such material becomes highly relevant. Ion implantation of metallic elements in semiconductors may result in the precipitation of metallic NPs in the implanted layer. It can occur either during the implantation [6] or the subsequent annealing stage [7]. Regarding high-dose Zn implantation in Si, it has been demonstrated that nanoscale hexagonal Zn and ZnO inclusions can be precipitated directly during implantation at room temperature [8].

For elements like Zn with combined diffusion mechanisms (interstitial/substitutional), the depth concentration profiles after annealing will be influenced by implantation induced defects. The ion implantation process will generate a high defect density in the Si crystal substrate and can amorphize it at doses above $10^{15}cm^{-3}$. The interaction of these defects and diffusing implanted atoms and also the recrystallization of the amorphous Si layer will have a strong influence on the diffusion of the implanted elements during annealing steps. Beside this the precipitation of the supersaturated solid solutions and intermediate compounds may occur at structural defects [4, 5].

Therefore a study of the properties of such material becomes highly relevant. Here we present a Rutherford back scattering (RBS) investigation of damaged layer and secondary ion mass-spectrometry (SIMS) investigation of Zn implanted profile in Si substrate, and examine how these profiles change during annealing.

2. Experimental Technique

A single crystal wafer of n-type Cz-Si(001) with thickness of 350μm and electron concentration of $3\times10^{15}cm^{-3}$ was implanted with $^{64}Zn^+$ ions at an energy of 100keV and ion dose $2\times10^{14}cm^{-2}$. The ion beam current density of $60nA/cm^2$ and a 2mm diameter was scanned over the substrate surface at an angle of 7°. The samples were subsequently annealed in a nitrogen atmosphere at 400°C for 60min and in an argon atmosphere at 700°C for 10min. Irradiation defects in a test sample were investigated by RBS of 2MeV He ions using ion channelling.

Secondary ion mass spectrometry (SIMS) was performed using a CAMECA IMS-4f at 150nA. Primary O_2^+ions with energy 10keV were used and mass-spectral resolution was 4000. The focused primary beam was rastered over an area of $200\times200\mu m^2$, with detection of ions from a circular area of 60μm diameter at the raster centre. The sputtering rate was approximately 0.5nm/s.

3. Results

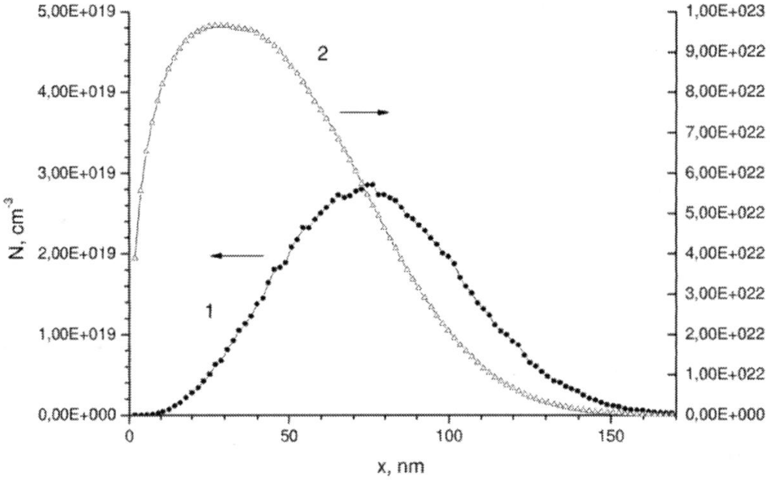

Figure 1. SRIM soft simulation results of distribution of implanted s $^{64}Zn^+$ ion (1) and Si target displaced atoms (Frenkel pair) (2)

According to SRIM simulation [10] the projected length for Zn ions is about 76nm for the given implantation conditions (figure 1, line 1). The $^{64}Zn^+$ ion produces about 2300 displacements of Si matrix atoms (radiation induced Frenkel pairs). Its distribution maximum is located at a depth of about 30nm (figure 1, line 2).

The experimental and simulated random and experimental aligned RBS spectra of He^+ ions for test samples are presented in figure 2. On the right side of the aligned RBS spectra there is a peak. As for the implanted sample the value of this peak in the aligned spectrum does not match that of a random spectrum, so the damaged layer is not completely amorphous. During annealing the value of this peak and its width decrease. If we assume that the damaged layer is completely amorphous, then it is possible to define a superficial density of the interstitial atoms of a Si matrix and this layer thickness [11]. It has been as a result found that after implantation the damaged layer thickness is 78nm, and the surface density of the interstitial atoms of a Si matrix is $3.9\times10^{17}cm^{-2}$. After preliminary annealing at 400°C for 60 minutes in a nitrogen atmosphere the damaged layer thickness decreases to 56nm and the surface density of the interstitial atoms decreases to $2.8\times10^{17}cm^{-2}$. After the final high-temperature anneal at 700°C for 10minutes in an argon atmosphere the damaged layer almost completely disappears and its thickness is only 2.0nm. Thus the associated density of interstitial atoms in the Si matrix is $1.0\times10^{16}cm^{-2}$. The aligned spectrum also shows peaks due to the presence of oxygen and carbon in the test sample. Their associated surface density is $N_O=4.4\times10^{17}cm^{-2}$ and $N_C=1.2\times10^{17}cm^{-2}$.

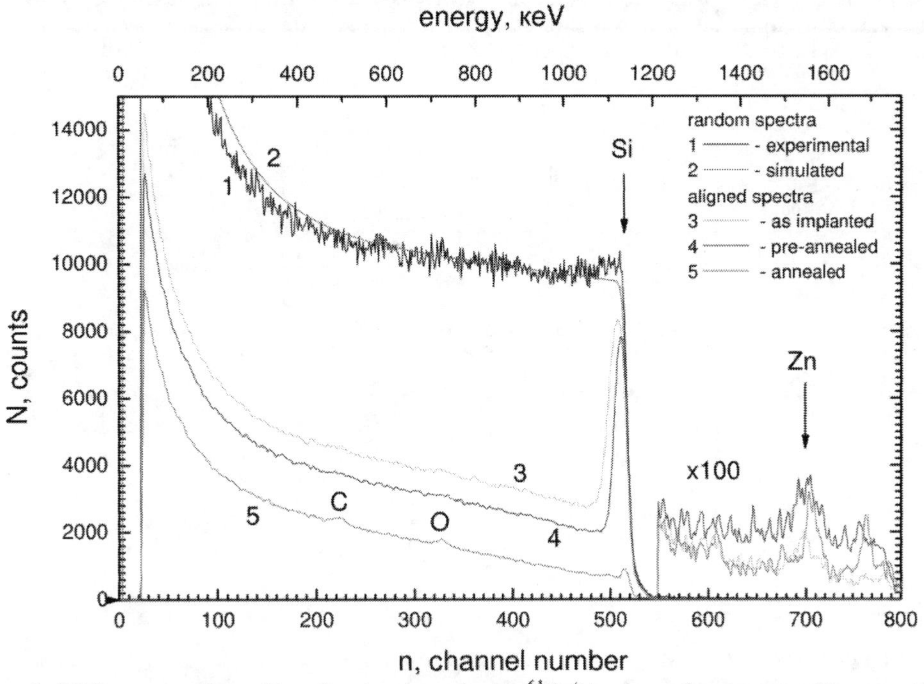

Figure 2. RBS patterns from Si wafers implanted with $^{64}Zn^{+}$ ions, random spectra (1) – experimental, (2) – simulated (on Fig.2, right they are not shown); aligned spectra: (3) – as implanted, (4) – pre-annealed at 400°C, (5) – annealed at 400°C + 700°C. Vertical arrows note channels of the particle scattering on the sample surface by ^{28}Si and ^{64}Zn atoms

Figure 3 shows the SIMS Zn depth profiles for the investigated samples. For the as-implanted sample the Zn distribution in depth profile (line 1) has a maximum at a depth of 75nm. line 2 shows SIMS spectrum of the implanted sample with pre-annealing at 400°C in nitrogen atmosphere for 60min. This also shows segregation to the right side of the concentration maximum. Particularly, we note that the value of displacement maximum has not changed, but its intensity has increased. Line 3 in figure 2 presents the sample annealed at 400°C in nitrogen atmosphere + 700°C in argon atmosphere. This curve shows two concentration peaks. The first curve maximum lies near the sample surface at depth less than 20nm. This peak is strong and consists of more than $5\times10^{19}cm^{-3}$. The second one occurs at a depth of 40nm near the projected range for the Zn ion implantation. Its intensity is higher than the limiting equilibrium concentration of Zn in Si.

4. Discussion.

In the as-implanted sample a damage layer with thickness about 78nm was formed near the sample surface. The radiation-induced point defects (Frenkel pairs) make the major contribution to the damage due to large amount of displacements per implanted Zn ion (about 2300). We note that as in the implanted sample, the radiation damage of the Si lattice and defect formation are determined by redistribution of radiation-induced point defects owing to secondary processes: annihilation of intrinsic interstitial Si atoms and vacancies, and a sink of interstitial Si atoms at the sample surface. One maximum is visible on the profile of Zn distribution in the depth profile for the as-implanted sample (figure 3, line 1). In this layer the Zn concentrations are considerably above the equilibrium limit of Zn solubility in Si namely $N_{Zn}^{lim}=6\times10^{16}cm^{-3}$ [13].

After pre-annealing at 400°C the damage layer thickness decreases to about 56nm. Because of high mobility of interstitial Si atoms and due to their sink towards a substrate surface the subsurface area is enriched by defects of vacancy type. Probably, these defects are clusters of intrinsic Si vacancies and/or complexes Si vacancy – Zn atom. Annealing leads to thermal activation of secondary processes and enrichment of a sample surface by vacancy type defects. This effect is observed as a reduction of the aligned RBS spectrum in a subsurface area.

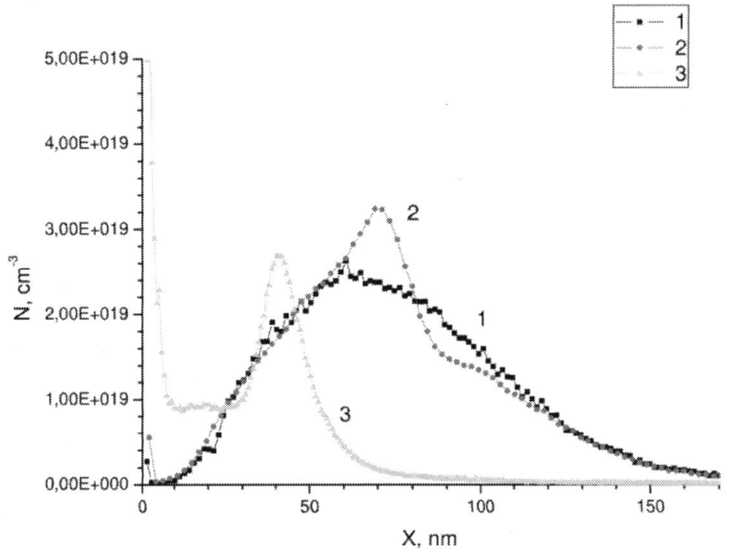

Figure 3. SIMS Zn distribution profiles on the depth, samples: as-implanted (1), pre-annealed at 400°C (2) and annealed at 400°C+700°C (3)

It is known that diffusion of Zn in Si proceeds by a mixed mechanism (interstitial/substitution) [13]. After pre-annealing at 400°C the feature in Zn concentration distribution at nearby 70nm can be caused by Zn diffusion towards the sample surface and into substrate depth, which interfere in two defect areas. One of them is enriched in intrinsic radiation-induced interstitial defects and is displaced at a depth nearby 30nm. Here, according to SRIM, there is a maximum of the displaced atoms (Frenkel pairs) distribution. Generated Frenkel pairs create damage in a crystal lattice, and the associated depth distribution is presented in figure 1, line 2. The second area is displaced to nearly 100nm, and has enriched so-called "end-of-range" defects. These defects are in a tail of impurity atom distribution behind its maximum, representing the small clusters of intrinsic radiation-induced interstitial defects, which are not yet annealed at 400°C.

After annealing at 700°C a reduction of point defect density and the complexes responsible for damage of the matrix crystal lattice is observed (figure 2, spectrum 5). The damage layer thickness has decreased to 2.0nm, and the density of the Si matrix interstitial atoms to $1.0\times10^{16}cm^{-2}$. Radiation-induced defects in the sample have almost all annealed out. Reconstruction of the radiation-induced damage layer has occurred from the intrinsic substrate and, apparently, most defects have diffused deep into the substrate. On the other hand the subsurface area has become enriched with Si vacancies due to the sink of interstitial Si atoms towards the crystal surface. In the damage layer near the interface with the intrinsic substrate there are located the weakly associated intrinsic interstitial Si atoms. RBS spectra for all studied sample temperature treatments demonstrate the same behavior (decrease of intensity and width) of the peak in figure 2 (spectra 4 and 5). Thus, the damaged layer thickness and interstitial Si atom density decrease during annealing in all samples. It means that the radiation-induced point defects in the damaged layer are relaxed through interaction with participating Si matrix interstitials and Zn impurity atoms, due to them clustering.

After annealing at 700°C there is an active redistribution of impurity Zn in the area of the damaged layer, therefore a Zn distribution in the depth profile (figure 3, curve 3) has undergone significant changes. Now in the SIMS spectra there are two concentration peaks: one located at the sample surface and another in the substrate depth. The first area thickness is less than 10nm, at the subsurface Zn concentration peak. On the substrate surface this depth is associated with a silicon oxide layer, which is an effective sink for Zn atoms. So this peak is connected with a Zn atom sink near the sample surface. The maximum near the sample surface corresponds to distortions of the crystal lattice, which are caused by association of radiation-induced point defects, which sink towards the crystal surface.

The second area is located in the subsurface region where the majority of Zn atoms are gathered. In this area another Zn concentration peak is located at depth of 40nm. After the second annealing the peak grows and moves closer to a sample surface (figure 3, comparing curves 2 and 3). This shows Zn atom clustering during precipitation from the super saturation solid solution of Zn in Si. High temperature annealing results in further growth of the Zn clusters. This peak is thus apparently caused by radiation-induced defects, which have formed clusters near the sample surface, which interfere with the fast Zn atom diffusion to the sample surface. The second peak corresponds to a depth where, according to SRIM simulations, the number of Si matrix atom displacements due to irradiation is a maximum. This peak occurs at a depth before the projected range, which is the most probable penetration depth for the implantation. For example, the projected range for Zn ion implantation is 75nm [10] and the second peak occurs at 40nm. Thus such obstacles do not exist at greater depths. A similar situation was observed in [4], where implantation and annealing were occurred under approximately similar conditions. This concentration peak transformation could also be due to the formation of Zn-silicide compounds ($ZnSiO_3$ type).

A third peak sometimes also occurs [4], deeper than the implant projected range. It is related to implantation defects at range end. At sufficiently high dozes this maximum is related to amorphous and damaged parts of the substrate.

During Zn ion implantation in the Si substrate and subsequent thermal annealing stages the initial process is characterized by Zn precipitation and formation of Zn-silicide (type of $ZnSiO_3$) compound due to HRXRD. The oxygen dissolved in Cz-Si can take part in this reaction leading to such formation. From IR-spectroscopy data the oxygen concentration in the initial Si wafer was about $N_O=6\times10^{17}cm^{-3}$.

5. Conclusions

PBS and SIMS spectral distribution in the as-implanted state is determined by redistribution of radiation induced point defects due to secondary processes (annihilation of Si intrinsic interstitial atoms and vacancies, a sink of interstitial defects towards the sample surface). The 78nm subsurface layer is disordered.

Heat treatment at 400°C for 1h leads to annealing of interstitial point defects and a thickness reduction of the damaged Si lattice layer to 56nm. This layer contains clusters or Zn- vacancy complexes. These complexes result from reactions between vacancies and implanted Zn atoms activated by thermal annealing. In the area enriched by Si interstitials the annealing stimulates Zn precipitation from Si<Zn> solid solution.

After annealing at 700°C for 10min the damage layer has reconstructed almost completely. A sharp redistribution of Zn was observed in this sample. The total amount of Zn in the layer decreased, which was probably caused by diffusion deep into the Si substrate. High temperature annealing contributes to further growth of the Zn clusters.

References

[1] Redel E. Transition Metal Nanoparticles. Sudwestdeutscher Verlag Fur Hochschulschrifte, 2009.
[2] Henry M O, Campion J D, McGuigan K G, Wthewalt M L and Lightowlers E C 1989 *Mater. Sci. Eng. B* **4** 201.
[3] Simov S, Kalitzova M, Karpuzov D, Yankov R, Angelov Ch, Faure J, Bonhomme P and Balossier G 1996 *J. Appl. Phys.* **79** 3470.
[4] Francois-Saint-Cyr H, Anoshkina E, Stevie F, Chow L, Richardson K and Zhou D 2001 *J. Vac. Sci. Technol. B* **19** 1769.
[5] Eryu O, Murakami K, Takita K and Masuda K 1988 *Nucl. Instr. Meth. Phys. Res. B* **33** 665.
[6] Lohner T, Angelov C and Mikli V 2008 *Thin Solid Films* **516** 8009.
[7] Kalitzova M, Simov S, Yankov R A, Angelov C, Vitali G, Rossi M, Pizzuto C, Zollo G, Faure J, Killan L, Bonhomme P and Voelskov M 1997 *J. Appl. Phys.* **81** 1143.
[8] Zollo G, Kalitzova M, Manno D and Vitali G 2004 *J. Phys. D: Appl Phys.* **37** 2730.
[9] Muntele I, Thevenard P, Muntele C, Chhay B and Ila D 2005 *Mater. Res. Symp. Proc.* **829**, paper B.2.21.
[10] http://www.srim.com/
[11] http://www.simnra.org/
[12] Feldman L C and. Mayer J W Fundamentals of surface and thin film analysis. North-Holland, NY-Amsterdam-Londom 1986.
[13] Milnes A. Deep Impurities in Semiconductors. Wiley, NY - London 1973.

AUTHOR INDEX